D1325813

ZERO DEGREES

CHARLES W. J. WITHERS

ZERO
DEGREES

GEOGRAPHIES OF THE PRIME MERIDIAN

 Harvard University Press

CAMBRIDGE, MASSACHUSETTS ◦ LONDON, ENGLAND ◦ 2017

Library of Congress Cataloging-in-Publication Data
Names: Withers, Charles W. J., author.
Title: Zero degrees : geographies of the Prime Meridian / Charles W. J. Withers.
Description: Cambridge, Massachusetts : Harvard University Press, 2017. |
Includes bibliographical references and index.
Identifiers: LCCN 2016040039 | ISBN 9780674088818 (alk. paper)
Subjects: LCSH: Prime Meridian—History. | Meridians (Geodesy)—History. |
Geographical positions—History.
Classification: LCC QB224 .W58 2017 | DDC 527/.2—dc23
LC record available at https://lccn.loc.gov/2016040039

FOR ANNE

CONTENTS

MAPS AND ILLUSTRATIONS

ZERO DEGREES

Longitude of a place is the difference of such place counted on the Equator, from some Meridian which passes through the place you reckon or count your longitude from; which Meridian is as it were the Land-mark of the whole sphere, being the Bounds from whence the Longitude of any particular place is accounted quite round the globe. Now this meridian from whence we begin to Reckon the Longitude, has been differently assign'd by several nations.

WILLIAM ALLINGHAM, *The Nature and Use of Maps* (1703)

PROLOGUE

IN NOVEMBER 1883 a sixty-seven-year-old English sailor addressed a letter to the world's geographical societies. There are, he began, "some things that belong to all mankind." What William Parker Snow had in mind was nothing less than a single prime meridian, a global base point or line to be used by all nations. From such an agreed point of terrestrial measurement, he reasoned, navigators and geographers could plot their voyages and align their maps, and perhaps astronomers could even chart the heavens, all using a standard point of earthly reference. One prime meridian for the world and only one, stressed Parker Snow, would be of "vast benefit to Science and Humanity."[1]

His appeal for a uniform base of measurement was rooted in personal experience. Born in 1817 in Poole, England, William Parker Snow was a sailor, an author, a sometime Patagonian colonist, a historian of the American Civil War, and, important to his own self-image, a veteran of Arctic exploration. He was also reputed to be psychic and lived, as one contemporary remarked, "on the edge of the Fourth Dimension." Parker Snow's supposed otherworldly powers, revealed in a dream in January 1850, convinced him (but few others) of the location of the ill-fated Franklin expedition that was lost in the Arctic in 1847–1848. As a young man, Parker Snow had himself nearly perished at sea during a storm when his vessel had a close encounter with another ship that had derived its mid-ocean position and course from an initial meridian different from his own.[2]

Parker Snow's entreaty was timely. In the late nineteenth century, there was no such thing as a single prime meridian. There was no shared global 0° base point from which the world's dimensions were reckoned in common—no one site of calculative origin from which geographers, navigators, astronomers, and timekeepers could fix their respective systems of measurement. Nor had there ever been. By the 1880s moves were certainly being made in several quarters to adopt a single prime meridian for the world. Parker Snow was aware of them. Yet during his time, the story of the prime meridian—its history, its geography, and its use by different nations and communities—was one of persistent national difference. The entry "meridian" in the 1825 *London Encyclopaedia*, for example, makes the nature of the problem clear: "The first meridian of a country is that from which its geographers, navigators, and astronomers, commence their reckoning of Longitude; and, the meridians having nothing in themselves to distinguish them from each other, the fixing upon any one for this purpose is quite arbitrary; hence different persons, nations, and ages, have commenced their longitudes at different points, which has introduced no small confusion into geography." For that author, at least, there was no likelihood that these differences and the resultant confusion would ever be reconciled: "National and even scientific jealousies are too strongly prevalent for us to hope that the world will at any early period fix on a common first meridian."[3]

Parker Snow lived in a world where more than twenty prime meridians were at work. Several European nations had long used the island of Ferro, the westernmost of the Canaries, as a prime meridian. This choice of geographical origin was an intellectual inheritance from the ancient Greeks. By 1825 and for some time before, however, as the *London Encyclopaedia* further reported, individual nations had commonly employed their own initial meridian: "Each considerable country now usually adopts the meridian of its own capital as the first." France, for example, used two—Ferro and from about 1720, Paris, centered upon the observatory there. After 1776 the United States did as well, employing Washington, DC, for astronomical purposes and Greenwich, England, for geography and navigation: this twin primacy was enshrined in U.S. law in 1850. In Britain from the later 1760s, the prime meridian most commonly used for navigational purposes was that based upon the Royal Observatory at Greenwich.[4]

Parker Snow's 1883 proposal for "one INTERNATIONAL PRIME MERIDIAN"— his typography emphasizing his case—was "at a spot on the Globe seemingly, to me, placed there by Nature for the purpose." His chosen spot was Saint Paul's Rocks (today known as Saint Peter and Saint Paul Archipelago, or Saint Peter and Saint Paul Rocks), a group of small islands close to the equator in the Atlantic and about 630 miles northeast of Cabo do Calcanhar on the Brazilian mainland. "If this spot were made a general PRIME MERIDIAN," he argued, "it would remove all difficulties as to interchange of longitude at Sea, and also do away with National or individual predilections for any particular locality." Intended as a new beginning for the world's benefit and measurement, Parker Snow's proposition bore in part the stamp of the old and of one nation's prime meridian above others': "This spot, if adopted at my suggestion, [is] to be called 'NEW GREENWICH, PRIME MERIDIAN.'"[5]

Saint Peter and Saint Paul Rocks today boasts a lighthouse, built in 1995, and a Brazilian scientific station, built in 1998, used for satellite communication. Parker Snow's plans for New Greenwich as the originating point of the world's measurement were similarly rooted in concerns over communication and safety at sea: "I do not confine myself to making these Rocks a mere *locale* for an imaginary line of Meridian. My plan embraces the *use* of them, and other Ocean Rocks, Islands, and Headlands, for more *Scientific* and *humane* purposes besides."[6] His idea, which he had been promoting in talk and text for over thirty years, was for a series of "Ocean Relief Depots." Most recently articulated in *Chambers's Journal* in November 1880, the scheme involved an immense network of cables and moored lightships, a "floating telegraph over the ocean world," with New Greenwich, the world's single prime meridian, as its o° heart.[7]

The idea behind the creation of a single prime meridian was to solve the geographical confusion caused by lasting and different national practices. Having one first meridian could save lives, standardize maps, unite various scientific communities, and provide a base point from which to regulate the world's measurement in space and in time. It would not, however, be easily achieved. To create a single prime meridian for the world required nothing less than undoing centuries of established national practice. Nothing less, that is, than abolishing the ages-old customs of numerous nations and communities that used—and used differently—several prime meridians in the course of their work in the world. Geographers and topographical surveyors could and commonly did work from an arbitrary o°, a measured prime meridian, according to their particular needs and circumstances in making maps. Astronomers calculated an observed prime meridian from the o° of astronomical observatories. As Parker Snow experienced, almost to the cost of his own life, sailors of different nations (and even those within the same nation) calculated their o° point of navigational departure in yet other ways—sometimes from astronomical observatories and sometimes from landmarks or other long-established customary first meridians.

Within a year of William Parker Snow's letter, though certainly not in direct answer to it, geographers, astronomers, navigators, and politicians from over twenty countries met in Washington, DC, at the International Meridian Conference in order to propose a single prime meridian for the good of the world. Of Parker Snow there is no sign. No trace of his proposal exists in the concerns voiced in the meeting. New Greenwich was not among the several candidate locations discussed there or being entertained by others elsewhere. There is no mention of Parker Snow or New Greenwich in the conference's printed proceedings. Yet in Washington one prime meridian was proposed above all others as the world's originating point of measurement for use in mapping, navigation, astronomy, and timekeeping. This was determined not by what nature provided, as Parker Snow had hoped, but through political and scientific debate over where and why and how one prime meridian among many should serve as the world's 0°.

Introduction

One Line to Rule the World

THE PRIME MERIDIAN IS the line and the point at which the world's longitude is set at 0°. It does not exist in any strict material sense, yet through maps and clocks, the prime meridian governs the life of every human on Earth. Both longitude, on the one hand, and time's measurement, on the other, are based on the prime meridian. Since 1884 it has been fixed at the Royal Observatory, Greenwich, in the United Kingdom. Global space has been universally and cartographically regulated from that moment in time and universal coordinated time set from that point in space.

Greenwich's position as the 0° baseline for the world's measurement in space and time is the result of the International Meridian Conference, held in Washington, DC, in October–November 1884 at the invitation of the president of the United States. In his opening remarks, Rear Adm. C. R. P. Rodgers, chair of the conference, welcomed "delegates renowned in diplomacy and science" on behalf of the president and reminded them of the aim of the meeting: "To create a new accord among the nations by agreeing upon a meridian proper to be employed as a common zero of longitude and standard of time throughout the world." The scientists and governments of the world, he stressed, now understood "that it is desirable to adopt a single prime meridian for all nations, in place of the multiplicity of initial meridians which now exist." One month later Rodgers closed the meeting following delegates' near unanimous agreement upon seven recommendations—"resolutions," as they termed them—over a global prime meridian and universal coordinated time. Key among them was the acceptance of "a single prime meridian for all nations" (Resolution I), the decision that Greenwich Observatory in the United Kingdom should be the world's "initial meridian for longitude" (Resolution II), an agreement that there should be a universal day (Resolution IV), and the ruling that "this universal day is to begin for all the world at the moment of mean midnight of the initial meridian" (Resolution V).[1]

For contemporaries the fixing of a single prime meridian for the world at Britain's Royal Observatory at Greenwich in 1884, and with it the global establishment of universal time, was a momentous event—or, at least, it had the promise to be so. For modern scholars, the Washington conference, Greenwich, 1884 and All That, has been seen as a vital moment in global science and cosmopolitan politics, a "touchstone for the world" in the shaping of modernity.[2] The story of Greenwich's metrological primacy is usually told as a seemingly inevitable consequence of fin de siècle British scientific and political authority. Earlier circumstances and the perspectives of other nations (when they are considered at all) are read as leading almost inexorably toward the "triumph" of Greenwich in the face of foreign competition: the global solution reached in Washington was a victory for British pragmatism, imperialism, maritime and commercial power, and the empirical sciences.[3]

This book presents an interpretation of the prime meridian question as something altogether more complex. It does so in relation to different national geographies; diverse private and public actors; issues of comparative metrology; and especially in the late nineteenth century the internationalization of science, space, and time. This is a book about a global problem in geography, astronomical science, and navigational practice—namely, the fact that there were so many prime meridians before 1884. It is about the implications of these many initial meridians. It is about the means adopted to solve this problem—that is, the Washington meeting. And it is about the longer-term and geographically uneven results both the problem and the solution proposed—what I call Washington's "afterlife." In the face of numerous alternate possibilities, how did the prime meridian, a singular world-ruling feature, come to be sited in Greenwich? Why was the world's prime meridian settled upon only in 1884, and what relationship did this decision over space have to the regulation of time? How are we to interpret this solution to the world's metrology, understand and explain the problems that prompted it, and fathom the fact that the solution proposed was put into effect only slowly and unevenly? These and other matters constitute what we may identify as and what I call here the "prime meridian question."

The Prime Meridian Question

All meridian lines are lines of longitude: geographical and mathematical conventions designed to delimit and to measure the earth. The positioning of the prime meridian, or of any of several initial meridians, is in one sense wholly arbitrary: it may reflect an individual's choice or an agreed-upon national political decision. In another sense, as delegates to Washington knew and as was understood for centuries beforehand, the choice of the "prime," "first," or "initial" meridian (the terms were interchangeable), was a far from arbitrary matter. There is an important initial distinction to be noted, however. There are two principal conceptions of prime meridians: the *cartographic* or *measured* and the *observed*. The cartographic or measured prime meridian is that meridian, either point or line, marked on a map as 0°, zero degrees of longitude. These initial meridians are entirely arbitrary and may become convention with repeated usage, or they might be replaced by the adoption of another cartographic first meridian reflecting a different purpose. An observed prime meridian is based upon an astronomical observatory and is associated with that observatory's publication of an ephemeris, an astronomical calendar in which the predicted positions of the principal celestial bodies are listed as an aid to astronomy and navigation: "Observed prime meridians constituted the foundations of eighteenth-century navigation and of the initial determination of positions for mapping; they are very few in number."[4]

In the past this distinction was often blurred in practice since astronomers and surveyors might refer equally to the "local" meridian in making use of measurements from their initial base point. Similarly, navigators used the term loosely or interchangeably when they referred to it at all, and early modern geographers commonly referred to the "prime" meridian in positioning longitude as east or west of a certain point that the user was invited to assume had been determined by observation, when it had not. As we shall see, the distinction has often been blurred in modern scholarship.[5]

What follows makes clear the distinctions between prime meridians observed and measured and between different communities' practices in relation to the context and the prime meridians under discussion. What we are examining here as the prime meridian question may for convenience be divided into three parts. The first is the problem delegates in Washington sought to address: the "multiplicity of initial meridians." The second is the "solution": properly, the solutions proposed in Washington that Greenwich be the world's initial meridian for longitude, that it be the basepoint for the establishment of a universal day, and that the universal day should begin "at the moment of mean midnight of the initial meridian." The third relates to the fact that the recommendations at Washington were just that—recommendations. Because of this we may study the "afterlife" of the 1884 Washington meeting for what it reveals about the uneven consequences of attempts to regulate the world.

The Problem of Multiple Prime Meridians

The existence of numerous cartographic and observed prime meridians and the difficulties they caused was certainly apparent to John Senex, mapmaker and one-time Royal Geographer to Queen Anne. As he put it in his *New General Atlas* (1721):

> Though the Western Nations agree to fix it [the prime meridian] in the West of our Continent, they don't agree on the Place where. Ptolemy, and the Ancients, fix'd it at one of the Fortunate Islands, now generally suppos'd to be the Canaries. Some of the Arabians follow'd him, and others plac'd it as Hercules Pillars, or the Streights of Gibraltar. Some Moderns would fix it at Tercera, others at the Isles of Cap Verd or Cape Verd it self, and some at the Pike of Teneriff, one of the Canaries. The Spaniards would have it at Toledo; the Portuguese at Lisbon; and, in short, every Nation may fix it at their own Capital if they please; but as Ptolemy has been follow'd by most, tis so likely to continue, especially since Lewis XIII of France did, by the Advice of the ablest Mathematicians, publish an Order of April 23, 1634, that it should be fix'd by his Subjects at the Isle of Fero, the most westerly of the Canaries.
>
> It is now become usual to count the Longitude westward as well as eastward from the Place where Geographers fix their first Meridian.
>
> The Difference among 'em [Senex here means different geographers, and he is largely referring to cartographic meridians], about fixing this Meridian has made great Confusion in their Maps, and occasion'd much difficulty in finding the Longitude of Places. . . . The only way to remedy this at present is, to give an account of the different Places where they fix their Meridians, and of their Distance from one another.

The Spaniards, since their Conquest of the West-Indies, place their first Meridian at Toledo; and from thence, contrary to all other Europeans, account their Longitude from East to West.

Bleau, the Dutch Geographer, and most of his Countrymen, place it as the Pike of Teneriff, one of the Canaries.

The French, as we have heard already, generally fix theirs at the Isle of Fero, and some of them at Paris.

Our English Geographers, as Camden, Speed, and others, fix it in the Azores Islands; some at the Isle of Corvo; and others, which is most follow'd at the Isles of St. Michael's and later ones place it at London.[6]

Senex's testimony over national differences makes clear the longevity of the problem as well as its complexity. In the ancient world, classical geographers and some Islamic scholars took the prime meridian to be the Canary Islands. The French formally acknowledged this position by royal edict in 1634, defining that nation's prime meridian from Fero (sometimes Ferro, Sp. Hierro), the westernmost of the Canaries, and did so on the basis of mathematical advice. Yet, Senex tells us, the French also reckoned from a different prime meridian, Paris. The Dutch likewise based their prime meridian on the Canaries but from the "Pike" (peak) of Teneriff (Tenerife). The British differed again, taking their prime meridian either from islands in the Azores or from London (which usually meant St. Paul's Cathedral). Although Senex had in mind the difficulties caused in geographically understanding and representing the world by using these different prime meridians, the problem was compounded by variations among the astronomical, geographical, and maritime communities over which prime meridians were used and for what purpose.

By the late nineteenth century and for some considerable period before, contemporaries everywhere encountered and lived with numerous prime meridians, each a reflection of the political authority and established scientific practices of a certain nation. Commentators knew this situation was unsustainable. The measurement of geographical space and of time in the world to new and agreed-upon standards was vital for the progress of science, international politics, and a rapidly modernizing world. Consider in this context the arguments of Sandford Fleming, a Scots-born Canadian railway engineer and a member of Britain's four-man Washington delegation in 1884. Fleming knew that advances in science and communication required the standardization of the world's time and longitude. In a series of papers from the mid-1870s, Fleming argued for the standardization of time and a single universal prime meridian: for this, he has been considered the "Father of Universal Time."[7]

Modern scholars, looking back on the nineteenth century's encounter with telegraphy, the railways, the cultural and technological effects of speed, and the closure of global space in an age of high empire know this as "time-space convergence."[8] To Fleming and others then experiencing a world of unregulated time and geographical confusion arising from the use of twenty or so prime meridians, modernity required a common prime meridian. As Fleming said, "The establishment of an initial or prime meridian as the recognized starting point of time-reckoning by all nations, affects the whole area of civilization, and conflicting opinions may arise concerning its position. Its consideration must therefore be approached in a broad cosmopolitan spirit, so as to avoid offence to national feeling and prejudice."[9]

These "conflicting opinions" constituted the problem that the International Meridian Conference sought to address. The development of national consciousness and the entrenchment of customary practices among scientists and other user communities within individual nations can explain the presence of different prime meridians. Some degree of unanimity existed throughout the early modern period, at least in Western Europe, around a prime meridian sited on Cap Ferro in the Canaries: Senex's evidence broadly confirms this. Different prime meridians appeared, from the later seventeenth century and especially in the eighteenth century, as a reflection of nations' relative political and military strength; as part of Enlightenment scholars' interests in geodesy and the power of reason and mathematics to delimit space; and as baselines for terrestrial and celestial measurement to meet the requirements of geographers, navigators, and astronomers. For historian of cartography Matthew Edney, the choice of these prime meridians serves "as a marker of distinct phases of cartographic practice between the Renaissance (various Atlantic islands), Enlightenment (Tenerife, Ferro, London), and Modernity (Greenwich, Paris, Cadiz, etc.)."[10]

But these phases were not always distinct. There were differences within as well as among nations and within and between various scientific users over which prime meridians to use. Geographers and cartographers depicted and worked with different 0° on their maps (some even showed several prime meridians on the same map). Geodesists and astronomers had to constantly recalibrate their measurements to take into account the multiple base longitudes used. Sailors would sometimes have to adjust their longitude at sea based on data from navigators aboard passing vessels. Or they would determine their position either by using a timepiece or by astronomical observation, calculating against different geographical features such as the last land sighted or the dome of St. Paul's—features which would, for convenience, then be taken as the originating prime meridian. Topographical surveyors found their mathematics in need of constant reappraisal: countries seemed to change their dimensions relative to another nation's prime meridian or even from the various prime meridians in use within a single nation.

The problem and subsequent confusion was not confined to the differences among nations, books of geography, and navigational or representational practices. Well into the nineteenth century, different prime meridians were employed even within nations. In 1882, as international concern for a single global prime meridian was reaching its height and invitations were being issued to come to Washington, Don Juan Pastorin, hydrographer to the Spanish Navy (and a delegate in Washington), spoke of his experience on these matters: "It has always seemed to me very lamentable that there should exist such a plurality of Meridians, and, while in the classes of the Naval College I could not understand why the unscientific plurality of our reckonings of longitude, condemned openly both by the Professors and the books we studied, should be persisted in." Persistent error in books and in navigational teaching was bad enough. The presence at one time or another of multiple prime meridians in his own country exacerbated the problem. Spain, continued Pastorin, "has counted the longitude from the Meridian of the Straits of Gibraltar, Toledo, the ancient College of Marine Guardo de Cadiz, San Fernando (in two different citations those of the two observatories, the ancient one and the present one), Ferrol, Cartagena, Plaza Major of Madrid, observatory of the same capital, Coimbra, Lisbon, (in three distinct places according to the successive observatories), the Cathedral of Manilla, the island of Heirro (in different points, some doubtful)—and, to-day, it is proposed heedlessly to give another Meridian of reference."[11]

The problem of different prime meridians highlighted the fact that there was no universal standard point against which to measure space and time. As the need for one became ever greater, a fact expressed most clearly in the nineteenth century but with distinct antecedents, the problem was not just that disparate scientific, nautical, geographical, and other parties used different prime meridians in different ways. The lack of a standardization of time compounded the issue. There was no universal time or agreed-upon civil day until the recommendations of 1884. Furthermore, two metrological standards were at work, the imperial and the metric, especially in linear and areal measurement. Here, too, long before Washington in 1884, during that meeting, and even afterward, appeals were voiced over the need for a single global standard for the world's measurement. Debates set national perspectives against proponents of a common cosmopolitan good. The problem of the prime meridian was that it was centrally implicated in all these matters. Establishing a single prime meridian was about choosing the appropriate place and the appropriate methods against which to regulate time and space in a world of difference.

The Prime Meridian Solution: Washington 1884 and a Single Global Prime Meridian

Greenwich's selection as the world's metrological baseline was far from inevitable. Delegates in Washington debated the respective merits of several leading candidates for the world's prime meridian. Berlin, Greenwich, Paris, Washington, and even one in the Bering Strait—which was termed Greenwich's "anti-meridian"—were each given serious consideration. The result intended was "a new accord" among nations. Delegates professed "absolute neutrality" as the scientific criterion driving their motives. Yet their voting patterns upon the seven resolutions sometimes became obscured as in other conference sessions they sharply advertised firmly held differences of opinion regarding individual nations' scientific and political interests. Washington, Greenwich, Berlin, and Paris may have been the principal candidates for the world's prime meridian at the 1884 meeting, but several other sites were mentioned outside the conference or were discussed at earlier scientific meetings without being completely discounted.

Over twenty prime meridians were in use or had been proposed by 1884: Jerusalem, Beijing, Philadelphia, Rome, Cap Ferro on the Canaries, Oslo, New Orleans, Madrid, Mecca, Kyoto, St. Paul's Cathedral in London, Greenwich, Paris, Berlin, Pulkova, Bethlehem, and the Great Pyramid of Giza, to name only a few. In several countries, the United States included, multiple prime meridians were in use for different scientific purposes. In his welcome, Rear Admiral Rodgers appealed for a globally useful outcome: "We seek only the common good of mankind, and gain for science and for commerce a prime meridian acceptable to all countries." Not the least of the difficulties facing Rodgers and the conference delegates as they debated a single global prime meridian was how to establish a basis for proceeding. What were the principles on which the meeting should work? What was the evidence against which delegates would decide? How would national practices be reconciled to effect international, even global, accord? These and other questions are addressed in what follows. Whereas the complexities of the prime meridian question are revealed in assessing the conduct and the content of the 1884 meeting, other issues also emerged from the meeting. The seven resolutions over space and time and global measurement proposed in Washington were recommendations only. As resolutions they were not binding upon the governments of the participating nations, far less those countries not represented. The 1884 International Meridian Conference was not the ultimate solution to the problem of the prime meridian.

The Prime Meridian's Afterlife

Greenwich was recommended as the world's 0° in 1884, but it was not taken up everywhere or at once. Several countries continued to employ prime meridians other than Greenwich on their topographical maps and nautical charts after 1884 or continued to use different local times. Yet neither should we see the Washington conference as a meeting without effect.

Explanations for these variations in uptake rest not only in the fact that the resolutions could not be enforced upon the nations of the world. They also lie in understanding that each resolution meant something different to the separate communities involved—the astronomical, the geographical, and the nautical—as well as to the public. Consideration of the prolonged and varied afterlife of Greenwich's proposed adoption is a vital element of the prime meridian question for what it reveals about the geographical, institutional, and social scale of the outcome. Plans to achieve global unanimity and remove long-running national differences produced varied responses among scientific communities within and across national boundaries. This was especially so not for Resolution II, that Greenwich be the world's "initial meridian for longitude," but for Resolution VI, which concerned the proposed unification of the astronomical and nautical (and civil) day. This was a world where different regimes of public timekeeping existed. The civil day began at midnight and had a duration of twenty-four hours. It was often divided into two twelve-hour sections (a.m. and p.m.), but the twenty-four-hour-long astronomical day began at noon after the start of the civil day. The nautical day was effectively equivalent to the civil day but in Britain only from the early nineteenth century and in the United States only from the late 1840s. In a world where standard time in relation to the running of the railways was a late nineteenth-century invention and where men like Sandford Fleming campaigned for the global use of a single system of timekeeping—"universal," "cosmic," or "cosmopolitan" time, as he variously called it—it is easy to see that the "solution" offered in Washington was quite far from consisting of just one answer or axiomatically suiting everyone.

The Prime Meridian as an Object of Study

What is here examined as the prime meridian question—the problem of multiple prime meridians, the solution proposed in Washington, and the afterlife of the 1884 meeting—has to date been addressed unevenly and in various ways.

John Senex's remarks of 1721 exemplify the widespread interest during the eighteenth century and earlier in the prime meridian as a geographical and navigational issue. This is evident in the attention given to the prime meridian as a practical subject in geographical and navigational texts as well as through the practice of cartography and ships' navigation. But with the exception of one or two important moments in the Enlightenment when the prime meridian was linked to questions of national geographical self-determination, the prime meridian as a subject of more scholarly study, with its own history and historiography that "speaks" to bigger issues of metrology, global modernity, and the authority of science, emerged only in the late nineteenth century. The prime meridian became the object of serious study at the moment its many national expressions became the source of widespread scientific concern.[12]

More recent scholarship has tended to address the prime meridian within national frames of reference. In Britain, Greenwich was established as the baseline for that nation's mapping and astronomical work from the later 1760s, following the endeavors of Nevil Maskelyne, the Astronomer Royal, who produced new and accurate ways of calculating longitude at sea in his *British Mariner's Guide* (1763) and *Nautical Almanac and Astronomical Ephemeris* (1767) and who in doing so took the Royal Observatory at Greenwich as his 0°. With this taken as the foundational moment for what would later be a proposed global primacy, Britain "ruled the world" from the 0° longitude of Greenwich—except in France. And in many other countries. The French, with their base referents of Ferro and Paris, ruled their world differently. For their geodesists, astronomers, and even French kings, it was axiomatic that measurements using the Paris meridian were the most accurate and during the eighteenth century in particular were considered the most scientifically authoritative. These were facts that French delegates to Washington were keen to point out and reluctant to concede. In the United States, competing views were held decades before 1884 over the prime meridian. Some authorities, notably in Washington, held to the view that an American prime meridian was essential to adequately symbolize America's political, scientific, and geographical postrevolutionary distance from Britain. Others argued the opposite—namely, that America should adhere to Greenwich's ruling authority as a global prime meridian for science and humanity's universal benefit.[13]

As the evidence of Don Juan Pastorin and Sandford Fleming indicates, the questions of the prime meridian concerned more than national differences, important though these were. Rather than see the prime meridian as an expression of British imperial empiricism, French truculence, American indecision, and errant Spanish pedagogy, it should be understood as a question of scientific authority, closely connected with the emergence of systems of timekeeping and, especially in the nineteenth century, modernity itself.

These issues, variously expressed, feature in others' discussions of the prime meridian. In his *One Time Fits All: The Campaign for Global Uniformity* (2007), the historian of time Ian Bartky assesses the prime meridian question as an important element in the move to standardize systems of timekeeping in the world, especially in the nineteenth and twentieth centuries and in the United States. As Bartky illustrates and as Sandford Fleming so clearly understood, the problem of time not being standard—of its not being "universal" or "cosmopolitan," as Fleming had it—was by the mid-nineteenth century vital to commerce, international communication, and the pace of life. As Bartky shows, the prime meridian question was bound up with developments in the regulation of time in Britain and Europe; with telegraphy; and in North America in particular with the regulation of railway timetables. As technological change made the world smaller, the need for a common point of spatial and temporal measurement became greater. As delegates in Washington came to realize, scientific principles had to be tempered in the interests of daily social life, while national interests had to be recast as questions of global unity.

Historian of science Peter Galison and literary scholar Adam Barrows both pay attention to the prime meridian in their respective accounts of temporality and modernity in the late nineteenth and early twentieth centuries. In *Einstein's Clocks, Poincaré's Maps* (2003), Galison examines the years immediately preceding the 1884 Washington conference. In several international meetings that considered the prime meridian, he sees a growing concern for the standardization of time as part of the regulation not just of science but of global civic life itself. More than any other period, the later nineteenth century was distinguished by rapid growth in the number of international scientific associations, by developments in scientific communication, and by emergent networks across national boundaries among disciplinary practitioners: these in combination marked the period as one of "international science in the age of nationalism."[14]

The prime meridian was debated in these terms in a sequence of scientific meetings—geographical congresses, mainly—beginning in Antwerp in 1871. Questions of a common metrology and the regulation and standardization of time were vital components of these meetings and the connecting threads between them. For years prior to Washington, from Antwerp in 1871, to Paris in 1875, to Venice in 1881, and to the International Geodetic Association meeting in Rome in 1883, the problems of the prime meridian were discussed as a particularly clear expression of a growing internationalism in science. Such internationalization, in science and in the regulation of time, was also understood to be a crucial element in political control: the coordination of clocks that regulated theoretical physics and drove communications technology was driven by "national ambitions, war, industry, science, and conquest." Adam Barrows's *The Cosmic Time of Empire: Modern Britain and World Literature* (2011) is similarly attentive to the regulation of time and its place in conceptions of modernity. His focus, however, is less the problem of multiple prime meridians than various literary representations of modernity in the wake of the 1884 resolution over Washington.[15]

In addition to its direct practical importance, the prime meridian was also a vital topic in broader discussions of accuracy and intellectual authority within science, even within particular subjects. The prime meridian question was also, I hope to show, bound up with the notions of epistemic and social authority that came from being accurate, or claiming to be accurate. For the French, particularly from the 1720s when the Paris meridian was recalculated and found to have been "misplaced," fixing the prime meridian was a *longue durée* project of eliminating error, repositioning the nation's baseline, and reevaluating the claims of scientific practitioners and their political patrons regarding accuracy. So too, from the late 1770s—notably, during the first thirty years of the nineteenth century and as a narrative that continued in later periods—were Britain's and France's joint attempts to fix the longitudes of the Greenwich and the Paris observatories and so better determine the absolute and relative position of their respective prime meridians. Following Norton Wise and others, we may take "precision" to be associated with standards of comparison and "accuracy"; broadly, to be about the degree to which one or a series of measurements corresponds to what is taken to be the true and agreed value. In Britain, France, and America, these questions were expressed in words and in numbers as reports and papers were prepared on different prime meridians and as the individuals and the institutions concerned sought credibility for themselves and validity for their claims in the language of mathematics.[16]

In the past the question of the prime meridian was exemplified by its presence in different nations and reflected nations' political authority as a measure of territorial sovereignty. Yet the authority over space and time, of which the prime meridian was both expression and symbol, was not rooted in common standards of measurement. The story of weights and measures, metrology, in the eighteenth and nineteenth centuries is a narrative of bewildering diversity with a huge range of measures in customary usage. Most evidently, metrology was a narrative of conflict resolution between, on the one hand, adherents to the British and imperial system of weights and measures and, on the other, proponents of the French metric system. As Ken Alder has shown, attempts to standardize and regulate the meter foundered in the face of geographical and political difference: the story of the meter is a story of error, political and epistemological tolerance, scientific authority, and national pride. Things were no different in Britain: British units of standard measurement were calibrated more than once in the eighteenth and nineteenth centuries.[17]

This matters because the question of which units—British or French, inches or meters—should be used to measure the world featured centrally in discussions on the prime meridian, especially in the 1860s and 1870s when the Great Pyramid of Giza was identified as a possible global prime meridian. This ancient monument, its advocates (bizarrely) claimed, enshrined and embodied British imperial units. Because it did so, the Great Pyramid should be the base point for the globe's metrology, and in the future all nations should be ruled by British measurements, which were themselves divinely ordained. Looked at from a modern perspective, this view and the opposition to the metric system that sustained it—the idea that the metric system was revolutionary; inaccurate; unworkable; and above all, French— seem odd. But for contemporaries the tension between imperial and metric units was a recurrent feature of eighteenth- and nineteenth-century scientific life and political debate, and the prime meridian—whether proposed for Giza or elsewhere—cannot be understood without attention to it.[18]

These facts have several implications. If, as has been claimed, the term "international science" was not fixed in its meaning, caution must be exercised about what "international" and "science" meant, or were held to mean, for contemporaries, whether in the later nineteenth century or for earlier periods.[19] The internationalization of science by the later nineteenth century was apparent not just in the growth of subject-based transnational associations. It was also evident in the ways in which scientific questions were seen, at least among Western nations, as a means to political, social, and cultural reform within and beyond national boundaries and in the idea of science as the means to and a form of the common public good.[20] Analyzing science's content *over* time while paying attention to scientists' and politicians' concerns *with* time raises questions about social context, disciplinary configuration, institutional agency, and individual influence. Understanding the prime meridian further requires consideration of whose political interests were served by seeing it as variously a matter of territorial sovereignty, mathematical accuracy, and metrological authority; the subject of specific usage by disparate scientific communities; and the expression of a universal "common good." This book draws from and extends the work of others in developing these claims. I have ignored "false leads" in which the "prime meridian" in question turned out to refer to works of poetry or prose or where historians used the idea of an "imperial meridian" to review certain moments in the imperial history of European nations.[21] Where the emphases of Bartky, Galison, Barrows, and others have been upon the prime meridian, modernity, and temporality, my principal concerns are with the prime meridian, metrology, and spatiality—with the geographies of the prime meridian.

The Prime Meridian Considered Geographically

The question of the prime meridian is profoundly geographical in several related senses. Not the least of these is the fact that the prime meridian is a principal feature of the world's measurement. It is one key manifestation of human attempts to "read," measure, and rule the planet's dimensions. What is now a single prime meridian for the world, at Greenwich, was in the past a geography of numerous prime meridians, each speaking to the interests of different nations. The solution to the problem of multiplicity was proposed in one setting, in one city, and yet had uneven consequences for the world as a whole. Understanding the prime meridian question is in this sense, importantly, a matter of paying close attention to geographical scale. The prime meridian was simultaneously something with global reach, interor transnational expression (in terms of the many international scientific meetings in which it was debated, for example), national importance, and different local meanings. It was also firmly identified with local settings and particular institutional, even individual, interests.[22]

The prime meridian question is geographical in terms of practice—that is, in terms of those processes—oceanic navigation, trigonometry, terrestrial survey, telegraphy, and astronomical observation—through which the world was put to order and in which the placement of 0° was either a constituent part or the object in view. The prime meridian was made real geographically in various forms: as a subject of inquiry in works of geography; educational texts and atlases; and, importantly, maps. The prime meridian was commonly, sometimes literally, at the center of map and atlas production; its delineation as measured or observed a base point for the mapmakers in question and, occasionally, a symbolic expression of the geopolitical reach of the nation in question.

The prime meridian question is geographical in terms of the particular sites—the physical locations—in which each prime meridian was located and the different social spaces—the intellectual sites—in which it was addressed as a problem or, as in Washington, a solution. In regard to the sites and settings in which the prime meridian question was articulated—In which of several astronomical observatories should the prime meridian be located?, Which international scientific meetings were influential in articulating the problem and proposing a solution?, Whose scientific voice did the most to shape the prime meridian?, and so on—my use of the term "geographical" is here strongly realist. That is, it is my contention that the nature of the prime meridian question was shaped by the places and the social spaces in which the question was addressed and by the places for which the solution was proposed. Where some commentators have seen the language of geography in the explanation of science's making and dissemination to be merely nominalist, believing that matters of place and space are not explanatory in themselves, what follows argues for the constitutive significance of certain sites and social and scientific meetings in explaining the prime meridian question.

These issues—of measurement, scale, geographical practice, geographical product and its representation, site and social setting, accuracy, social and political interest, and the differences among diverse interested parties in what the prime meridian meant—are all consistent with a range of recent work that has addressed the production and reception of science geographically. Because what was held to be science varied in its form and meaning from place to place, different subjects of knowledge had distinct character and cognitive content according to setting. Seen in these terms, the prime meridian was also, I suggest, one element in the emergent science of geography as it evolved as a subject over time and in different places— whether in Paris, either as the French king ruled upon Ferro in 1634 or in the Académie des Sciences in the 1720s; in London in the late 1770s; in early nineteenth-century Boston and Philadelphia; or at the moment of its "solution" in Washington in 1884. At the same time, the problem of multiple prime meridians, debated as it was in the social and epistemic spaces of conference venues; as the subject of individuals' interests; as the symbol of nations' identity; as part of competing metrologies; and as a matter of technological and cultural change in periods of high nationalism, scientific internationalism, and global modernity, illustrates particularly clearly the geographical dimensions of science.[23]

The Narrative Structure of the Book

Zero Degrees is about the prime meridian as a definitive geographical and world-ruling feature, its creation, and its long-run intellectual history and significance. It is about the geographies of science and the science of geography; about science and politics, globalization, accuracy, metrology, epistemic and social authority, national identity, and the work of individual scientists and national scientific communities in periods of growing internationalization in science. It is about how a single line in space and a point in time came to have global authority.

The book is divided into three parts. Broadly, these reflect the main elements of the prime meridian question: the problem of multiple prime meridians, the solution of 1884 that was Greenwich, and the continuing afterlife of the prime meridian question after 1884. I have already discussed the "answer," the proposal of Greenwich, and will return to it in Chapter 5. Otherwise, the structure of the book is mainly chronological, addressing the prime meridian from the work of classical geographers to its expression in one form or another in the twentieth century, although I end, briefly, with reference to the prime meridian in the early twenty-first century. Principally, what follows examines the prime meridian question in the 250 years between 1634 and 1884 before turning in Part III to the 1884 Washington meeting and its aftermath. The themes outlined above—accuracy, credibility, national identity, geographical practice, the internationalism of science, and so on—are discussed within this structure. While it would have been possible to examine each of these themes as a chapter, one consequence of such an approach would have been to arrive, as it were, at Washington, Greenwich, and 1884 several times over. Early chapters cover the evidence of several centuries, often at the scale of the nation. Later chapters descend into a necessary detail, even documenting the content of individual scientific papers, imparting the rules governing the scientific meetings at which the prime meridian was discussed, and scrutinizing the day-by-day, session-by-session working through of the issue in Washington.

In beginning Part I, "Geographical Confusion," Chapter 1 examines the problem of multiple prime meridians in the centuries before c. 1800, notably from the 1634 Paris meeting, which in intent if not in outcome may be seen as an early modern precursor to the 1884 Washington conference. The prime meridian question is studied in relation to its different national expressions; as part of textual traditions and practices within geography, astronomy, and navigation; in relation to the "longitude problem" in the later eighteenth century; and as a matter of accuracy and authority in international attempts to "fix" by triangulation the prime meridians of Greenwich and of Paris. Chapter 2 focuses upon the prime meridian in the United States in the century before 1884, or rather upon America's several prime meridians. As I show, it is in the United States that we find perhaps the clearest expression of multiple anxieties over the prime meridian in terms of national identity, political self-determination, scientific authority, and differences between scientific communities. In Chapter 3, the first of two chapters that make up Part II, "Global Unity?," the prime meridian is examined in relation to questions of metrology in science and politics in order to place it in a wider intellectual context. Chapter 4 looks at the prime meridian question as a geographical and political concern within international geographical meetings from 1871. Here in particular there is an evident "sharpening" of focus as I turn to the workings of several scientific meetings and to the words of influential commentators in directing the form and the content of the prime meridian question.

The theme of the prime meridian as a solution is returned to in Chapter 5, the first of two chapters making up Part III, "Geographical Afterlives." This chapter examines the verbatim report of the Washington meeting and contemporaries' reactions to it as delegates' deliberations were reported upon in the popular press and in scientific periodicals. Chapter 6 examines the afterlife of Washington and Greenwich in different settings and material forms, with further attention to detail a necessary feature. Chapter 7 reviews the principal features of the prime meridian question and looks at the several ways in which Washington 1884 and the processes discussed there have been the subject of later popular commemoration.

We know that the 1884 International Meridian Conference happened and what it proposed. This book is about why: What was the problem to which this was the "solution"?

PART I

GEOGRAPHICAL CONFUSION

1

"Absurd Vanity"

The World's Prime Meridians before c. 1790

IN INTRODUCING HIS BOOK, *A New System of Geography*, in 1762, Anton-Friedrich Büsching, the German professor of philosophy and teacher of geography at the University of Göttingen, made clear exactly what the prime meridian was: "By the first Meridian we understand that particular one, among the other innumerable Meridians, from which we begin to reckon the degrees on the Equator from west to east." The prime meridian, continued Büsching, was a work of human artifice and a matter of national difference, not one of nature's disposition: "Nature has indeed fixt no particular Meridian for this purpose, all of them having an equal right to this honour so that it is left entirely to our choice to fix upon any one of them for the first Meridian." As he further remarked: "The *Hollanders* and many others fix their first Meridian at *Pico* on the island of *Teneriffe*; as, on the contrary, the *French*, ever since the year 1634, by order of *Lewis* XIII. draw the first Meridian through the island of *Ferro*, and in this they are generally followed by modern geographers; particularly by the Cosmographical Society at *Nuremberg*, and by the authors of the Berlin Sea-Atlas published in 1749. The *Swedes* draw their first Meridian through *Upsal* [Uppsala]."

Knowing *what* the prime meridian was did not extend to common agreement over *where* it was. Büsching was impatient with this lack of consensus: "It were to be wished that all geographers were agreed in this point." Many contemporaries were of like mind. In his *A New System of General Geography, in which the Principles of that Science are Explained* (1780), Ebenezer MacFait, a geography teacher in Edinburgh, observed how "British geographers, for the most part, reckon longitude from the meridian of London, or rather, from the meridian of the Observatory at Greenwich" and how the "diversity of meridians, in globes, maps, and books of voyages, is apt to perplex beginners; and, at any rate, costs some pains and time to reduce them from one method of reckoning longitude to another." "'Tis pity," MacFait pronounced, "they did not all agree to reckon longitude from Teneriff." The English historian Edward Gibbon was characteristically forthright in remarking upon the matter in 1789:

> By the Greek and Arabian Cosmographers the first meridian was loosely placed at the Fortunate or Canary Islands: the true position of the Isle of Ferro has been determined by modern observation; and the degrees of longitude proceed with singular propriety from the western limit of the old Hemisphere. The absurd vanity of the Spaniards and Dutch, the French and English has variously transformed this ideal line from a common and familiar term: the longitudes of Madrid and Amsterdam of Paris and Greenwich must now be compared, and the national diversity of speech and measure is aggravated by a new source of perplexity and confusion.

To William Faden, geographer to King George III, the situation was easily stated: "Each Nation, having chosen its first Meridian, has occasioned no small confusion in Geography."[1]

As Gibbon hinted, the prime meridian that the ancients used and understood was held to mark the western edge of the known world, from which longitude was marked eastward, rather than being a central feature of the globe from which longitude was calculated both east and west. Yet by the early modern period, this common understanding had disappeared, even within nations. In France, for example, a royal decree in 1634 fixed Ferro (Sp. El Hierro) in the Canaries as its prime meridian. But by the 1720s, the French had shifted the meridian to Paris, a choice that would prove a source both of national pride and considerable anxiety, occasioned by the demands and presumptions of accuracy.

The geographical confusion that resulted from the use of different prime meridian lines—Gibbon's "absurd vanity"—became still more complicated by the later eighteenth century. By the last decade of the century, each professing claims of unprecedented accuracy, British and French authorities were hard at work trying to fix, by trigonometry and astronomy, the exact geographical location of the observatories at Greenwich and in Paris. Their claims and activities reflected a desire to join the observed and the geographical prime meridian, in those two places at least: fixing observatory sites allowed what we might think of as national "sight lines," base points for an ever-more accurate mapping of one's respective national space, to be determined. But the role of the Greenwich and Paris observatories in the 1790s as twin sites within an international plan that aimed at the accurate determination of two separate 0° prime meridians was, in effect, only a particular refinement of a long-standing more complex situation. Büsching, MacFait, Gibbon, and Faden lived in a world where the Dutch had one geographical baseline of global reckoning, the Spanish another, the British yet another, and the French different ones still.

This chapter traces the geography of the prime meridian between its positioning in the worldview of classical geographers and the establishment of the prime meridians of Greenwich and of Paris in the late eighteenth century. The chapter is in three parts. The first examines the multiple prime meridians at work, chiefly in Europe and in the European geographical imagination, looking, in particular, at how the different prime meridians in use came into existence. I discuss the appearance in the Western intellectual tradition of one "category" of prime meridian, the magnetic or agonic prime meridian and, by the mid-seventeenth century, its virtual disappearance. The second part explores the connections between national identity, mapping, scientific usage, and the role of the prime meridian in the century following the establishment, in 1667, of the Paris Observatoire and, in 1675, of the Royal Observatory, Greenwich, England. Here, I consider the prime meridian in relation to the "longitude problem" and its solution, at sea and in print, in the late eighteenth century. Against this background of national differences over the prime meridian in later eighteenth-century Europe, the third part considers plans made, from the 1770s, to connect the observatories of Paris and Greenwich trigonometrically. Taken together, the chapter is an account of why, for Portuguese and Spanish navigators, Dutch mapmakers and German philosophers, English geographers and French kings, and students of astronomy and geography everywhere, different prime meridians produced such "confusion in Geography."

Ruling Lines in Space and Time: The Prime Meridian before c. 1667

The historical geography of the prime meridian before the establishment of astronomical observatories in Paris and in Greenwich has several related features. One was the enduring legacy of the prime meridians used in the classical world. In medieval and early modern Europe, various astronomer-mathematicians moved beyond these classical references to take new and different 0° terrestrial baselines for their astronomical observations. From 1492, as the result of European voyages of global discovery and exploration, European mariners and navigators drew upon the ancients' authoritative lines and, in part, thought that they had found a mid-Atlantic prime meridian. Heading to the Americas, the empirical encounter with what appeared to be magnetic constancy in the mid-Atlantic suggested to Europe's sailors that, in fact, there was a natural baseline for a 0° longitude from which to effect the earth's measurement.

Classical Lines and Points of Origin

In the history of European geographical practice, theoretical attempts to measure the earth, particularly its latitude and longitude, began with the ancient Greeks. The first to use a prime meridian as a starting point for global measurement was, as far as we know, the geographer-mathematician Eratosthenes (c. 276–c. 196 BCE). He took Alexandria, his place of residence, as the prime meridian in order to determine the length of the classical ecumene, the known world of the Mediterranean. The Rhodes-based Hipparchus of Nicea (c. 190–c. 120 BCE), took Rhodes as his prime meridian in criticizing Eratosthenes's measurements (Eratosthenes had mistakenly assumed that his Alexandrian prime meridian passed through Rhodes). These attempts were flawed because few reliable observations of latitude had been made and because there existed no practical means of reliably deriving longitude, the east–west measurement of the earth. This twin problem would beset agreement upon the prime meridian for centuries.

The most significant Western European classical authority with respect to the prime meridian is Claudius Ptolemaeus, known as Ptolemy (c. 90–c. 198 AD). Ptolemy's importance rests in his choice of the most westerly land then known, what he called the Fortunate Isles, or in modern terms the Canary Islands. This was the result of decisions he made in estimating (in fact, exaggerating) the length of the Mediterranean and, from that, the size of the world as a whole. In traditional Islamic and South Asian societies, a number of prime meridians were recognized. Islamic geographers adopted Ptolemaic notions in similarly placing their prime meridians to the west of the known world but without always agreeing upon Ptolemy's use of the Fortunate Isles. The ancient Indians used a central meridian based on Laṅkā (Ceylon) for their calculation of planetary motion and may have used this as their base point for longitude and terrestrial measurement. From the mid-second century AD, the principal prime meridian in South Asia was based upon the ancient observatory at Uijain, in Madhya Pradesh in modern-day India. In some early Arab tables of longitude, reference is made to another prime meridian in the East, based on an uncertain locality called Jamāgird or Kangdiz. Cultures in the Far East based their prime meridians at significant cosmological sites—Kyoto for early Japan, for example—and in China the prime meridian was placed at Beijing.[2]

For the ancients and for Europeans before the fifteenth century, the Canaries were the western edge of the known world. Ptolemy's choice of prime meridian at Ferro was a global edge, not a centrally positioned line of longitude. But because his works of geography survived into the early modern period and were subject to critical scrutiny during the European Renaissance, the prime meridian situated with reference to the Canary Islands came to have twin significance by the end of the fifteenth century. It was understood and adopted by Europe's geographers and astronomers and, as a classical geographical precept, it could be tested by European voyagers. Despite the discovery in the 1430s of the Azores farther west of the Canary Islands; of the yet more westerly Cape Verde Islands from the 1450s; and, from the 1490s, of the Americas, most geographers, navigators, and mapmakers initially stuck with Ptolemy's choice as their prime meridian. However, they increasingly recognized its arbitrariness in a world whose geographical dimensions were enlarging, to the west and to the east, and so required new mapping and measurement. And they began, concurrently, to question other Ptolemaic ideas about, for example, the width of the Atlantic and the earth's symmetry and size.[3]

For astronomers, as opposed to sailors and navigators, the Canaries had never been canonical. Islamic and Christian astronomers alike tended to use observatories as their base zero. They took the observed meridian of different observatories as their respective points of origin for the ease of calculating ephemerides, the printed tables of the position of astronomical objects at regular intervals of time from which to measure celestial phenomena and position features on the earth. The Alfonsine Tables, for example, drawn up under the patronage of King Alfonso of Castile and first employed on January 1, 1252, used Toledo as their baseline, and so in medieval and early modern Iberia, Toledo was the prime meridian commonly used. In the later fifteenth century, the astronomer Johannes Müller (or Mueller), posthumously known as Regiomontanus, used Nuremberg for his base meridian. Johannes Kepler based his ephemeris, the so-called Rudolphine

Tables named after the Holy Roman emperor Rudolf II, upon Ulm. Tycho Brahe, the Danish astronomer, used his Uraniborg observatory on the Baltic island of Hven as a prime meridian for his astronomical observations from the 1570s. By the late sixteenth century, most astronomers followed Nicolas Copernicus in reading the heavens as a heliocentric system, rather than as a system centered upon the earth, as Ptolemy had. But since none of the several ephemerides then in use was particularly accurate, the Alphonsine Tables, with their Toledo baseline, remained dominant into the seventeenth century before Kepler's Rudolphine Tables replaced them.[4]

Led Astray by the Compass: Points of Magnetic Constancy, c. 1492–c. 1650

Island groups in the Atlantic and on the American continent were not the only geographical features that oceanic navigators encountered as they tested Ptolemaic conceptions of the world and extended European commerce and geographical knowledge. At several points of longitude close to the Azores, ships' crews reported that their compasses showed very little or no magnetic variation from north. Christopher Columbus encountered this in mid-September 1492 on his first voyage to the Americas. Sebastian Cabot experienced it in 1498. Off the Azores in the 1580s, the Elizabethan navigator John Davis did too. Davis argued that the first meridian should be based on Saint Michael's Island, one of the Azores: "The reason why the accompt of longitude doth begin at this Ile, is, because that there the Compasse hath no variety, for the Meridian of this Ile passeth by the Poles of the world and the poles of the Magnet, being a Meridian proper to both Poles." It seemed that nature had indeed provided a line of what appeared to be magnetic constancy, a maritime zero of use in determining global longitude. Similar thinking lay behind Mercator's famous world map of 1569 in which he established the projection that bears his name by showing navigational lines of constant bearing and, broadly, the shape of terrestrial features but not their relative size. Mercator drew a prime meridian through the Cape Verde Islands on the basis of several reports that there the compass showed no variation. Although others employed the magnetic prime meridian before, and after, 1569, Mercator's map is the most important example, a position achieved in part by its use in other influential works of European cosmography and cartography in the mid-sixteenth century.[5]

Nature proved inconstant with respect to the magnetic prime meridian. Different navigators cited the same phenomenon—an observed lack of magnetic variation as the needle held to north—but at different places.

Some navigators found it near the Azores; some at the Cape Verde Island group. Some mariners reported that there was no point of longitude and latitude at which the compass did not vary. Geographical and mathematical texts record this diversity of views. In the 1594 edition of *Mr. Blundeville His Exercises,* for example, the English mathematician Thomas Blundeville affirmed the view that "late Cosmographers of these dayes doe make the first Meridian to passe through the islands called Azores, which islands . . . are situated more Westwards from the foresaid Insulæ fortunatæ [Canary Islands] by five degrees." His reason was that "the Mariners Compasse . . . wil never incline to the true North pole, but when they saile either by the Ile S. Mary or S. Michaell [in the Azores]." By the 1622 sixth edition of *His Exercises,* Blundeville was more circumspect: "Dieurs [diverse] learned Pylots . . . have found the variation of their compasse to be as much there, as else where, and not rightly to point to the respective point, which is supposed to be upon the earth." As Blundeville cautioned, "Without such a certain Meridian, no true account from any Latitude can be made."[6]

What early modern navigators had encountered but not fully understood was the phenomenon of geomagnetism, or terrestrial magnetism. The earth has a magnetic field, but it is not constant in its strength or alignment. The fact that the compass needle seemed unwaveringly to point north suggested that there was a location, somewhere in the Atlantic, where a 0° baseline might be situated: at that point there was no variation or declination from north. In modern terms this is the agonic line of zero variation, or the agonic meridian. As studies of terrestrial magnetism have shown, however, the agonic zero varies over time and by longitude as the earth's magnetic field fluctuates. Isogonic lines of magnetic variation—lines connecting points of the same magnitude—curve as they near the poles and so are not constant in their position. To early modern cosmographers and navigators, the Azores, both Saint Michael's (modern-day São Miguel) and the Corvo Islands, were taken as prime meridians because there nature's hidden forces seemed steady. In his map of 1594 and in his 1612 globe, for example, the Dutch mapmaker Petrus Plancius has the prime meridian running through the Azores. With others, he believed that the absence of magnetic variation indicated that these islands were the location of the world's 0° and therefore a natural baseline for computing the world's longitude. Several early modern European cartographers show the world as two hemispheres, the "Old World" and the "New World," touching where nature's magnetic forces seemed not to favor either (Figure 1.1). Between c. 1508 and c. 1688, this agonic prime meridian was the subject of much attention from ships'

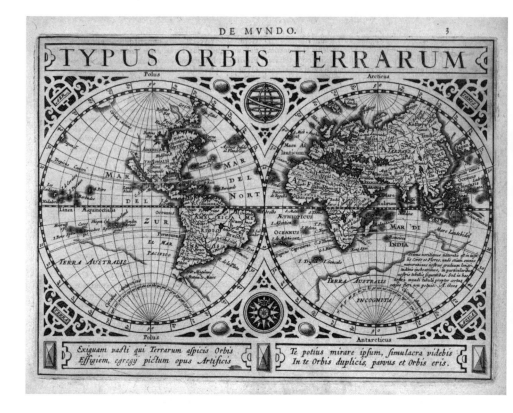

FIGURE 1.1 This map of 1628 shows the Eastern and Western Hemispheres of the globe touching at the equator at what was presumed to be the agonic meridian, the line of zero magnetic declination. Here, the agonic meridian is based upon the westernmost island of the Cape Verde group. *Source: Atlas minor Gerardi Mercatoris al. Hondius plurimis aenis tabulis auctus et illustratus* (Amsterdam: Ex officina, 1628). Reproduced by permission from the National Library of Scotland.

captains and Europe's mapmakers, but there was no agreement as to its exact position because there was no one position to be exact about.[7]

Some even thought that the agonic meridian was the basis for that dividing line that separated the colonial reach of Spain and Portugal in South America. The distinguished nineteenth-century geographer Alexander von Humboldt interpreted the actions of the Spanish-born Pope Alexander VI, under whose aegis South America was divided and ruled along a line 370 Castilian leagues west from Sant Antão on the Cape Verde Islands, as being based on an understanding of magnetic variation. To Humboldt, the pope had rendered "an essential service to nautical astronomy and the physical science of terrestrial magnetism." The truth is more prosaic: This line of continental demarcation is the result of imperial politics, not papal authority. It is the result of the Treaty of Tordesillas in 1494, at which the Spanish and Portuguese delimited their zones of influence in the New World. Because different Cape Verde Islands were chosen as the baseline for the earth's magnetic zero—sometimes Sant Antão, sometimes Fogo or Bonavista—the supposed 370 Castilian leagues could not be measured accurately. The ruling "line" that resulted from the Treaty of Tordesillas reflects the early modern Iberian geopolitical imagination more as a zone of intention than as a fixed line of global politics rooted in longitudinal certainty. However much the compass seemed unwavering near the Azores or Cape Verde Islands, the problem of accurately replicating on maps and globes the observations taken from various places meant that nature's inconstancy provided no single secure base for global measurement.[8]

These issues—Ptolemaic conceptions of the prime meridian as marking the world's western margin; the seeming invariance of magnetic declination, which, although practically encountered, was difficult to fix longitudinally; and the positioning of the prime meridian as a matter of users' convenience—were recognized and summarized in 1622 by the leading Dutch cartographer Willem Blaeu:

> Although the prime meridian is arbitrary, nevertheless it seemed good to the ancients to place it in the west because there a boundary of the world existed, whereas no expedition to the east could ever discover one. And for that reason Ptolemy (to whose zeal and industry all owe the soundness of geographical science though they do not all admit it) started with the furthest known limit of the west, called the Fortunate Islands in the Atlantic Ocean, and placed there his prime meridian, which disputable starting point almost every one has kept out of respect for his authority. But in our days a good many think this starting point ought to be based on nature itself, and have taken the direction of the magnetic needle as their guide and placed the prime meridian where that points due north. But that these are under a delusion is proved by that additional property of the magnetic needle through which it is no standard for the meridian, for itself it varies along the same meridian according as it is near one land mass or another. But those who agree that on account of its instability the magnetic needle is of no use, disagree about the prime meridian. Wherefore, in order that, for the greater convenience of Geography,

one certain meridian may be kept and observed as the first commencement, we, following in the steps of Ptolemy have chosen the
same islands and in them Juno, commonly called Tenerife, whose
lofty and steep summit covered with perpetual cloud, called by the
natives El Pico, shall mark the prime meridian.[9]

In effect, European voyages of global discovery between the 1430s and
the 1590s established both the importance and the difficulty of longitude's
measurement, particularly at sea, but did nothing to determine where the 0°
for such measurement should be. Astronomical observation and the related
ephemerides that court astronomers and others produced used different first
meridians. Such variations in the terrestrial start point did not present individual practitioners with any difficulty. However, when used comparatively,
they were the root cause of that constant recalculation and of the geographical and astronomical confusion over different 0° that was such a distinctive
geographical feature of the early modern world.

By the later sixteenth and early seventeenth centuries, at least four
principal geographical prime meridians were in use in the minds, texts,
and maps of European mathematicians, geographers, and mapmakers. The
prime meridian at Cap Ferro, or Ferro, the westernmost of the Canaries, was
a Ptolemaic conception. For Ptolemy, this was the world's western margin.
If we may take Blaeu literally, Ferro continued to be used in deference to
Ptolemy. Tenerife, also in the Canaries, was understood and used as a prime
meridian, not least because the volcanic peak of El Pico (today, Teide) was
recognizable at a distance by sailors. And on the basis of the seeming constancy of terrestrial magnetism, other navigators used either Saint Michael's
or the Corvo Islands in the Azores or Sant Antão, the westernmost of the
Cape Verde Islands. At the same time astronomers used diverse other prime
meridians as the basis for the various printed ephemerides. Before the
mid-seventeenth century, Europe's mapmakers employed different prime
meridians at different times on different maps according to either their
adoption or rejection of Ptolemy; their acceptance or not of navigators' reports of magnetic variability; and because, simply, there was no common
understanding, either in natural philosophy or through political agreement,
by which any one first meridian, geographical or observed, might be deemed
"better" than another (Table 1.1).[10]

Seen in relation to this wider context, King Louis XIII of France's decision in July 1634 to ratify Cap Ferro in the Canaries as the prime meridian
of choice for France might be interpreted as a political decision born of

Table 1.1 The prime meridian in the work of selected European cosmographers and mapmakers, c. 1507–c. 1688

Date and author of work		Location of prime meridian
1507, c. 1515–c. 1528	Quirini Green Globe[a]	Cape Verde Islands
1508	João de Lisboa, map	Azores
1538, 1554	Gerardus Mercator, maps	Canary Islands (Ferro)
1547	Fernandez de Oviedo, map	Azores
1564	Abraham Ortelius, map	Tenerife
1569	Gerardus Mercator, map	Cape Verde Islands
1570	Abraham Ortelius, map	Azores, Cape Verde Islands
1594	Thomas Blundeville, *His Exercises*	Azores
1601	Jodocus Hondius, globe	Azores
1612	Petrus Plancius, globe	Azores
1622	Willem Blaeu, atlas	Tenerife
1656	Nicolas Sanson, map	Ferro
1679	Pierre du Val, map	Ferro
1688	Vincenzo Coronelli, globe	Ferro

Source: W. G. Perrin, "The Prime Meridian," *Mariner's Mirror* 13 (1927); Horace E. Ware, "A Forgotten Prime Meridian," *Publications of the Colonial Society of Massachusetts* 12 (1908–1909); Art R. T. Jonkers, "Parallel Meridian: Diffusion and Change in Early-Modern Oceanic Reckoning," in *Noord-zuid in Oostindisch perspectief*, ed. J. Parmentier (The Hague: Walburg), table 1, 13. On the Quirini Green Globe, see Monique Pelletier, "Le Globe Vert et l'Oeuvre cosmographique du Gymnase Vosgien," *Bulletin du Comité Français de Cartographie*, 2000, 163, 17–31.
[a]Author unknown. Schöner or Martin Waldseemüller?

geographical necessity. Ancient authority was being affirmed, as it were, in the face of inconsistent contemporary empirical evidence. Louis XIII's decision followed a meeting in April 1634 of naval men and mathematicians, under the direction of Cardinal Richelieu. At the meeting, the French authorities also took note of the views of García de Céspedes, since 1596 the Cosmographer Major to Spain's Council of the Indies and an advisor to both Phillip II and Phillip III of Spain on proposals to determine longitude at sea, who remarked: "Que la ligne du vrai méridien devoit passer par les Canaries et particulièrement par l'île de Fer" (the line of the true meridian passes through the Canaries and in particular the Isle of Ferro).

In fact, the decision of 1634 was rooted in the politics of war with Spain, not in Ptolemaic authority or in the words of a Spanish courtier-astronomer. The French declaration regarding Cap Ferro was part of contemporary diplomacy over maritime conflict in relation to the Thirty Years' War (1618–1648), with the decree stating that French ships were not to attack Spanish or Portuguese shipping in waters that lay to the east of this first meridian and north of the Tropic of Cancer. This view over the demarcation line for maritime hostilities reflected contemporary intellectual debates over the legal jurisdiction of the oceans. In his *Mare liberum* (*The Free Sea*) of 1609, the Dutch legal philosopher Hugo Grotius developed the idea of the "open ocean," with all seas free to all nations. Others disputed this, including the English philosopher John Selden, whose 1635 *Mare clausum* advanced arguments for limitations upon the sea, as for land-based territory. King Louis XIII's decision in 1634—"the King in consequence forbids all pilots, hydrographers, designers or engravers of maps or terrestrial globes to innovate or vary from the ancient meridian passing through the most westerly of the Canary Islands"—was the result not of geographical consensus but the pragmatic approach of absolutist kings.[11]

Against this evidence of French realpolitik and the positioning of the prime meridian in the work of early modern cosmographers, we must also weigh the practices of oceanic navigators. Studies in the intellectual history of geomagnetism have made use of ships' logbooks to reveal onboard practices with regard to longitude and the different 0° baselines employed. Such navigational practices provided the means to test—and, over time, to disprove—belief in an Atlantic agonic meridian. They also varied from nation to nation. An examination of 536 logs from Dutch ocean-going vessels in the period 1598–1800, including ships of the Dutch East India Company, admiralty vessels, merchants, traders, and whalers, shows that later Dutch mariners tended not to use the agonic prime meridian that Mercator and other Low Country cartographers favored (see Table 1.1). The principal prime meridian used by the Dutch maritime community was not the Azores but the Canaries. Given Blaeu's observations, this shift may have begun as early as 1622. It was certainly taken up, more or less uniformly, by about 1675. For the historian of geomagnetism Jonkers, "This was not some belated honorary recognition of Ptolemy's work, but due to purely practical considerations. Tenerife's Teide volcano [the "El Pico" in Blaeu's commentary of 1622] rises an impressive 3,718m above sea-level, making it by far the highest peak for hundreds of miles around. . . . In addition, the Canaries had the advantage over the other two groups [the Azores and Cape Verde Islands] in being most easily reached from Europe, favoured by northeast tradewinds and Canaries Current, and ideally located as waystation for Atlantic crossings to north America and the West Indies, African destinations, and the routes to south America and the East Indies."[12]

For the French, the effect of their king's unilateral declaration regarding Cap Ferro was variable. Contrary to the view of one commentator that Louis XIII's 1634 declaration "had the effect of encouraging the use of Ferro," there were marked differences over time within the French maritime community. Little is known of what baseline French (or other) ships used in the early seventeenth century, but an examination of 468 French logbooks in the period c. 1670–1789 shows that it was not Ferro that predominated in French oceanic navigation before about 1750 but Tenerife (evident in 103 French logbooks in this period) and the Cape Verde Islands (from which 135 French ship's logs took their prime meridian).

It was ultimately the ship's itinerary, not the chart or the logbook, that determined the choice of prime meridian. The evidence of ships' logs points to differences in customary usage regarding the prime meridian within as well as among national maritime communities. Logs show that French ships bound for the East Indies favored the Cape Verdes; those on triangular voyages across the Atlantic—from France, to West Africa, to the Americas, and back to France—used the Canaries as their base meridian. In French cartography, by contrast, as Figure 1.2 shows, the Ferro prime meridian remained the dominant choice as an initial meridian even decades after Paris had been adopted as the nation's longitudinal baseline in the early 1720s. On board French ships, although Mercator charts and Dutch sea atlases were available, "no single prime meridian ever reached the coveted status that Ferro was supposed to have, but never attained. In this respect, French navigational practices thus differed thoroughly from the Dutch example, up to the 1750s." The transition in French maritime culture away from Tenerife or the Cape Verdes to the use of Paris as the baseline was not complete until the second half of the eighteenth century.[13]

Two hundred and fifty years after his pronouncement upon Cap Ferro, Louis XIII's words were to seem belatedly prophetic. French delegates at the 1884 International Meridian Conference in Washington, DC, cited the 1634 French royal decree as a precedent for a neutral international prime meridian (see Chapter 5). Even as they did so, however, they overlooked the overtly political context of the king's ruling, ignored the navigational practices of French mariners and others, and made light of the problems that Louis XIII's decision caused later generations of land-based French courtly astronomers and geodesists working to determine Cap Ferro's geographical and longitudinal position vis-à-vis the prime meridian of Paris.[14]

FIGURE 1.2 This inset of a map of Europe in 1757 by the leading French cartographer, D. Robert de Vaugondy, clearly shows the line of the "Premier Méridien" as adjudicated by France's King Louis XIII in the edict of 1634. By this time Paris had been adopted as France's prime meridian, based on the Paris Observatoire. *Source:* Robert de Vaugondy, *Atlas universel* (Paris: Quai de l'Horloge du Palais, 1757). Reproduced by permission from the National Library of Scotland.

Longitude, National Identity, and the Prime Meridian, c. 1667–c. 1767

In the century between the building of the Paris Observatoire in 1667 and the solution to the longitude problem proposed by Britain's Astronomer Royal, Nevil Maskelyne, in his *Nautical Almanac and Astronomical Ephemeris for the Year 1767*, the different prime meridians ruled and used by astronomers, geographers, and practical navigators became more directly and distinctively associated with national identity. Discussions concerning the prime meridian and the solutions proposed to the longitude problem became more evidently "modern." The prime meridian was studied within formal scientific institutions, commonly observatories, academies, and learned societies. Debates were couched in the mathematical and rhetorical languages of observational astronomy, topographic mapping, and scientific cultures aiming at precision in method, plainness of expression, and accuracy of outcome. The prime meridian became, too, a subject of debate in less formal spaces, such as coffee houses, and of satire in verse and newsprint.[15]

In these terms the historical geography of the prime meridian in the century after 1667 is one illustrative element within what later scholars term the "Scientific Revolution." The prime meridian, and its close cousin, longitude, was a matter of scholarly disputation and public comment as well as of evident differences in regard to tacit expertise and practical usage. The prime meridian also became intimately associated with the seafaring and earth-ruling colonial activities of the eighteenth-century fiscal-military state. Its different positioning and use was examined by verifiable experimentation and repeatable calculation, and its uses were treated in printed works claiming scientific authority. Questions of accuracy and precision were central to geodetic inquiry. The depiction of different prime meridians on maps and charts was bound up with what Edney calls "modes" of Enlightenment cartographic and metrological culture concerned with topographical and longitudinal accuracy.[16] These elements of context are important in explaining the two themes explored here: the emergence, in the first half of the eighteenth century, of Paris as the site of France's observed and geographical prime meridian and the role of different national prime meridians in disparate countries in providing a solution to the increasingly important longitude problem.

Knowing France, Fixing Paris

The establishment in 1666 of the Académie des Sciences (at founding, the Académie Royale des Sciences de Paris) under the direction of King Louis XIV and Jean-Baptiste Colbert and the building of the Paris Observatoire a year later constituted a new beginning for natural philosophy in France in general and for the prime meridian in particular. Astronomical and topographical studies were advanced, initially under the guidance of Colbert and from 1671 until 1712 under the direction of the Italian mapmaker and mathematician Giovanni Domenico Cassini, usually known as Jean-Dominique Cassini or Cassini I. Related astronomical and terrestrial measurements by Jean-Félix Picard, author of *Mesure de la Terre* in 1671, underlay the annual production in France, from 1679, of the *Connaissance des temps*, the Paris-based ephemeris containing tables giving the calculated position of the sun, moon, and other heavenly bodies and textual commentary and other observations by mathematicians and astronomers. From the early 1680s, topographical measurement in France was held to ever-stricter standards using coordinated plans of astronomical calculation, trigonometry, and triangulation. The work of Philippe de la Hire, Picard, and Cassini I and the later mapping of France under the work of the Cassini "dynasty"—for so we may describe the aggregative work of Cassini I; his son Jacques Cassini (Cassini II); his son, César-François Cassini de Thury (Cassini III), known also as Cassini de Thury; and Jean-Dominique, comte de Cassini (Cassini IV)—exemplifies better than for any other country during the Enlightenment the power of mapping to shape national identity through territorial representation.[17]

This new beginning for science and for France—for the idea of France as a territory and a polity produced through topographical science—at once encountered two problems. The first was that the initial calculations emerging from this enlightened cartographic national self-determination "shrank" France by several hundred square miles, notably, on the nation's western and Mediterranean coasts. The second was the persistent difficulty of accurately fixing the prime meridian baseline against which France might be measured and of calculating where in longitudinal terms Paris stood in relation to Cap Ferro. Colbert, Picard, and others cartographically produced, and reduced, their nation, and in so doing they were the first to show Paris as France's first meridian (Figure 1.3). For these measurements to be accurate, Paris needed to be fixed in relation to Cap Ferro, France's existing prime meridian. But where were the Canaries as an island group, relative to France? Where, exactly, was Cap Ferro? Answers

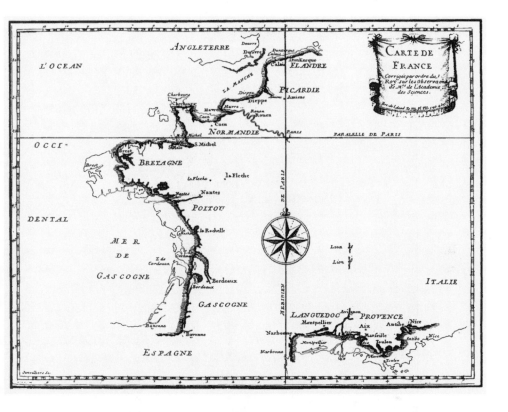

FIGURE 1.3 France, geodetically reduced. This map, engraved in 1693, shows (*thin and faint lines*) the extent of France as delineated by the French mapmaker Nicolas Sanson (1660–1667) and (*bolder lines*) the extent of the nation following its geodetic measurement by Jean-Baptiste Colbert and other members of the Académie des Sciences. Source: "Carte de France," *Mémoires de l'Académie Royale des Sciences depuis 1666 jusqu'à 1699, tome VII* (Paris, 1729). Reproduced by permission from the National Library of Scotland.

to these questions were evident in work that topographically refashioned France, produced new national limits, and prompted a distancing from the edict of 1634.

This edict, as academicians and others came to understand, had created a clear problem. Based as it was upon one imprecisely located classical prime meridian and upon a royal whim determined by politics, not precision, the edict offered no basis for accurate astronomical observations or mathematical calculation of terrestrial limits. Because the Ptolemaic prime meridian of Cap Ferro was not fixed with accuracy, longitudinally speaking,

FIGURE 1.4 Enlightenment France, regulated and ordered. It is realized and represented in a series of measured triangles with the nation's baseline and prime meridian clearly showing the "méridien de l'Observatoire Royal de Paris."
Source: César-François Cassini de Thury, "Nouvelle Carte . . . de la France," 1744. Reproduced by permission from the National Library of Scotland.

it could not be France's 0° in anything other than a symbolic way. The essential distinction between an observed astronomical prime meridian and a measured or geographical prime meridian could not be properly made without reference to a known site, the observatory in Paris, as the basis for the nation's measurement. Paris, and not an island or an island group somewhere to the west of France, had to be the nation's baseline. But how should one "fix" Ferro in relation to Paris and Paris in relation to the rest of France?

In addressing these issues between the late seventeenth century and the mid-1740s, France's astronomers, mathematicians, and geographers produced the first modern corpus of material specifically focused upon the prime meridian as a means to promoting national identity through science. It is evident in a remarkable cumulative genealogy of observational recordings, correspondence, debates, maps, and texts. In addition to Picard's work and the map of 1693, it includes inquiries in 1702 by the philosopher-mathematician Jean-Joseph La Montre, who distinguished between the geographical and astronomical meridians in terms of their relative importance and accuracy, and Philippe de la Hire's 1704 *Description et explication des globes.* It includes a lengthy memoir of 1722 from Guillaume Delisle, who held the title of Géographe de l'Académie from 1702, and, from 1718, was Géographe du Roi to the young King Louis XV, during which he proposed a figure of 20° of difference between Cap Ferro and the Paris Observatoire. These positions of civic authority and royal patronage lent Delisle's geographical views considerable weight.

The work of Louis Feuillée in 1724 is a further vital element in this corpus. His voyage to the Canaries to determine firsthand the longitude of Cap Ferro is the first geodetic expedition aimed at determining the position of the prime meridian; contra Delisle, he computed the difference in longitude to be only a little over 19°. The culmination is Cassini de Thury's *La méridienne de l'Observatoire Royal de Paris, vérifiée dans toute l'étendue du royaume par de nouvelles observations* of 1744. Known usually as *Méridienne vérifiée,* this is an encyclopedic inquiry running to 292 pages, with a further 250 pages of supporting tables of astronomical and longitudinal calculations and maps depicting France's "emergence" through the medium of trigonometry (Figure 1.4). As a result of all of this and other work, Paris between c. 1693 and c. 1744 moved to and became the center of France's increasingly self-confident national calculation; its prime meridian not just *le meridian,* one prime meridian among the geographical and astronomical many, but *la méridienne, the* prime meridian—the nation's ruling line and, quite possibly, the world's.[18]

Asserting Accuracy

Beneath these expressions of accuracy and national self-determination lay a further problem. Statements about accuracy also express an intolerance of error. Because of its convenience for calculation, it was Delisle's figure of 20° for the longitudinal difference between Paris and Cap Ferro—based as it was on "calculation and good sense"—that was used as the difference between these two prime meridians and not Feuillée's more accurate figure of 19°55′3″ west of Paris. This continued to be the case even into the nineteenth century, despite information provided after 1720 by later expeditions and the work of the Dépôt des cartes et plans de la Marine. For example, revised figures following the work in 1742 of the distinguished astronomer Pierre Le Monnier—that "the actual figure" was 20° 2′ 30″—were overlooked in favor of the expediency of national calculation using only an approximate figure.

In an important sense, the authority of the Paris prime meridian, *la méridienne,* lay in the scientific claims of those astronomers, geographers, and natural philosophers who professed it. In Enlightenment France, their advocacy of accuracy was an epistemological and even a political necessity. Achieving accuracy in the measurement of the earth was a cumulative process involving fieldwork, sedentary reflection, complex computation, and trust, both in the instruments used and in the authoritative figures doing the work. Accuracy—specifically, the claim to have "accurately" positioned Paris in relation to Ferro—was almost a form of moral judgment expressed in numbers. "Fix" the country's prime meridian relative to another prime meridian, and France's geographical limits would be firmly known, its capital astronomically determined, the Académie des Science's position as a state-sponsored institution strengthened, and its members' standing as authoritative men of science enhanced. For France between about 1720 and 1744—the second date marking what one authority has seen as a "monumental year" in the history of Enlightenment geodesy—this interpretation is lent support by the parallel endeavors of French academicians then engaged in terrestrial measurement in South America as part of expeditions testing competing theories about the shape of the earth.

Far away in the viceroyalty of Peru, as they sought to measure a degree of latitude at the equator, Charles de la Condamine, Pierre Bouguer, and others used mathematics as the language of choice to reduce the world to empirical order. Trigonometry and cartography were the particular medium of representation.[19] Fixing France's prime meridian was one expression of that nation's geographical self-determination, a principal chapter within a broader intellectual history of geodesy. The political value of the prime meridian, and this is a point to which I return in different contexts throughout this book, accorded considerable significance to accuracy. This was despite the difficulty of actually achieving it—concerns over the accuracy of Paris's position continued long after 1744—and the awkward fact that sailors still tended to be guided by pragmatism rather than precision.

Determining Longitude in Britain

In Britain the study of the problem of the "prime meridian" has, to a degree, been overshadowed by that of the longitude problem, notably, in the intellectual and institutional history of the British navigational sciences in the century before the 1767 publication of Maskelyne's *Nautical Almanac and Astronomical Ephemeris*. It is, of course, specifically because fixing precise longitude at sea is so vital but so difficult that the locations of different prime meridians continued as a matter of geographical confusion, within and among nations, communities of chart- and mapmakers, oceanic navigators, astronomers, and geographical authors.

We know that the adoption of Greenwich as Britain's effective observed prime meridian for oceanic navigation as well as significant advances in text-based and instrumentally mediated astronomical calculations solved the longitude problem in the later 1760s. We know that solving the longitude problem featured within the Royal Society's work from its foundation in 1660, as it did with the Royal Observatory at Greenwich from its foundation in 1675. In this, there are obvious parallels with France. But there is a danger, too, that in our knowing the ultimate answer to the longitude question in Britain—Greenwich, the conjunction there from 1767 of one geographical and observed prime meridian, and the authority of the *Nautical Almanac and Astronomical Ephemeris*—we discount the role that prime meridians other than Greenwich played as the problem was worked out. There is, additionally, the danger that we overlook quite how contemporaries, both at sea and on land, dealt with and understood the practical problems arising from the existence of multiple prime meridians. Nevertheless, with these dangers in mind, it is important to understand the nature of the quest to determine precise longitude.

In order to calculate longitude and so know one's position west or east of a given position on the earth's surface, it is necessary to compare the local time with that of a given meridian. Until the invention of the telegraph in the nineteenth century and its use in determining longitude and calibrating the accuracy of different prime meridians—topics addressed in later chapters—there were four ways to do this. The first was by observing eclipses: the eclipses of the satellites of Jupiter were preferred, given their frequency, although eclipses of the moon had long been known about and used. The second was by observing the position of the moon in relation to the stars, principally, through lunar occultation—that is, as the moon comes between the observer and more distant celestial objects. The third was by observing the moon's passage across the local given meridian. These three astronomical methods, each theoretically possible, in practice required consistent observations, complex computation, and some form of tabulation or written register—a predictive ephemeris or equivalent catalog of the stars and tables of the moon's motion—produced for the chosen meridian. Difficult enough to do on land, each of these three methods was seriously impracticable at sea. The fourth method was by knowing the time of the given initial meridian by means of a clock that either never varied or varied only at a known rate. This method was practicable at sea provided a timepiece could be invented that kept regular time as it, and the ship on which it traveled, moved from place to place. Dava Sobel has stressed the instrumental expertise of the clock maker John Harrison in producing, in improved versions between 1730 and 1773, what became the marine chronometer, capable, once set to a given meridian, of keeping time across long sea voyages. Harrison's "seawatch," or timekeeper, a technical solution to the longitude problem and calibrated to Greenwich mean time, represents the work, according to Sobel, of a "lone genius who solved the greatest scientific problem of his time."[20]

More recent accounts of the solution to the longitude problem reveal it to be more complicated: an iterative and accumulative process involving a German astronomer-geographer, French geodesists, testing voyages to the Caribbean, and British printed tabulations—each different "instruments of science," to be sure, but collective testimony to a more complex solution than individual genius and the marine timepiece alone. The process began with the building of the Greenwich observatory in 1675 and the appointment of the Astronomical Observator and first Astronomer Royal, John Flamsteed. Flamsteed, who knew Picard and de la Hire's work, conducted experiments there to ascertain whether the earth rotated at a constant speed (a vital assumption in calculating longitude by celestial observation) and established Greenwich time as the temporal base for the terrestrial measurement of astronomical phenomena.

Even as he was doing so, and as the French were working to site and authorize Paris as their prime meridian, Henry Oldenburg, secretary to the

Royal Society in London, received correspondence in August 1675 from Juan Cruzado, professor of mathematics in Seville and "captain in chief" to Spain's Council of the Indies. As part of a letter on longitude and other shared interests, Cruzado proposed an altogether different prime meridian for global use in order "to bring about agreement among geographers." His proposal centered upon an island near Brazil. According to Cruzado, it lay exactly on the equator, 9° west from the meridian of the peak of Tenerife and 42° west from Uraniborg: "In it there is to be found everything conducive to [the determination of] the prime meridian, as I judge. . . . It is proper, for precision, that the location of this prime meridian should be small, so that the numerical value of the longitude may be expressed more exactly [that is, should be accurately determined]." Cruzado's motives were directed at the collective good but perhaps also at furthering his own reputation: the island in question, formerly named Abroxos, he had named Cruzado, apparently from its cruciform shape but perhaps for himself. "I should be very grateful," he entreated Oldenburg, "if your worship would lay my opinion and discovery before a meeting of the Royal Society so that, if it is judged useful to them, there may afterwards be universal agreement among geographers and I may congratulate myself on having brought something acceptable to the distinguished Royal Society." Flamsteed, much pressed "on account of lack of time," wholly ignored Cruzado's remarks on the prime meridian, commenting, and then via Oldenburg, only on the Spaniard's astronomical calculations: the Royal Society was not called upon to debate Cruzado's alternative option.[21]

By the early eighteenth century, and because in Britain "finding the longitude" was of public as well as of scientific concern, commissioners of longitude were appointed in 1714 with a reward, on a sliding scale, to be given to those who could fix time at sea with accuracy: £10,000 was available to "the First Author or Authors, Discoverer or Discoverers . . . of the said Longitude to One Degree of a great Circle, or Sixty Geographical Miles," with £15,000 to those who could calculate accuracy to "two-thirds of a degree" and £20,000 for accuracy "to within half a degree." The work of this body—which, in the 1760s, became the Board of Longitude—has recently been the subject of detailed study.[22]

In Britain several prime meridians other than Greenwich were put to work in a variety of proposals as contemporaries debated how best to solve the longitude problem. Where these proposals have been noticed at all, they have, until recently, been dismissed because they advocated solutions to the longitude problem that were, for one reason or another, practically unworkable.

But if our focus is upon the prime meridian rather than longitude, such works offer not error but insight into how, why, and by whom certain prime meridians were used: the point is not in the fact that such "bizarre" schemes would not have worked but in the intellectual legitimacy afforded different prime meridians as they featured in the solutions advanced.

Consider in this context the proposal put forward in 1714 by William Whiston, Lucasian Professor of Mathematics at Cambridge between 1703 and 1710, and Humphry Ditton, master of the Royal Mathematical School in Christ's Hospital, London, in their *A New Method for Discovering the Longitude both at Sea and Land*. Their scheme involved a combination of sight and sound and was prompted in part from having witnessed and heard the fireworks displays marking the Peace of Utrecht in 1713, which ended the War of the Spanish Succession. It depended chiefly upon the firing of rockets from a single elevated point and a network of ships from which to launch and observe these projectiles—"large Shells," as they put it—being fired to a height of exactly 6,440 feet above sea level and at exactly midnight. "Fire or Light 6440 Feet high," they argued, "will be visible, in the night time, when the Air is tolerably clear about 100 measured, or 85 Geographical Miles: i.e., one whole degree, and 25 minutes of a great Circle, from the place where it is, even upon the surface of the sea." Compass bearings would give the direction of the respective ships. "Hulls of Ships, without Sails or Rigging" would be moored at regular intervals, even in deep water, thus producing a network of observation points that were also firing points. The distance of the observer from the lightship would be obtained either by noting the difference in time between seeing the rocket's glare and hearing the explosion or by calculating the elevation of its highest point. The "Hulls or ships" would be "fixed in proper Places as to Latitude . . . and as to Longitude, by Eclipses of the Sun, or Moon, or of Jupiter's Planets, or by the Moon's Appulse to fixed Stars; or rather by an actual Mensuration of Distances on the Surface of the Sea by Trigonometry, just as Monsieur Picard and Cassini measur'd the Length of a Degree of a great Circle on the Land." The several advantages listed for this scheme began by emphasizing that "this Method requires no Depth of Astronomy, no Nicety in Instruments, and but seldom any Celestial Observations at all, either as to the Latitude, or the Hour at the Ship; and so is to even the common Sailors the *most Practicable*."

What distinguished Whiston and Ditton's "big bang" longitude proposal was their choice of the peak at Tenerife as its 0° baseline: "It is farther humbly propos'd to the Learned, Whether it may not be proper for all Nations, upon this Occasion, to agree upon one *first Meridian*, or beginning of

Longitude, for the common Benefit of *Geography?* And whether it may not be proper, in that Case, to fix it to the *Pike of Tenariff,* as the most noted Place already; and as the place whence the Highest and most generally useful Explosion must in this Method be made every Midnight continually for the Discovery of the *Longitude* itself."[23]

The fact that Tenerife was a commonly understood and used prime meridian, chiefly by the Dutch, was clearly relevant. It is interesting that a classical prime meridian should be appealed to instead of Flamsteed's or Cassini's observatory sites. At least a prime meridian was given: other schemes, such as the anonymous *An Essay Towards a New Method to Shew the Longitude at Sea* of 1714, had no such base point. This scheme proposed the establishment of a series of large lighthouses, lit by fires in their lower floors, from which, through sets of mirrors, a tower of light would be projected into the sky to reflect off the clouds: "If the Light would reach two Miles high, with sufficient Strength, the Orbicular Form of the World would allow its visibility about 200 [miles]." Shutters drawn over the light at agreed intervals would allow sailors to determine their longitude "if the Sailor knows the time of Night where he is."[24] Unworkable as this and Whiston and Ditton's schemes were, the idea of points of light, moored ships, and networks of light from a fixed prime meridian as the basis to global navigation would be returned to, over 160 years later, in the words of William Parker Snow, whose own too-close encounter at sea with ships on a course derived from different originating prime meridians prompted, as we have seen, a similar never-realized scheme for global navigational safety.

For the scientific writer Jane Squire, the solution to longitude lay in a series of celestial star charts, which she termed "cloves," that navigators would memorize. Her chosen prime meridian was the manger at Bethlehem: "Beginning our Account of Longitude at *Bethlehem,* and in that Part of the Heavens which was Zenithal to it, at the *Time Our Lord Jesus Christ vouchsafed there to be born for us,* will indulge our Christian Gratitude, and by fixing our Æra of Place, with that of our Time, facilitate the Calculation of all Observations from it." Squire's proposal—a "simple Easy Method," as she put it—appeared first in 1731 and, in expanded form, in 1742. In its proposition regarding Bethlehem as the "Terrestrial first Meridian," for which she proposed a Bethlehem ephemeris, her work is consistent with a particularly Christian interpretation of universal history and scriptural chronologies characteristic of works of Enlightenment historical inquiry. Her proposal would have echoes in later years: discussions at Washington and elsewhere a century and a half later addressed the merits of Jerusalem as the world's prime meridian (see Chapter 6). In her own time, Squire's proposal was almost wholly ignored, partly for its impractical nature and partly because as a woman Squire was outside the "legitimate" circles of those men puzzling upon longitude's solution. What matters is not that it was ignored or that, in its dependence upon the memorization of thousands of star charts, Squire's scheme was as unworkable as that of Whiston and Ditton's Tenerife-based son et lumière. The importance of these schemes rests in the evidence they present of different prime meridians being used to ameliorate that confusion in geography of which they were in part the root cause.[25]

Solutions to the longitude problem came through the international combination of words, mathematics, and at-sea instrumental testing with one particular prime meridian in mind. The simultaneous invention in 1731 of the double-reflection quadrant, forerunner to the sextant, permitted the accurate observation of celestial bodies in relation to one another and in relation to the horizon, from on board moving ships. In 1755 Tobias Mayer, the German astronomer, lecturer in geography, and mapmaker, sent notice of his lunar tables, first produced in Göttingen in 1752, to the British Admiralty, from which body they reached the Board of Longitude in 1756. Nevil Maskelyne tested Mayer's lunar tables, using the quadrant, in a voyage to Saint Helena in 1761. He achieved an accuracy of longitude of better than 1°; that is, he was fewer than sixty nautical miles out in his calculation of linear distances. In 1763 Maskelyne published *The British Mariner's Guide*, which laid out the principles of lunar-distance observations. In the guide Maskelyne calculated all the longitudes "from the meridian of the Royal Observatory at Greenwich, which is only five minutes thirty-seven seconds of longitude to the east of St. Paul's Church in London" (Figure 1.5). Using a second set of revised and more accurate tables from Mayer, Maskelyne tested the German's lunar tables and the accuracy of John Harrison's H4, the fourth version of Harrison's marine timekeeper, on a voyage to Barbados in 1763, all as an official trial for the Board of Longitude. From 1765 Maskelyne drew together proposals for a nautical ephemeris based upon a range of observations at the Royal Observatory. *The Nautical Almanac and Astronomical Ephemeris for the year 1766* was finally assembled and available for sale in January 1767 (Figure 1.6). This was a key moment in the solution to the longitude problem, especially when used at sea in combination with a marine timekeeper. *The Nautical Almanac and Astronomical Ephemeris* gave tables of lunar distances with predicted angular distances from the center of the moon to selected stars or to the center of the sun. Although the longitudinal calculations to be undertaken were not simple, they were now reduced in

[XXIX]

A TABLE

Containing the Longitudes of Places that have been determined by Aftro-nomical Obfervations, reckoned from the Meridian of the Royal Ob-fervatory at *Greenwich* ; and alfo their Latitudes.

Names of Places.	Con-tine.	Country.	Coaft or Province.	Latitude. ° ' ''	Longitude. In Degrees. ° ' ''	Longitude. In Time. H. M. S
Abbeville	Eur.	France	Picardy	50 7 1 N	1 49 45 E	0 7 19
Abo	Eur.	Finland	Baltic Sea	60 27 10 N	22 13 30 E	1 28 54
Achem	Af.	N. W. Pt. Ifl. Sumatra	Indian Ocean	5 22 N	95 34 E	6 22 16
Agra	Af.	India	Moguls	26 43 0 N	76 44 0 E	5 6 56
Aix	Eur.	France	Provence	43 31 35 N	5 26 15 E	0 21 45
Alby	Eur.	France	Languedoc	43 55 44 N	2 31 15 E	0 10 5
Alexandretta	Af.	Syria	Mediter. Sea	36 35 10 N	36 20 0 E	2 25 20
Alexandria	Af.	Ægypt	Mediter. Sea	31 11 20 N	30 16 30 E	2 1 6
Amiens	Eur.	France	Picardy	49 53 38 N	2 18 0 E	0 9 12
Ancona	Eur.	Italy	Mediter. Sea	43 37 54 N	13 30 30 E	0 54 2
Angers	Eur.	France	Orleanois	47 28 8 N	0 33 45 W	0 2 15
Angoulême	Eur.	France	Orleanois	45 39 3 N	0 8 45 E	0 0 35
Antibes	Eur.	France	Mediter. Sea	43 34 50 N	7 8 30 E	0 28 34
Antwerp	Eur.	Flanders	River Scheld	51 13 15 N	4 24 15 E	0 17 37
Archangel	Eur.	Ruffia	White Sea	64 34 0 N	38 5 50 E	2 35 40
Arica	Am.	Peru	South Sea	18 26 38 S	71 11 0 W	4 44 44
Arles	Eur.	France	Provence	43 40 33 N	4 38 0 E	0 18 32
Ifl. of Afcenfion	Af.	Angola	S. Atl. Ocean	7 57 0 S	13 59 0 W	0 55 56
Athens	Eur.	Turkey	Archipelago	38 5 0 N	23 52 30 E	1 35 30
Auch	Eur.	France	Gafcony	43 38 46 N	0 30 0 E	0 2 0
Aurillac	Eur.	France	Lionois	44 55 10 N	2 27 0 E	0 9 48
Auxerre	Eur.	France	Burgundy	47 47 54 N	3 34 15 E	0 14 17
Avignon	Eur.	France	Provence	43 57 25 N	4 48 30 E	0 19 14
Avranches	Eur.	France	Normandy	48 41 18 N	1 22 45 W	0 5 31
Antie. Babylon	Af.	Mefopotamia	Riv. Euphrates	33 0 0 N	42 46 30 E	2 51 6
Bagdad	Af.	Mefopotamia		33 21 0 N	43 46 30 E	2 55 6
Balafore	Af.	India	Bay Bengal	21 20 0 N	86 0 0 E	5 44 0
Bayeux	Eur.	France	Normandy	49 16 30 N	0 42 45 W	0 2 51
Bayonne	Eur.	France	Bay Bifcay	43 29 21 N	1 30 0 W	0 6 0
Great Bear Ifl.	Am.		Hudfon's Bay	54 34 N	79 56 0 W	5 19 44
Beavais	Eur.	France	Ifl. of France	49 26 2 N	2 4 45 E	0 8 19
Berlin	Eur.	Germany	River Elbe	52 32 30 N	13 26 15 E	0 53 45
Befancon	Eur.	France	France Compte	47 13 45 N	6 2 30 E	0 24 10
Beziers	Eur.	France	Languedoc	43 20 41 N	3 12 30 E	0 12 50
Cape Blanco	Am.	Patagonia	Atl. Ocean	47 20 S	70 5 0 W	4 40 20

FIGURE 1.5 The role of the Royal Observatory at Greenwich as a later eighteenth-century center of geodetic calculation is clear from this table providing the relative longitudinal positioning of places in the world "determined by Astronomical Observations." *Source:* Nevil Maskelyne, *The British Mariner's Guide* (London: printed for the author, 1763). Reproduced by permission from the Centre for Research Collections, University of Edinburgh Library.

Apparent Times of the external and internal Contacts of VENUS with the Sun's Limb at several Places, as they may be expected to happen on June the 3d 1769.

	1st ext. cont.	1st int. cont.	2d int. cont.	2 dext. cont.	Latitude.	Supposed Longitude from Greenw
	h. m.	h. m.	h. m.	h. m.		h. m. f.
Greenwich — —	7 6	7 25	—	—	51 29 N	0 0
Edinburg — —	6 53	7 12	—	—	55 58 N	0 13 13 W
Dublin — —	6 41	7 0	—	—	53 20 N	0 24 54 W
Tornea — —	8 43	9 2	14 56	15 15	65 51 N	1 36 48 E
Kittis — —	8 43	9 2	14 56	15 15	66 48 N	1 36 48 E
Attengaard — —	8 39	8 58	14 51	15 10	69 59 N	1 32 27 E
Wardhus — —	9 12	9 31	15 24	15 43	70 35 N	2 5 36 E
North Cape — —	8 51	9 10	15 3	15 22	71 23 N	1 44 48 E
Bear Island — —	8 12	8 31	14 24	14 43	74 32 N	1 5 36 E
Spitzbergen, Bell-Sound	7 57	8 16	14 8	14 27	77 15 N	0 51 2 E
Petersberg — —	—	—	15 21	15 40	59 56 N	2 1 20 E
Tobolski — —	—	—	17 53	18 12	58 12 N	4 32 51 E
S. John's, Newfoundland	3 40	3 59	—	—	47 32 N	3 31 13 W
Quebec — —	2 29	2 48	—	—	46 55 N	4 39 36 W
Hudson's Bay — —	0 49	1 8	6 54	7 13	58 56 N	6 19 40 W
Boston — —	2 26	2 45	—	—	42 25 N	4 42 29 W
Williamsberg — —	2 3	2 22	—	—	37 20 N	5 6 20 W
Jamaica, Port Royal	2 4	2 23	—	—	18 0 N	5 7 2 W
Mexico — —	0 19	0 38	6 14	6 33	20 0 N	6 54 40 W
Cape Corientes — —	23*48	0 6	5 44	6 3	20 50 N	7 25 40 W
Cape St. Lucar —	23*31	23*50	5 28	5 47	23 15 N	7 41 40 W
Cape Conception —	22*34	22*53	4 33	4 52	35 30 N	8 37 40 W
Fernambuca, Brazil	4 50	5 9	—	—	8 13 S	2 20 8 W
Conception, Chili —	2 24	2 43	—	—	36 43 S	4 50 44 W
Bombay — —	—	—	18 9	18 28	19 18 N	4 47 48 W
Madrafs — —	—	—	18 41	19 0	13 13 N	5 20 9 E
Calcutta —	—	—	19 14	19 33	22 30 N	5 53 43 E

N. B. The Contacts marked with Asterisks belong to the 2d Day of June 1769, according to Astronomical Time.

FIGURE 1.6 *The Nautical Almanac and Astronomical Ephemeris* was a work for future reference. It presented the predicted positions of the heavenly bodies for the coming year and could also be used for recurrent but infrequent astronomical events, such as the transit of Venus (here for 1769). Note the "supposed longitudes" of places listed with reference to Greenwich. *Source: The Nautical Almanac and Astronomical Ephemeris, for the Year 1769* (London: Commissioners of Longitude, 1768). Reproduced by permission from the Centre for Research Collections, University of Edinburgh Library.

DIFFÉRENCE DES MÉRIDIENS
entre Paris & les principaux lieux de la Terre.

LA Table qui donne les différences des Méridiens entre Paris & les principaux lieux de la Terre, est le résultat de toutes les observations que les Astronomes font depuis un siècle pour perfectionner la Géographie.

Les situations des lieux de la Terre se déterminent par latitudes & longitudes; la *latitude* est la distance d'un lieu de la Terre à l'Équateur, comptée depuis l'Équateur en allant vers le nord ou vers le sud; la *longitude* géographique ou terrestre est la distance comptée d'occident vers l'orient depuis le premier Méridien; & dans les Cartes françoises, le premier Méridien passe par les îles Canaries. Supposons à Paris 20 degrés de longitude.

Mais comme c'est à la méridienne de Paris que toutes nos observations se rapportent, & que toutes nos Tables y sont assujetties, nous nous sommes contentés d'indiquer la différence entre le Méridien de Paris & ceux des principaux lieux de la Terre.

Lorsqu'on a l'heure qu'il est sous le Méridien de Paris, & que l'on cherche l'heure sous un autre Méridien; s'il est à l'orient de Paris, il faut ajouter la différence des Méridiens avec l'heure de Paris; si c'est à l'occident de Paris, il faut la retrancher.

Les latitudes & les différences des Méridiens où il y a des étoiles* ont été déterminées par les observations de l'Académie; celles où il y a des croix † ont été déterminées par d'autres Astronomes; celles où il n'y a rien de marqué sont fondées sur l'estime, sur le rapport des Voyageurs, ou sur des observations moins certaines que les autres. Cette Table est suivie d'une autre que m'a procuré M. d'Après de Mannevilette: elle contient la position des principaux lieux de son Neptune oriental, & a été rédigée par l'Auteur même; ainsi il est à présumer que les Navigateurs lui sauront gré de la peine qu'il a bien voulu prendre à cet égard.

TABLE DE LA DIFFÉRENCE
des Méridiens en heures & degrés, entre l'Observatoire Royal de Paris & les principaux lieux de la Terre, avec leur latitude ou hauteur de Pole.

NOMS DES LIEUX.	Différ. des Méridiens		LATITUDES ou Hauteurs du Pole.
	en Temps.	en Deg.	
	H. M. S.	D. M.	D. M. S.
Abbeville.	0* 2. 1. oc.	0. 30.	50* 7. 1.s.
Abo. *Finlande*.	1† 19. 30. or.	19. 52.	60† 27. 7.
Agra *du Mogol*.	4† 57. 36. or.	74. 24.	26† 43. 0.
Aix *en Provence*.	0* 12. 25. or.	3. 7.	43* 31. 35.
Alby.	0* 0. 45. oc.	0. 11.	43* 55. 44.
Alep *de Syrie*.	2 20. 0. or.	35. 0.	35† 45. 23.
Alexandrete.	2* 16. 0. or.	34. 0.	36* 35. 10.
Alexandrie *Egypte*. . . .	1* 51. 46. or.	27. 57.	31* 11. 20.
Alger.	0 0. 29. oc.	0. 7.	36* 49. 30.
Amiens.	0* 0. 8. oc.	0. 2.	49* 53. 38.
Amsterdam	0 10. 36. or.	2. 39.	52* 22. 45.
Ancone.	0* 44. 42. or.	11. 11.	43* 37. 54.
Angers.	0* 11. 35. oc.	2. 54.	47* 28. 8.
Angoulême.	0* 8. 45. oc.	2. 11.	45* 39. 3.
Antibe.	0* 19. 14. or.	4. 49.	43* 34. 50.
Anvers.	0* 8. 17. or.	2. 4.	51* 13. 15.
Archangel.	2* 26. 20. or.	36. 35.	64 34. 0.
Arles.	0* 9. 12. or.	2. 18.	43* 40. 33.
Avignon.	0* 9. 54. or.	2. 29.	43* 57. 25.
Avranches.	0* 14. 51. oc.	3. 43.	48* 41. 18.
Aurillac.	0* 0. 28. or.	0. 7.	44* 55. 10.
Auch.	0* 7. 20. oc.	1. 45.	43* 38. 46.
Auxerre.	0* 4. 57. or.	1. 14.	47* 47. 54.
Barcelone.	0. 0. 28. oc.	0. 7.	41† 26. 0.
Basle.	0 21. 0. or.	5. 15.	47 55. 0.
Bayeux.	0* 12. 11. oc.	3. 3.	49* 16. 30.
Bayonne.	0* 15. 20. oc.	3. 50.	43* 29. 21.
Beauvais.	0* 1. 1. oc.	0. 15.	49* 26. 2.

FIGURE 1.7 France's overall aim of "perfecting geography" by using accurate longitudinal measurements centered upon the 0° of the Observatoire de Paris is made clear in the "Différence des Méridiens" and the "Tables de la Différence des Méridiens." *Source: Connaissance des temps, pour l'année commune 1777* (Paris: Imprimerie Royale, 1776). Reproduced by permission from the Centre for Research Collections, University of Edinburgh Library.

duration (at least supposedly) from over four hours to about thirty minutes and based upon a predictive text. Crucially, Maskelyne based his ephemeris on, and Harrison set his chronometer to, the 0° meridian of Greenwich.[26]

As significant as these events were to the longitude problem, there is not a straight and directly consequential line between Maskelyne's *Nautical Almanac and Astronomical Ephemeris*, the choice of Greenwich as its base meridian for navigational purposes, and the adoption of Greenwich as the world's prime meridian in 1884.[27] The evidence tells a more complex story. British mariners before the mid-eighteenth century differed from their Dutch and French counterparts in their use of a prime meridian. As the author of the 1703 nautical manuscript *Sayling by the True Sea Chart* noted, "The first meridian may be indifferently placed, anywhere, ther being nothing in nature on the Earth consider'd with relation it might have with the heavens, which obliges us to place it in one place rather than in another." This being so, "Our sea fareing men generaly speaking make the meridian they parted from the first meridian." If, as Jonkers remarks, English oceanic practice regarding the use of a prime meridian reflected "the more utilitarian principle of taking the last land sighted," there were differences, too, between the ships of the East India Company and the Royal Navy. The latter adopted a prime meridian based on London earlier and more consistently than did their East India counterparts. Works such as the Admiralty's *Regulations and Instructions relative to His Majesty's Service at Sea* (1731) make no mention of Greenwich or any single base meridian as the required start point to a ship's log, merely noting that differences in practice and that the reckoning of longitude should form part of the crew's journal keeping. Only from about 1780, as Harrison's marine timekeeper became more widely available, did Greenwich become commonplace in the British navy. Even then, the availability of instruments should not suppose their widespread usage. Thomas Brisbane, a military officer on an East India Company ship bound in 1795 for the West Indies, observed how "in that immense fleet there were perhaps not 10 individuals who could make a lunar observation." Such distinctions in practice help explain why no single prime meridian rose to early prominence in Britain. They also illuminate for Britain, as for the Dutch and the French, the differences within and between communities of practitioners before and after the appearance of Maskelyne's 1767 text.[28]

If we consider *The Nautical Almanac and Astronomical Ephemeris* in terms of its book history, two further points are noteworthy. The first is that practical navigation texts published in Britain in the 1770s, and many in the 1780s, do not make mention of *The Nautical Almanac and Astronomical*

Ephemeris or of the lunar-distance method. Some texts simply remark, "Meridian is where you take your departure from." In his *Synopsis of Practical Mathematics* (1771 and 1779), Alexander Ewing recommended taking the Lizard, in Cornwall, as the initial meridian, which would have made using *The Nautical Almanac and Astronomical Ephemeris* effectively impossible. The second is to note that Maskelyne's nautical almanac, which depended in part upon Mayer's tabulations for its structure, was not the first such national ephemeris. Greenwich's authority from 1767 as the site of Britain's prime meridian may be seen as a national achievement born of, and itself prompting, international cooperation, even in France. There, Abbé Nicholas Louis de Lacaille had demonstrated the value of lunar-distance observations and had incorporated them into the *Connaissance des temps* for 1761 using the prime meridian of Paris for his tables. From 1774 to 1788, the *Connaissance des temps* incorporated Maskelyne's lunar-distance tables, based on Greenwich, together with other, Paris-based, tables (Figure 1.7). The 1789 edition even used the British tables converted to the Paris meridian. The Greenwich tables were similarly included in the Spanish *Almanaque náutico* from 1792 and in other comparable national works. If the first point is to call for caution over *The Nautical Almanac and Astronomical Ephemeris*'s immediate effect in terms of navigational practices regarding the use of Greenwich as Britain's prime meridian, the second concerns that text's rather "hybrid" nature. We will see in Chapter 2 on the prime meridians in the early United States how Maskelyne's work was again used, in amended form, to structure debates over a new prime meridian for that newly emergent nation.[29]

Differences in National Regulation, c. 1767–c. 1790

Across Europe, on maps, in geography books, and in maritime navigation, numerous prime meridians existed in parallel as, respectively, the establishment of the Paris Observatoire and the publication of *Connaissance des temps* and the Royal Observatory at Greenwich and *The Nautical Almanac and Astronomical Ephemeris* helped secure the prime meridians of Paris and of Greenwich. In eighteenth-century maps and atlases of Russia's empire, for example, Ferro, Paris, and Greenwich all appear as topographic base meridians. Between 1727 and the mid-1740s, that empire's prime meridian was placed at its westernmost islands of Dagö and Ösel in what is today Estonia: this is a late parallel of Ptolemy's and some Islamic geographers' use of the prime meridian as marking the western edge of the known world. In Spain, where as many as fourteen prime meridians appear on topographical survey maps at

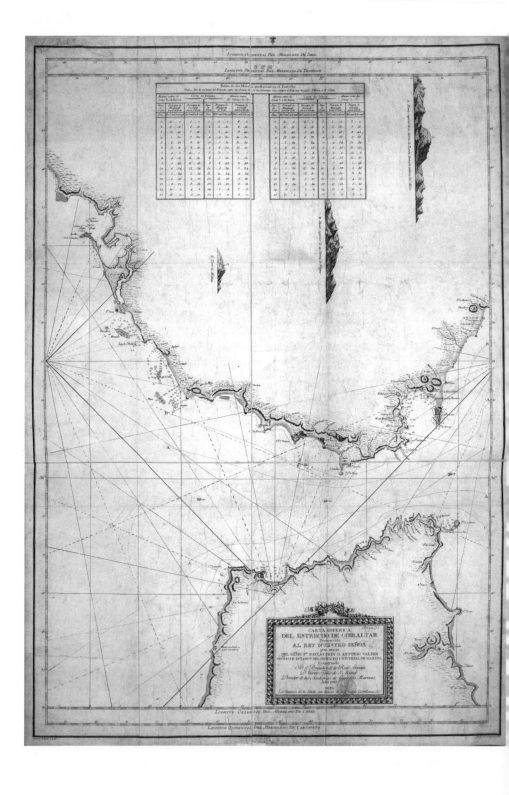

one time or another, coastal maps dating from 1768 show longitudes east or west of Paris, Cadiz (the site of an observatory established in 1753), Tenerife, and Cartagena; this last a reflection of its status as a major seaport. In 1762 the Spanish cartographer Tomas Lopez noted that "we Spaniards regularly count from the Pico de Teide, 12° 37′ west of Madrid"—that is, from the peak of Tenerife. In his *Principios de geografía* (1775), Lopez presented longitudinal tables of the differences between Tenerife and other prime meridians then in common use. Coastal charts from the 1780s "fixed" Spain's national outline with reference to the prime meridians of Paris, Tenerife, Cadiz, and Cartagena (Figure 1.8). Tenerife and Cadiz both feature as 0° meridians in Antonio Cavaniles's *Geografía del reyno de Valencia* (1797). For the Dutch the long dependence upon the prime meridian of Tenerife was shifting from the 1750s as their mariners used Greenwich more and more. The Dutch longitude committee, established in 1787, recognized this mixed usage of prime meridians and the fact that no other seafaring nation, with the exception of Spain, used Tenerife, and Spain did not use Tenerife alone. In 1826 the Netherlands formally adopted Greenwich as their prime meridian for navigational purposes: there, "the rules were adopted to practice, not the other round."[30]

In France, coastal charting mirrored terrestrial topographic surveys in aiming to produce more accurate delimitations of the nation. Different prime meridians appear on topographic and hydrographic maps as a reflection of the diverse points used by mapmakers and sailors (Figure 1.9). Such maps could depict as many as six prime meridians: on Figure 1.9, from north to south, these are the Azores, the western side of Ferro, "from the Pike [peak] of Tenerife," from the Lizard, from London (by which was meant St. Paul's Cathedral, not Greenwich), and Paris. This information was important with respect to navigation in coastal waters, although it is unlikely that such maps were actually carried on board and used by ships' officers. These meridians are, nevertheless, largely all cartographic or geographic prime meridians rather than observed (astronomical) and geographical prime meridians: on this map, only Paris and Cadiz were prime meridians on the basis of observed astronomical and geographical reckoning.

FIGURE 1.8 As the map frame shows, it is possible from this late eighteenth-century Spanish coastal chart of the Strait of Gibraltar to calculate longitude from any one of four different prime meridians. *North to south:* Paris, Tenerife, Cadiz, or Cartagena (Spain). © The British Library Board, *Carta esférica del Estrecho de Gibraltar* (Cadiz, 1786), Maps*18440. (11.).

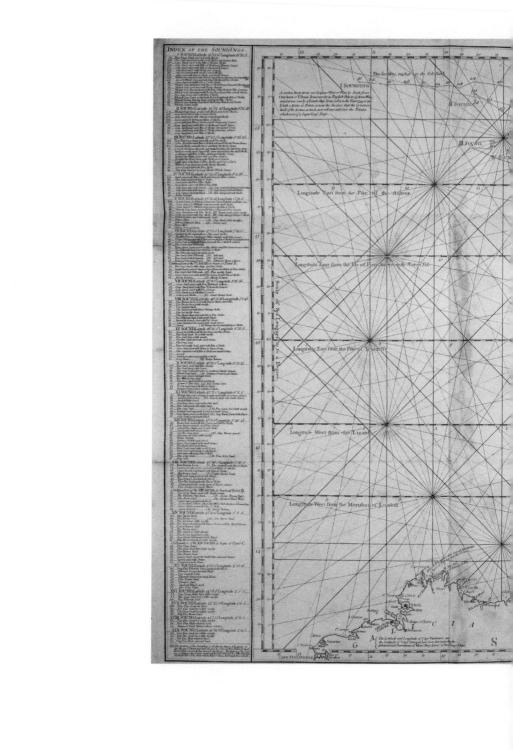

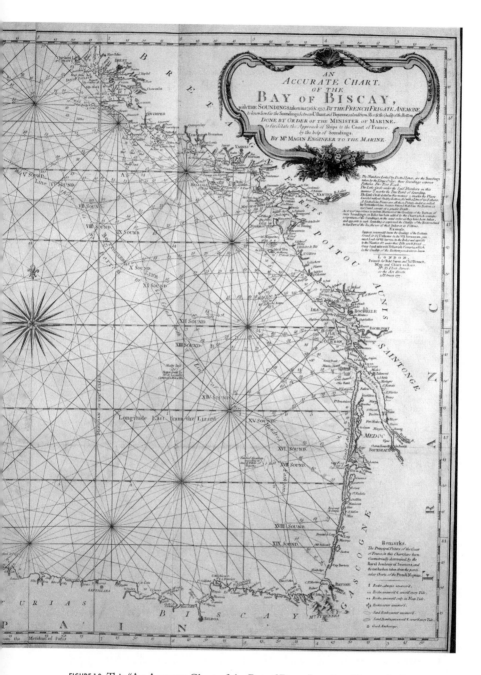

FIGURE 1.9 This "An Accurate Chart of the Bay of Biscay" produced in 1757 indicates six different 0° baselines. They are, *north to south:* the Pike [peak] of the Azores, Ferro, the Pike of Tenerife (today, Teide), the Lizard (in southwest England), London (St. Paul's Cathedral), and the meridian of Paris. © The British Library Board, Thomas Jefferys, *Neptune Occidental: A Complete Pilot for the West Indies* (London: Jefferys and Sayer, 1778), Maps. C.6.d.6.

In Britain, the first known chart with 0° longitude drawn through Greenwich dates from 1738. Some later eighteenth-century evidence seems to deny the primacy of Greenwich as the nation's definitive baseline. In the posthumously published *West-India Atlas* (1775) by the leading mapmaker Thomas Jefferys, geographer to King George III, Jefferys's editor Robert Sayer stated their reference points for the atlas:

> For a First Meridian we have made use of that of the island of *Ferro*, the most westerly of the Canaries, marking besides, at the top of each Chart, the difference of longitude from the meridian of London. Here, in England, we unhappily have not been sufficiently agreed upon a first meridian, which some reckon from St. Paul's Church at London, others from the Royal Observatory at Greenwich, and several from the Lizard-Point in Cornwall; these differences in meridians produce a perplexity in the reductions we are obliged to make in comparing Charts with one another. We could make a large catalogue of meridians which caprice and national pride have led different authors to make use of.

For Sayer and the late Jefferys, it seemed likely that the meridian at Ferro would dominate: "It is at present generally adopted in Europe." In his texts of navigational instruction, Samuel Dunn, mathematician, astronomer, and, from the mid-1770s, examiner for the East India Company, made it clear that there were differences not only in the customary usage between the first meridians of London and of Greenwich, as we have seen of the French for the Cape Verdes and the Canaries and the Dutch for Tenerife and Greenwich. He also taught that fact, and the longitudinal difference between the two, in his *Nautical Propositions* (1781). In his *Theory and Practice of Longitude at Sea* (1786), Dunn devoted a section—and, we may presume, a taught session—to "First Meridians." His instructions outlined the "Principles, geographical and hydrographical, concerning the prime or first Meridian, from which Longitude begins to be reckoned, and how it is formed." He advised his readers and students that while "the first meridian hath been supposed by the British Nation to be that of London for long time past, probably because it is the Metropolis of the Kingdom," more recently, prime meridians were to be based upon particular sites of a nation's science: "Since Astronomical Observatories have been erected at different places throughout Europe for the express purposes of improving Geography and Navigation, these first Meridians have been supposed to begin at several of

SITUATION des principaux endroits de Paris, de Londres & des environs, où l'on a fait des Observations astronomiques.

PARIS.	DIFFÉR. des Mérid.		LATITUDE.		
	M.	S.	D.	M.	S.
Façade sept. de l'Observatoire royal...	0.	0,0	48.	50.	14
Observatoire du collége Mazarin.....	0.	0,1 or.	48.	51.	29
Coupole du palais du Luxembourg....	0.	0,1 or.	48.	51.	0
Place du Palais royal.............	0.	0,2 oc.	48.	51.	46
Observ. de la Marine. Hôtel de Clugny...	0.	1,8 or.	48.	51.	14
Collége Royal.................	0.	2,2 or.	48.	51.	7
Observ. de la cour des Capucins......	0.	2,3 oc.	48.	52.	3
École royale militaire.............	0.	7,6 oc.	48.	51.	9
Tour de Châtillon..............	0.	14,0 oc.	48.	47.	49
Hôtel de Pasly, près la Muette......	0.	14,5 oc.	48.	51.	37
Observatoire de Colombe..........	0.	20,3 oc.	48.	55.	28

LONDRES.	H.	M.	S.			
Spital-square, où est l'obs. de M. Canton..	0.	9.	33 oc.	51.	31.	16
Newington, village où M. Bevis a observé..	0.	9.	35	51.	33.	43
Coupole de Saint-Paul de Londres.....	0.	9.	38	51.	31.	0
S. John's square.................	0.	9.	39	51.	31.	33
Red lyon street.................	0.	9.	39 $\frac{1}{2}$	51.	31.	30
Observ. de M. Graham dans *Fleetstreet*..	0.	9.	40	51.	31.	2
Surrey street, Observatoire de M. Short..	0.	9.	42	51.	30.	54
Convent garden, Observat. de M. Bird..	0.	9.	44	51.	31.	0
Malboroug house, S. James parch.....	0.	9.	47	51.	30.	37

FIGURE 1.10 The longitudinal difference between Paris and London and the relative situation of places near the Observatoire de Paris is clear from this table in the *Connaissance des temps.* Note, however, that whereas the longitudinal position of the cupola of St. Paul's Cathedral in London is calculated, together with that of private observatories in London, the longitude for the Royal Observatory at Greenwich is not provided. Source: *Connaissance des Temps, pour l'année commune 1777* (Paris: Imprimerie Royale, 1776). Reproduced by permission from the Centre for Research Collections, University of Edinburgh Library.

these Observatories respectively, and the Longitudes of places have been reckoned from them accordingly." By the time Dunn came to publish his *New Atlas of the Mundane System* in 1788, he had convinced himself of the defining role of Greenwich: "The English astronomical tables are adapted for the Meridian of Greenwich, from which last mentioned place the longitudes of places are now usually reckoned."[31]

In short, while different prime meridians were commonplace in Europe at the end of the eighteenth century—a matter of "caprice and national pride" and the cause of geographical confusion—it matters where you look for an explanation of these facts and of their consequences. Books of geography and astronomy taught that there were multiple prime meridians (Figure 1.10). The study of these books also shows that it was only slowly, by the 1780s, that "national" prime meridians were accepted: Greenwich in Britain; Paris in France; and, in Spain, the trio of Cadiz, Toledo, and Madrid. Such texts taught what the prime meridian was—a practical proposition to be solved via instruction within often rather formulaic treatments of geodetic, astronomical, and longitudinal principles. At sea, expediency ruled, with single prime meridians coming to dominate only in the 1750s in France, perhaps the 1780s in Britain, and the 1820s in the Low Countries. On maps and charts, and for astronomers, numerous topographical and navigational prime meridians continued to be used. For different practitioners, basing a map or an observation or a calculation at sea at a given $0°$ made operational sense. But for comparative purposes, the diverse usage of different baselines necessitated constant recalculation. For map readers and users, for students, and for anyone wanting to measure where places were in relation to one another or, more grandly, to delimit the globe with any degree of accuracy, the multiplicity of prime meridians continued to make geography, astronomy, and navigation awkward and perplexing.

Fixing Greenwich and Paris, c. 1776–c. 1790

By the late 1770s, astronomers and geographers in Paris and in Greenwich were acutely aware that knowing the world depended on where you started from and that whatever start points were chosen for longitude's verification, they should be accurately determined. Cassini de Thury's 1744 *Méridienne vérifiée* and earlier work had established new dimensions for France. It also had the effect of exposing inaccuracies in Picard's astronomical measurements for the prime meridian of Paris. Small though the error was, about five *toise* or ten meters, the error would have had major implications for France, for astronomy, and for the placing of the Parisian prime meridian had it been compounded for the nation as a whole. Cassini de Thury's "Nouvelle carte de la France" of 1744 (Figure 1.4) presented in its eight hundred principal triangles and nineteen trigonometrical baselines a view of France that was, essentially, abstract and geometric. It was, moreover, without detailed reference either to local topographical features or to human settlements. For these several reasons, the map did not satisfy France's King Louis XV. The king instructed Cassini III to map the nation anew, and from 1756 France began once more to emerge in maps unparalleled in their precision, detail, accuracy, and standardization. Upon Cassini de Thury's death in 1784, the project was continued by his son, Jean-Dominique, comte de Cassini (Cassini IV). By then, royal authority was waning. During the French Revolution, the Cassinis' map project was effectively "nationalized" and, technically, never completed. A further complicating factor was that the standards of measurement used in depicting France, such as the toise, were anything but standard and were shortly to be replaced by that new and revolutionary unit of metrology, the meter (a topic addressed in Chapter 3).[32]

In undertaking his revisions after 1744, Cassini de Thury extended his sight lines beyond France: into Holland and the Low Countries in 1748 and toward Vienna after 1762. As part of intentions to unite, trigonometrically, Paris, Vienna, and London, he also sought to extend his lines of Enlightenment earthly measurement toward Dover and Britain. In that respect he recalculated London's latitude, observing as he did so that British astronomical calculations were inaccurate—"an uncertainty of 15″." This had implications for the accuracy of the longitude of the Greenwich observatory. Sir Joseph Banks brought Cassini de Thury's 1776 paper on these matters to Nevil Maskelyne's attention in April 1785. In a paper read before the Royal Society in February 1787, Maskelyne responded to the "doubts entertained by the late Royal Astronomer of France concerning the latitude and longitude of this Royal Observatory."

The details of Maskelyne's paper need not concern us, based as they are on astronomical calculations by James Bradley, Maskelyne's predecessor as Astronomer Royal; discussions of instrument type and observational procedure; the problems caused by atmospheric refraction; and Cassini de Thury's reliance on research by Nicolas Louis de Lacaille in 1752. But Maskelyne's remarks upon "the difference of meridians of Greenwich and Paris, in reply to the late M. Cassini's doubts on the subject" do merit examination. Bradley's figures of a longitudinal difference of 9′ 20″ (nine minutes and twenty seconds) between Greenwich and Paris, reported Maskelyne, had stood since

1749. From 1763, on the basis of a sequence of observed transits of Mercury across the sun, astronomer James Short advised that the difference was 9′ 16″. Drawing upon a further dozen observers' reports, including, in Paris, those of astronomer Charles Messier, and calculations over the differences between the two meridians made in the period between November 1773 and November 1785, Maskelyne reported variations in longitude ranging from 10′ 46″ in March 1776 ("Air hazy" in Paris) to 8′ 54″ in August 1783 ("Air very clear" in Greenwich). Overall, he calculated a mean of 9′ 31″. An additional set of forty calculations, taken between February 1775 and December 1786, was presented showing the differences between the meridians at the observatory at Greenwich and the Hotel de Clugny in Paris, 2″ east of the Paris Observatoire: an assessment of their mean put the difference of the meridians at the two observatories at 9′ 20″ (exactly what Bradley had taken it to be in 1749). Maskelyne summed up the evidence: "Hence the difference of meridians of the two Royal Observatories, by the observations made in the Royal Observatories themselves, is 9′ 30″; and by the observations made by M. Messier, at the Hôtel de Clugny, and reduced to the Royal Observatory is 9′ 20″. The mean of both results is 9′ 25″. But if greater weight be given to the latter determination than to the former in the ratio of 2 to 1, on account of the series of M. Messier's observations being the most complete, the difference of meridians will be 9′ 23″."

Drawing upon further evidence—including from astronomer-surveyor Pierre Méchain, the editor of the *Connaissance des temps* who, later, would be centrally involved in plans for establishing the metric system in revolutionary France—Maskelyne proposed a compromise for the two observed prime meridians: "For the present, I infer, we may take the difference of meridians 9′ 20″, as being within a very few seconds of the truth, till some more occultations of fixed stars by the moon, already observed, or hereafter to be observed, in favourable circumstances, and carefully calculated, shall enable us to establish it with the last exactness. To collect and calculate such observations I have not leisure at present; but the field of calculation is equally open to the celebrated astronomers of Paris, the observations made at this place [Greenwich] being now published accurately."[33]

The harder and more often observers looked, it seemed, the more their assessments differed over where, exactly, Paris and Greenwich lay in relation to each other. We should recall that the French were already living with compromise: their siting of Paris was based upon Delisle's judgment of a 20° difference between Paris and Cap Ferro in the face of others' more precise estimations. Accuracy in the 1770s and 1780s over these two prime meridians was similarly an exercise in tolerance, the "last exactness" only ever a working figure. The degree of difference is not the important feature here, but its causes are telling: on "the field of calculation," different methods produced different results, as did local weather and instrumental practices. What is important was the fact of difference, and Maskelyne's acceptance of it, until such time as new work might better close the gap between Paris and Greenwich and between error and "the truth."

Nevil Maskelyne's response to the doubts expressed by Jean-Dominique Cassini de Thury in 1776 over the exact location of the observatories of Paris and of Greenwich, and thus of the geographical position of those centers of Enlightenment calculation, took the form of mathematical computation. The man who from Britain's perspective addressed the problem in the field was William Roy, the Scottish military surveyor and geodesist: in 1791 Roy would become the founding father of Britain's Ordnance Survey, the country's national mapping body. In 1784 Roy was at work on Hounslow Heath, establishing a baseline for Britain's trigonometrical survey. Maskelyne recognized that this work, if extended across the Channel, might act to "determine the difference of meridians of Greenwich and Paris to great exactness." Roy's 1790 account of his trigonometrical operations to determine the distance between the prime meridians of Greenwich and Paris is a tour de force of late Enlightenment geodesy in keeping with Cassini de Thury's *Méridienne vérifiée* (1744) and Cassini IV's *Déscription géometrique de la France* (1783) (whose works Roy often cites). It is a key work behind the establishment of the Ordnance Survey.[34] Roy's narrative and that of his French counterparts, also published in 1790, afford insight into the geography of the two prime meridians as contemporaries looked to site them at the end of the eighteenth century.[35]

On September 23, 1787, Cassini IV, Pierre Méchain, and the mathematician-astronomer Adrien-Marie Legendre of the Académie des Sciences met with Roy and his fellow surveyors in Dover to begin a sequence of observations across the Channel. The aim was to position the two national observatories by trigonometry and so bring the two nations closer together, geodetically speaking. Making use of a new theodolite designed by the instrument maker Jesse Ramsden; a John Harrison marine timekeeper courtesy of the Board of Longitude; and, on the French side, Lenoir and Borda's repeating circle, a series of sight lines and triangles was drawn up along the South Coast of England—when the "tempestuous weather" permitted—and extended into France (Figure 1.11). Explosions on the respective coasts were used to effect alignment by sight and sound. Where possible, the angles

Pl. IX.

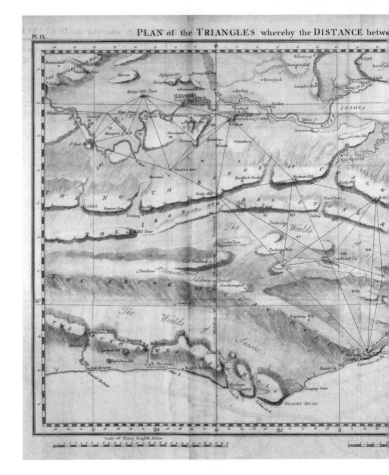

PLAN of the TRIANGLES whereby the DISTANCE betwe

Scale of Thirty English Miles.

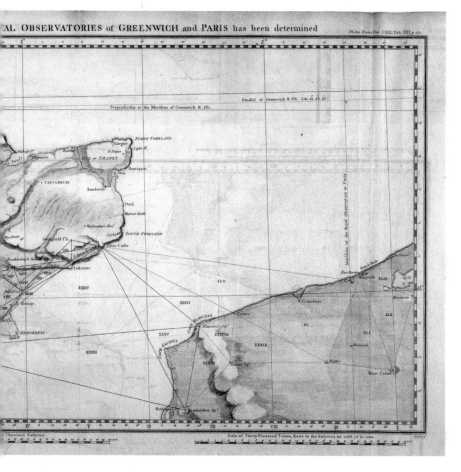

FIGURE 1.11 The intentions of collaborative international regulation through coordinated triangulation are evident in William Roy's "Plan of the Triangles"—but note too the recognition of the competing units of linear measurement: English miles (*left-hand corner*) and French toises (*right-hand corner*). *Source:* William Roy and Isaac Dalby, "An Account of the Trigonometrical Operation, Whereby the Distance between the Meridians of the Royal Observatories of Greenwich and Paris has been Determined," *Philosophical Transactions of the Royal Society of London* 80 (1790), 111–614, facing p. 272. Reproduced by permission from the National Library of Scotland.

and lines this new triangulation produced were compared both with Cassini IV's *Déscription géometrique* and with his father's *Méridienne vérifiée*. After forty-eight British and French triangles had been drawn up and compared, a reported difference of fifteen inches was testimony, wrote Roy, "to what a wonderful degree of accuracy operations of this sought may be brought when fine instruments are made use of, and great care bestowed in the application of them." Given that the work proceeded with precision, such errors as had occurred seemingly having "compensated for, or destroyed each other," they would, wrote Roy, "be enabled to determine the difference of longitude between the two Royal Observatories within a mere trifle of the truth."[36]

Roy's confidence was misplaced. The determination of the difference between the meridians of Paris and of Greenwich put the figure at "9′ 19″ nearly" (the figure given was 9′ 18.8″). What these inquiries appeared to suggest, assuming the British-based triangulation to be correct and those of Dover to Dunkerque to be accurate, was that those farther east and west were not: "It follows, that all the longitudes of the great map of France, the labour of more than half a century, will be considerably affected thereby, in proportion to the distances of the places, eastward or westward, from the meridian of the Royal Observatory at Paris respectively." The search for precision had uneven national consequences. Much as had been the case a century before (see Figure 1.3), the implication was that France in about 1789 was less certain in its shape and size nearer its coastal margins than along the line of its Parisian prime meridian. So the British claimed. For Roy the solution lay in the conjunction of different methods. These included trigonometry, the use of an accurate timepiece (Harrison's timepiece), and what he termed "the next best mode to angular measurement"—namely, "that of marking, by means of well-regulated clocks, as was done in the South of France, the repeated instantaneous explosion of light, observed at stations as far distant to the eastward and westward of the place of explosion as the circumstances will permit in practice, these distances having been for the purpose accurately settled by trigonometrical operation." Finally, locating the two prime meridians lay in repeated astronomical observation: "It will readily be conceived, that the object here in view is solely this; namely, that astronomers who live near those places, and have their time, that is to say the directions of their meridians very accurately ascertained, may, by their future corresponding observations (which should only be occultations of the fixed stars behind the moon's dark limb) compare the old with the new longitude, and thus be enabled to satisfy the curious world, which of the two comes nearest to the truth." Fixing these two prime meridians in space appeared only a matter of time: the solution would be arrived at via a potent mixture of mathematics, instrumental accuracy, patience, and gunpowder.

For their part the French placed great faith in the instruments used, especially Lenoir and Borda's repeating circle (its design using twin telescopes allowed multiple measurements of the same points), as the subtitle to their 1790 account makes clear: *Description et usage d'un instrument propre à donner la mesure des angles, à la précision d'une seconde* (Description and usage of instrument suitable for measuring angles to the precise second). Taking into account adjustments relating to the curvature of the earth, the calculations of this joint Franco-British venture produced two results to define the separation of the meridians of Greenwich and of Paris: 9′ 20.6″ and 9′ 18.6″. The latter was very close to Roy's figure of 9′ 18.8″, the former closer to that compromise figure of 9′ 20″ derived from Maskelyne, Cassini III, and others based upon astronomical calculations taken in Greenwich, in Paris, and farther afield between 1773 and 1786. But the fact of different figures left unresolved the question of the exact position of the two prime meridians. Additionally, the French, confronted by different units of length in their use of the toise, were again left unsure of their nation's dimensions.

The Franco-British triangulation project of 1787–1790 marked a phase of genuine international collaboration, although the scientists involved did not agree exactly where the respective prime meridians of Greenwich and of Paris lay. At century's end these and other prime meridians remained in use, each a testimony to the vanity of nations.[37]

o   o   o

Writing about geographical confusion is to risk producing it. Let me be clear, then, about the main issues discussed in this chapter and their significance for what follows. By about 1650, four main prime meridians were at work in the world in terms of European navigational practices: Ferro, Tenerife, the Azores, and the Cape Verde Islands. By the end of the eighteenth century, Paris and Greenwich were more commonly employed in addition to Ferro. Many other prime meridians existed, each the expression of topographic needs or, as in the case of Spain, a reflection of the authority of astronomical observatories or terrestrial institutions.

The existence of so many prime meridians meant, simply but universally, that there was no agreed-upon standard baseline for the world's geographical measurement. Each nation—and within each nation, different communities and users—measured itself against a different 0° and often against several first meridians. For a few countries—France from about 1667; Britain from about 1675 and, more evidently, after 1767; and for Spain, Cadiz from 1753—observatories facilitated an astronomically determined observed baseline. Even when this permitted an agreed-upon national measurement, the authority of one initial observed meridian in any one nation had to be understood in relation to the use of other 0° in the work of geographical authors, topographical surveyors, mapmakers, authors of navigational texts, and sailors.

There were several key features to the geography of the prime meridian between Ptolemy's adoption of Cap Ferro on the Canaries and the attempted verification of the Paris and Greenwich observatories in the late eighteenth century. The "discovery" between c. 1492 and c. 1508 of a seemingly magnetically constant compass bearing underlay attention to the agonic prime meridian in European navigation and mapmaking. This "category" of prime meridian was never agreed upon with certainty and featured only until about 1650. King Louis XIII of France's 1634 edict over Cap Ferro was an act of absolutist geopolitics, not accurate geodetics. Feuillée's voyage of the 1720s tested Cap Ferro's longitudinal position in the field. But his estimation of the island's position provided only an inconvenient truth—one replaced by the more easily usable figure of 20°, "calculated" and verified by Delisle, a person of geographical authority. Accuracy, always an end in view, was a negotiated compromise reached through accommodation of error. Cassini's *Méridienne verifiée* of 1744 and Roy's "Account of the Trigonometrical Operation" of 1790 provided through geodesy and triangulation what the *Connaissance des temps* and *The Nautical Almanac and Astronomical Ephemeris* provided for astronomy and navigation in 1679 and 1776: authority through observation and measurement, the textual verification of practical experimentation.

Different nations' and communities' use of different prime meridians caused widespread confusion. It is far from clear, however, given practices of use by navigators, geographers, mapmakers, and astronomers that we should think simply of distinct differences among nations or even of "national" prime meridians. To examine the question of multiple prime meridians only at the scale of the nation ignores social and epistemic distinctions in the use and significance of the different prime meridians. What may be interpreted, chronologically, as the emergence (and disappearance) of certain categories of prime meridian may also be interpreted, geographically, as the story of certain key locations—Cap Ferro, Tenerife, Greenwich, Paris. It is also a narrative of social and thematic difference, in relation to different professional communities and the several prime meridians they worked with.

2

Declarations of
Independence

Prime Meridians in America,
c. 1784–1884

IN THE UNITED STATES OF AMERICA in the late eighteenth and early nineteenth centuries, at least three prime meridians were in use: Philadelphia, New York, and Washington, DC. A fourth, New Orleans, was employed for a few years after the Louisiana Purchase of 1803 since that city, in lands previously held by the French, was by convention longitudinally sited with reference to Paris. America's philosophers, scientists, naval administrators, and politicians were dissatisfied with this lack of consensus and would, in different ways and places, consider all four of these options and others in debates about a compromise choice that would linger for decades.[1]

This range of options, albeit briefly described, is nonetheless important to recognize because when America's prime meridian debate is discussed at all by scholars, it tends to be depicted as a stark contrast between just two proposals: the establishment of a new and national prime meridian versus the use of an existing prime meridian that could be used universally.

Silvio Bedini has traced the lineage of presidential dictates from Thomas Jefferson onward regarding the importance of an American national prime meridian, usually proposed for the nation's capital. Matthew Edney, in complementary work, has outlined the relationship between the Washington, DC-based civil servant and astronomer William Lambert and the Philadelphia-based philosopher Benjamin Vaughan as a dispute between proponents of national and universal solutions. Thus understood, the prime meridian question for the United States in the century before the 1884

Washington meeting has been cast as a choice between independence from the scientific and political claims of other nations, notably Britain, and a more practical reliance on an already established o° baseline with a potentially global reach.[2]

The story of America's prime meridian is both more complicated and more significant than this simple binary suggests. Examination of the debates before 1884 among individuals, within scientific bodies, and throughout administrative circles in the United States over which of several prime meridians to use reveals a multiplicity of sides. The views expressed arose from contemporaries' concerns over accuracy, civic utility, and disciplinary and institutional authority. That these views were not easily reconciled is evident from the fact that in 1850 the United States formally adopted two separate prime meridians—Greenwich in England for navigational purposes and Washington for astronomical purposes and as a baseline for the country's topographical survey. This compromise was born out of an unresolved and seemingly unresolvable tension between national interests and the universal common good. In one sense the case of America before 1884 perhaps best represents that categorical distinction in the geographies of the prime meridian between the "observed" and the "cartographic" prime meridian: between Washington and/or Greenwich in the first category and other American cities in the second. In another sense, the American case blurs these distinctions since from the later eighteenth century we are shown a conflict between different prime meridians on the global scale as well as between different types of prime meridians within one nation's bounds.[3]

This chapter discusses these issues in three parts. The first examines debates over the prime meridian and American national identity from c. 1784 to the early nineteenth century. The second considers the political and the scientific deliberations over America's prime meridian between c. 1812 and the mid-nineteenth century, looking in particular at debates within the U.S. Congress and within the American Association for the Advancement of Science (AAAS) over choosing a national or a universal initial meridian. A key piece of evidence discussed is a report from 1849, intended as a reconciliation between competing views, which urged using a different prime meridian in the United States from those already at work. This 1849 report underpinned the decision of the U.S. Congress in 1850 to adopt two prime meridians. The third part addresses the place of the prime meridian and competing metrologies in American political and scientific discourse from 1850 to about 1884, a narrative theme that will be taken up again, in Chapters 3 and 4.

American Geographical Identity and the Prime Meridian, c. 1784–c. 1812

Before and even more strongly after political independence from Britain, the geography of the North American continent presented new challenges to established knowledge and to political administrators. The new Republic that was the United States after 1776 was slowly rendered visible by the gradual east–west movement of its surveyors and map makers. Following the expeditions of Meriwether Lewis and Samuel Clark and the Louisiana Purchase, America significantly expanded to the west and south. New on-the-ground information was scientifically and politically necessary, for map makers, geographical authors, and statesmen, if the emerging nation was to be fully understood and properly governed. In America, however, knowledge of both the content and the methods of geography depended on books from everywhere but America. Almost all of what was written about America was not founded on firsthand evidence. American authors accordingly turned to provide what has been seen as a new "print discourse of geography," or even "a geographic print revolution in early America," in which American books of geography become the means to know oneself and one's nation. Noah Webster promoted the Americanization of place names and sought to provide a national frame of reference for different spoken accents in his *Grammatical Institute of the English Language* (first published in 1783). Thomas Jefferson's *Notes on the State of Virginia* (1787) offered a critique of European views, particularly those of Buffon, about the limiting effects of America's climate and the alleged inferiority of American fauna. For the Connecticut Congregationalist minister Jedidiah Morse, a new American identity required new books of American geography.[4]

Morse's role as geographical midwife to the birth of the American Republic is well documented. In his *Geography Made Easy* (1784); *American Geography, or A Present Situation of the United States of America* (1789); and *Elements of Geography* (1795), Morse produced works for a society and polity only recently separated from Britain. Keen to subvert the textual hegemony of British and European books of geography, Morse reversed their usual text order in his geographies, putting America first and treating it at greater length: *Geography Made Easy* is the first book in Western print history to do this. Morse's "revolutionary" textual project (for such it was) should be interpreted in multiple ways. It was a deliberate subversion of the European publishing premises on which it simultaneously depended. It was an attempt to establish a national identity through and for American geography. And it was a form of cosmopolitan writing that saw geography as a means to national self-realization, a rejection of the past, and a moral blueprint for a new republican present and future.[5]

A New Beginning for a New Nation

Morse was insistent that America should have its own prime meridian, a point and line of origin against which America could distance itself from Britain, define its separate identity, and mark its westward progress as a nation. A map of the early United States that showed both the meridian of Greenwich and the meridian of Philadelphia, the latter intended as a symbolic beginning for the new Republic, prefaced his *Geography Made Easy* (Figure 2.1). For Morse, Philadelphia—home to the Liberty Bell—seemed an ideal foundational site for the new nation: "The meridian which passes through Philadelphia is fixed, in this work, as the first, because of the size, the beauty, the improvements, and the central situation of that city." America's baseline was not that of other nations: "The meridian of Philadelphia is the first for Americans; that of London for the English; and that of Paris for the French."[6]

Morse was not the first to see Philadelphia as America's point of origin. A Philadelphia prime meridian first appeared on maps in 1749 (and would do so as late as 1816) and was used by the map maker Lewis Evans for his "Map of the Middle British Colonies" of 1755. Like Morse, Evans considered Philadelphia's progress in "Letters, and the mechanic Arts, and the public Spirit of its inhabitants" as "Reasons sufficient for paying it the particular Distinction of making it the first Meridian of America." Evans thought the choice would among other things enable accurate self-knowledge: "And a Meridian here I thought the more necessary, that we may determine the difference of the Longitude of Places by Mensuration; a Method far excelling the best astronomical Observations; and we may be led into several Errors by always reckoning from remote Meridians" (by which he meant either Greenwich or Paris). A prime meridian for America based on the city of New York features on maps in 1776, and one based on Boston appears on maps of 1733 and 1775. In each case, as for Evans and Morse, this was for the less accurately defined cartographic or geographical prime meridian rather than an observed first meridian from an astronomical observatory. Accuracy and exact placement, despite his aims, proved problematic for Morse himself, even when—and perhaps because—he had a choice to work from. As one critic of Morse's plans noted, in words that foreshadow others' later

FIGURE 2.1 Jedidiah Morse's use of a Philadelphia prime meridian in his 1784 map is a cartographic declaration of geographical independence for the United States, showing the Philadelphia 0° (*bottom of the map*) and the longitude "West from London" (*top*). Note that the longitude is to London, not to Greenwich. *Source:* Jedidiah Morse, "Map of the United States of America" (Philadelphia, 1784). Photo courtesy of the Newberry Library, Chicago, Ayer Collection, 109.5.M7 1784.

concerns, "It is of no consequence where the first meridian is, provided it be fixed and in general use. A first meridian ought to be a precise point. But Philadelphia is an extensive city; and Mr. M. has not informed us, from what part of he reckons. . . . But whether Mr. M.'s plans be good or not, he frequently deviates from it. For Beside Philadelphia, he has three other first meridians; Washington, London, and the Observatory at Greenwich."[7]

From the early 1790s, the discussion of America's prime meridian centered not upon Philadelphia but upon Washington, DC. President Thomas Jefferson's plans for a national meridian to be based in Washington drew upon the work of the surveyor Andrew Ellicott, who from 1791 had been working to the president's directions in surveying the nation's capital. Ellicott undertook a "true meridianal line by celestial observation." This became the basis on which "the whole plan," that is, the laying out of the city of Washington, was established. Jefferson's plans for an American prime meridian to begin in the capital reflected his concerns, which Morse had voiced two decades earlier, about ensuring the nation's scientific and geographical development, about mapping the country to register and represent its future shape, and about furthering America's independence from other nations with respect to the prime meridian. Under Jefferson's aegis but directed on the ground by Isaac Briggs, surveyor general of the Mississippi Territory, and Nicholas King, a city surveyor in Washington, further measurements were taken for the baseline in Washington to run through the president's house. This work was brought together on October 15, 1804, in a report titled *Record of the Demarcation of the First or Prime Meridian of the United States.* In this report, the case is laid out for an American national prime meridian: a new defining line, foundational moment, and point of origin for America's history and geography.[8]

From late 1804 the management of the decision to delimit an American prime meridian in the nation's capital was the responsibility of William Lambert, a War Office clerk and amateur astronomer. Lambert had worked in the Department of State when Jefferson was its secretary and would correspond with him over the prime meridian question through the end of Jefferson's presidency in 1809—and afterward, as Lambert also took up the matter with Jefferson's successors, James Madison and James Monroe. Lambert's reports to Congress and his own publications provide valuable contemporary insight into the importance assigned to an American prime meridian for the young country's scientific and political identity.[9]

Lambert undertook astronomical observations and calculations in 1804—several from the presidential garden—and again in 1805, possibly in relation to the predicted solar eclipse of 1806 but principally with a view to fixing the position of the capital. These observations and computations he brought together in pamphlet form in 1805 (Figure 2.2).[10] This pamphlet became the basis for proposals he brought forward to Congress in December 1809 on the question of an American prime meridian. In his memorandum, Lambert is unequivocal as to the significance of a new prime meridian for America, his view that Washington should be the site for that American prime meridian, and the importance of accuracy in its determination:

> By the plan of the city of Washington, in the Territory of Columbia, the capitol in that city is intended as a first meridian for the United States of America; but, in order to establish it as such, the distance between it and some known meridian in Europe, or elsewhere, measured or estimated on a parallel with the equator, and referred to the centre of the earth under the respective meridians for which the computations may be made, should be ascertained on correct principles, and with due precision. As many of our mariners and geographers are still in the habit of taking their departure, or reckoning their longitude, from Greenwich observatory, in England, it will not, it is hoped, be considered as an instance of unpardonable presumption for attempting to extricate ourselves from a sort of degrading and unnecessary dependence on a foreign nation, by laying a foundation for fixing a first meridian of our own.[11]

Lambert's memorandum, including calculations incorporated in his 1805 publication, was referred to a congressional committee.

On March 28, 1810, the committee's chair, Representative Timothy Pitkin, reported to Congress. Recognizing that "the necessity of the establishment of a first meridian ... from which geographers or navigators could compute or reckon longitude, is too obvious to need elucidation," Pitkin's committee reiterated the problem and proposed a solution:

> It would perhaps have been fortunate for the science of geography and navigation, that all nations had agreed upon a first meridian, from which all geographers and navigators might have calculated longitude; but as this has not been done, and in all probability never will take place, the committee are of opinion that, situated as we are in this Western hemisphere, more than three thousand miles from any fixed or known meridian, it would be proper, in a national point of view, to

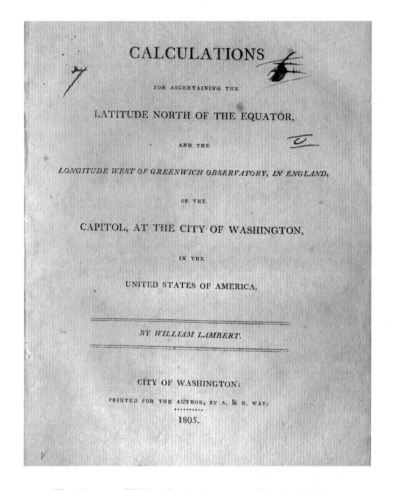

FIGURE 2.2 The title page of William Lambert's 1805 pamphlet, in which he argues on political grounds for a national prime meridian—Washington, DC—for the United States. Reproduced by permission from the American Philosophical Society.

establish a first meridian for ourselves, and that measures should be taken for the eventual establishment of such a meridian in the United States.

In the interim the committee had examined maps of America and found that "the publishers have assumed different places in the United States, as first meridians." This, they observed, in terms that echo those of eighteenth-century European commentators, "creates confusion, and renders it difficult, without considerable calculation, to ascertain the relative situation of places in this country. This difficulty is also increased, by the circumstances that, in Louisiana, our newly acquired territory, longitude has heretofore been reckoned from Paris the capital of the French empire."[12]

It is easy to see why geographical and astronomical debate around an American prime meridian was a political necessity in early nineteenth-century America. The country seemed not to know its own shape or where its principal places lay relative to one another and certainly did not know the extent of its territories to the west. In incorporating new lands within its political jurisdiction, America faced the challenge of adjusting not only to France's metrology but also to Britain's. Yet precisely because American natural philosophers and politicians were aware of Britain's ruling authority in the form of the Greenwich meridian—"a meridian with which we have been conversant"—they did not see adjusting maps and navigation tables to be a problem. It would not be difficult, the congressional committee argued, "to adapt all our maps, charts, and astronomical tables, to the meridian of such a place," that is, to adjust to a new American prime meridian: "And no place, perhaps, is more proper than the seat of Government" (Washington).

Congressman Pitkin proposed that the House take the matter forward. Even as he did so, however, and while paying testimony to Lambert's work, he sounded a cautionary note over the methodological procedures used to date, the need for precision in procedure, and the accuracy of the reported outcome:

It appears by the papers submitted to the consideration of the committee, that Mr. Lambert has calculated the longitude of the capitol in the city of Washington, from the royal observatory at Greenwich, by one of the most approved methods now in use for that purpose, viz: an occultation of a known fixed star by the moon. . . .

The committee would observe, that Mr. Lambert appears to be well acquainted with astronomical calculations; and that, so far as the committee have had time to examine them, they appear to be correct. In a question, however, of so much nicety, the correct decision of which depends so much on the accuracy of the observations made, and the goodness of the instruments used, and when the smallest error in the data will necessarily produce an erroneous result, full reliance ought not to be placed on calculations made from a single observation.

Indeed, in order to be certain of a correct result, it may be proper that more than one of the various methods of ascertaining longitude should be used. . . .

The committee, are, therefore, of opinion that, in order to lay a foundation for the establishment of a first meridian in this Western hemisphere, the President of the United States should be authorized to cause the longitude of the city of Washington, from the observatory at Greenwich, in England, to be ascertained with the greatest possible degree of accuracy; and that he also be authorized, for that purpose, to procure the necessary astronomical instruments.[13]

The necessary astronomical instruments included not just the appropriate devices—a telescope, a regulator or land-based clock, and transit and equal altitude instruments—but, crucially, an observatory.

This last is important. Lambert's work from 1805 to 1809 and the congressional discussions of 1809–1810 identified for America the key need for an observed prime meridian defined from an observatory rather than from a geographical meridian as laid down by Ellicott, Briggs, and King. It would not be soon established: the Naval Observatory, a governmental institution, was established in Washington only in 1844. But debates over its utility highlight the importance afforded an American prime meridian. As Bedini argues, "Lambert, like Jefferson, was concerned with establishing a prime meridian for the United States as part of an astronomical institution of the government." There is then a causal and consequential political logic connecting Lambert's advocacy of a prime meridian to matters of political and scientific independence from Britain and for an astronomical observatory to be the proper means to realize such independence by accurately delimiting an observed American prime meridian. America's sense of itself as a scientific nation depended, in Lambert's eyes, upon defining and delimiting its own first meridian. But achieving this involved more than a simple distancing from British rule, actual and metaphorical. Representative Pitkin's remarks about accuracy and the danger of working from a handful of observations, particularly his admission that he and his committee had neither sufficient time to consider the details of Lambert's proposal nor the expertise to judge its author's astronomical and mathematical credibility, were concerns shared by others.[14]

Dissenting Voices and an Appeal to Universalism

To the Philadelphia-based diplomat and political economist Benjamin Vaughan, Lambert's logic in advocating a national prime meridian was wholly wrongheaded. Vaughan outlined his objections in a never-published

1810 manuscript titled "An Account of Some Late Proceedings in Congress Respecting the Project for Establishing a First Meridian, with Remarks." This important if rather inchoate document presents a clear counternarrative to those proposing an American prime meridian and so is valuable in understanding the geographies of the prime meridian both in America and more generally.[15]

Vaughan used his manuscript to document the geography of the prime meridian in historical context. He discussed Ptolemy's ideas and the long-run significance of Ferro despite the fact that it had no associated ephemeris or observatory and that its longitude was not accurately delimited. The French, he reckoned, under Picard and de la Hire in the late seventeenth century, were "the first important violators" of their own 1634 ordnance. He cited the debates within the French geographical and astronomical community in the early eighteenth century over the presumed accuracy of the calculated distance of Ferro from Paris; the Dutch use of Tenerife; and, for British authors, the slow move away from a variety of first meridians to the wider acceptance of St. Paul's and of Greenwich. In short, Vaughan understood the importance of the prime meridian as a means to national and potentially to global measurement. He showed familiarity with recent events of global discovery, citing Cook's voyages, Macartney's 1792 embassy to China, and Pierre-Simon Laplace's and others' work in celestial physics. Most important—and drawing upon this and other work to substantiate his arguments—Vaughan offered a reasoned rebuttal of Lambert's claims over an American prime meridian for the use of America.

Vaughan's contrasting view was rooted in recognition of the debt owed to Britain given its scientific influence upon America, despite recent political differences, and an understanding of the importance and status of Greenwich, despite its relatively recent adoption in navigational and geographical terms. Vaughan's view is throughout global rather than national: "As it is plain, that the relative position of places within the United States may be correctly settled, without any attention to the meridian of Greenwich; a knowledge of that meridian must chiefly be desirable, for the sake of connecting our own geography with that of the world at large [underlining original to source throughout]; though this is not stated as the object here proposed." Vaughan recognized, presciently, that "the choice of a prime or fixed meridian includes the choice of a prime or fixed noon, to serve in reckoning longitudes." There was, he argued, "nothing unusual in choosing a foreign, as well as an extremely distant place, for a prime meridian," and he cited the examples of Ferro and the peak of Tenerife in support of this point.

This was, he averred, "the moment for fixing a new meridian at Washington, with any certainty of its being permanent, or with the slightest hope of its being adopted by other nations."

In Vaughan's eyes, American ambitions with regard to a Washington prime meridian threatened to impose an unwelcome metrological authority on other nations, in the same way that Lambert was claiming that Britain and Greenwich had done with respect to the United States: "Again," Vaughan argued, "if we design to form a permanent prime meridian for the whole Western hemisphere (as we are pleased to call it) we ought to recollect that the meridian of Washington is not likely to agree with the wishes of the other inhabitants of the two Americas, who are probably thrice as numerous as ourselves, and still less with the wishes of the nations of the other hemisphere." Vaughan appealed throughout to notions of universal utility, not national identity: "Thus if we have a pride let it be a useful pride, which will end in receiving glory from others, and not in cherishing pomp and vanity among ourselves." Aiming his broadside at Lambert, Vaughan had an even loftier final possibility in mind: "But allowing that we make no change at present in our own meridian, there is a grand change always to be kept sight of by ourselves, the British, the French, Spanish & other nations; which is, that of considering the whole globe as an unit, & having one prime meridian for the whole."[16]

Vaughan's critique of Lambert's reasoning was by implication a critique of the "official" view for an American prime meridian. It spoke to an altogether different, cosmopolitan, and enlightened vision for the sciences in America and for the perfectibility of terrestrial measurement—"the prime meridian's power to focus space for political reasons," as Edney has it. It was also born of a concern for accuracy and a shared epistemology, in the form of maps as representations of national space and of the authority of a universal prime meridian. Vaughan spoke of a future when, as he put it, "All maps speak an uniform language" and when he mentioned the need for "an uniformity of the mode of reckoning longitude" and how "a prime meridian fixed by the consent of a few civilized nations on judicious and impartial principles, may be expected to be gradually adopted by more of them."[17]

It is tempting to see Benjamin Vaughan as the unheralded prophet to Washington 1884 and the International Meridian Conference, but it is also wrongheaded. Vaughan's work was born of a particular moment in the status of American science and the country's geographical identity. At its core his opposition to Lambert's memorandum was rooted in an understanding of the historical and geographical problem that was the prime meridian, the confusion caused by different first meridians, and the admission of America's then-poor standing as a scientific nation: "As the citizens of the United States, for example, are without a fixed public observatory or an astronomical almanac (for a native of Great Britain resident among us merely reprints for us the nautical almanac of Greenwich, with some variations), our attempt to figure in the world of science by a measure so purely mechanical and so useless as that of changing our prime meridian must call to mind the many scientific and useful points which we have neglected to notice on this occasion." Vaughan's "An Account of Some Late Proceedings in Congress Respecting the Project for Establishing a First Meridian" is an appeal to a future scientific universalism and to its constituent shared practices, including the earth's measurement: "When a point of this kind [the prime meridian] also shall once be settled, it may lead to an actual uniformity in weights, measures, and money, or to the establishment of an ideal scale for each of these branches ... and as a still farther advantage, it may render the use of our modern calendar completely universal amongst Christian nations."[18] These sentiments about metrological uniformity and standards other than the prime meridian alone were very much of Vaughan's age, as will be discussed in Chapter 3, while his remarks about the universalism of a modern Christian chronology echo those of Jane Squire half a century earlier.

Other people took issue with Lambert's proposals over a Washington prime meridian, but for different reasons. To the distinguished Bostonian astronomer and mathematician Nathaniel Bowditch, Lambert's congressional reasoning was flawed in several respects. Unlike Vaughan's critique, which remained in manuscript form and may thus have been unknown to William Lambert, Bowditch's criticisms and Lambert's response took place in print in the public sphere. The exchange between the pair represents the first public airing in America of differences over the prime meridian as a line of national or global science.

Nathaniel Bowditch first took issue with Lambert's 1809 memorandum in an essay in October 1810. It was true, observed Bowditch, "that different first meridians have been marked on some of our maps and charts," but because most also showed Greenwich, it required "no more than one additional line on a chart, for each first meridian." Bowditch noted Morse's earlier use of the Philadelphia meridian in one of his books on geography as one of several "instances to the contrary" and hoped that this would be corrected in later editions "since no one at present supposes that Philadelphia will ever be the first meridian of the United States." Although

Bowditch did not specify a particular publication, he is probably referring to Morse's *American Geography,* his most widely printed and popular work. Particularly indefensible in Bowditch's eyes was Lambert's language, in which Lambert claimed that to count the longitude from an *English* observatory was "a sort of degrading and unnecessary dependence on a foreign nation" and an "incumbrance unworthy the freedom and sovereignty of the American people."

For Bowditch such rhetoric had no place in scientific discussion: "It is an address rather to the passions than the reason of our countrymen; and is an example of the introduction of national prejudices in matters of science, which cannot be too much reprobated." Instruments necessary to determine the longitude were not so expensive, but the expense of an observatory—"perhaps one of the objects Mr. Lambert had in view"—would be very great. Far more useful, Bowditch considered, would be a "complete and accurate survey of the sea coast of the United States, for the want of which, our navigators are daily suffering in approaching our shores." In this he echoed Vaughan's sentiments over the utility of mapping, for the good of American mariners and indeed the whole nation, to determine the country's correct shape and dimensions. Bowditch also endorsed the scrutiny given to America's maps by Pitkin and his congressional colleagues, although there is no evidence to suggest that the two men were in direct communication over the issue.

The great weight of Bowditch's criticism was reserved for the mathematical part of Lambert's work. This, reckoned Bowditch, "is for the most part a compilation, with many needless repetitions and palpable mistakes, evincing a great want of knowledge in the principles of the calculations." He concluded with a direct recommendation to Congress and America's administrators: "In our opinion the work of Mr. Lambert, both as respects its object and execution, is wholly undeserving the patronage of the national legislature; and we sincerely hope, that honourable body will discountenance this, and all future attempts to introduce innovations of this nature, which would only tend to 'create confusion,' and increase the labour in the calculations of geography and navigation."[19]

The counterreaction was soon in coming. Lambert was incensed at Bowditch's criticism of his plans for America's initial meridian—both at its tone and because it was in the public domain. Lambert saw this criticism as an attack upon his personal credibility, accusing Bowditch of "twistical cunning—ingenious quibbling" and of "zeal for the British nation . . . to the prejudice of our own."[20] Bowditch defended his comments in 1811 in a printed defense of his initial review of Lambert's memorandum. Bowditch considered Lambert's personal comments "beneath our notice." Because "the object" in question, an American prime meridian, "is of considerable national importance," it was necessary, argued Bowditch, that scientific argument give the lie to expressions of personal and national political sentiment: "We have thought it our duty to exhibit such proofs and authorities as will fully substantiate our observations in the mind of all scientifick men." Bowditch offered a lengthy critique of Lambert's astronomical evidence and "*proved him to be wrong in every instance,* and even in the calculation of the occultation at Washington, in which he more particularly challenged us to discover *one single mistake*—we have pointed out several."

Congressman Pitkin's cautionary remarks over the dangers of siting an American prime meridian on the basis of a limited number of observations looked well founded. Bowditch averred that a concern for accuracy and scientific authority motivated him:

> In doing this, we have been influenced not so much by a desire of vindicating ourselves from his impotent aspersions, as of lending our aid (so far as we are able) to the cause of truth: We shall observe, that the whole tenour of Mr. Lambert's papers bears strong marks of his wishing to make the business of his memorial a question of party politicks, than which nothing can be more improper and unworthy a man of *real* science. We trust that the good sense of the legislature of the United States, will prevent the adoption of a scheme that would be so injurious to the cause of science in our country.[21]

Like those of Benjamin Vaughan, Bowditch's criticisms amount to a defense of science over national sentiment. They were rooted in doubts over Lambert's scientific credibility, his political rhetoric, and his dependence upon an observational strategy whose limited results were inadequate for the purpose they sought to serve. They stemmed too from Bowditch's own position as an arbiter of scientific authority with respect to practical navigation—hence his remarks (echoing those of Vaughan) over the role of accurate coastal surveys rather than the expense of an observatory in determining an American prime meridian.

In 1802 Bowditch authored *The New American Practical Navigator,* a work whose many editions would secure him lasting recognition as the founding father of modern maritime navigation in the United States, especially following its adoption in 1868 by the United States Navy Hydrographic Office. Before 1802 there were in America several versions of Nevil Maskelyne's

Nautical Almanac and Astronomical Ephemeris (1767), including one edited by the British émigré John Garnett, and another, *The New Practical Navigator and Daily Assistant,* that was first published in 1772 by John Hamilton Moore, an Edinburgh-born teacher of navigation. Bowditch intended his 1802 work to be an extension of the third—and the first American-published—edition of Moore's *New Practical Navigator and Daily Assistant,* which had appeared in 1799. Whereas *The Nautical Almanac* is an annual publication, consisting primarily of data tables that change each year (with some description of their use), Hamilton, Moore, and others produced navigation manuals that were not published annually but did contain some data from Maskelyne's almanac. Based on *The Nautical Almanac and Astronomical Ephemeris,* Moore's book was commonly used by the American maritime industry for its longitudinal and latitudinal calculations. This, almost certainly, was the work that Benjamin Vaughan had in mind when he noted, as shown previously, how a British-born author had amended the Greenwich-based almanac for use in America. But in turning to amend the work for his own purposes, Bowditch found it "so erroneous in the tables, and faulty in the arrangement" that he resolved to address the problem afresh. To this end he turned both to an earlier and popular work of practical navigation, *The Elements of Navigation* (1750) by John Robertson; to Nevil Maskelyne's calculations of longitude and latitude in his *British Mariner's Guide* (1763); to the *Tables Requisite,* which were first published in 1766; and to *The Nautical Almanac and Astronomical Ephemeris.* In bringing together these several works—"singly neither of them was complete"—Bowditch found numerous errors. He corrected over eight thousand calculations in Moore's work and over two thousand in Maskelyne's. Bowditch noted that most of his corrections of Maskelyne's work were to the last decimal place: "The corrections would but little affect the result of any nautical calculation." In astronomical measurement, however—upon whose unchallenged authority politicians might determine national foundations—accuracy, or at least the claim to accuracy, was everything.[22]

Bowditch's actions and words are one expression of the differences within communities in America engaging with the prime meridian. For sailors, calculations corrected to within the last decimal place were meaningless in their day-to-day practical navigation. To astronomers, being correct to the last decimal place was about being correct for the right longitudinal place, especially when siting an observatory or sighting from one. In the thousands of mathematical corrections that justified Bowditch's claims, we find not only the basis for his criticism of Lambert's method but also an appeal to practical reason. This, like Vaughan's, was founded on a view of science as a universal common good and upon a belief in the national benefits of accurate coastal survey. Lambert's appeal for an American prime meridian, no less shaped by a belief in practical reason but with far less robust evidence, stemmed from the belief that in political, geographical, and scientific terms America should distance itself from Britain. To men such as Vaughan and Bowditch, who proposed retaining Greenwich as the prime meridian, America had no need of a new national baseline: science, not politics, should dictate claims to truth.

Bowditch was also the translator, in four volumes, of *Mécanique céleste*, by French astronomer and mathematician Pierre-Simon Laplace. As he was preparing the final volume, Bowditch looked back upon his public dispute with Lambert in the *Monthly Anthology and Boston Review:* "These two papers [Bowditch's articles of 1810 and 1811] were fatal to the proposed project; and, fortunately for the interests of science, Greenwich continues to be the first meridian for all who speak the English language."[23] The truth was more complicated.

Washington's Ruling Authority, c. 1812–1850

In July 1812, in a report to the House of Representatives, U.S. Secretary of State James Monroe returned to the issue of Lambert's memorandum and the American prime meridian. He made clear the distinction between the political will for an American prime meridian in the nation's capital and the issue of scientific accuracy upon which the meridian question rested:

That the principal object of the submission of these papers to the Department of State [Lambert's memorial of 1809], seems to have been to obtain from it a report as to the policy, in a national point of view, of establishing a first meridian in the United States at the seat of their Government, and not as to the accuracy of the observations and calculations already made, respecting such a meridian. To do justice to the latter, or scientific part of the subject, would require a profound knowledge of astronomy and mathematics, in the higher branches, to which the Secretary does not pretend.

Monroe was aware of the observational limits to Lambert's astronomical calculations, of Bowditch's emphasis upon more—and more accurate—readings, and of the political value of scientific precision. But to Monroe the importance of a first meridian to America was not in doubt:

The advantages of a first meridian are known, even to those who know least of the science on which it depends. . . . Scientific men agree that it would be of advantage to science, if all nations would adopt the same first meridian. . . . In admitting the propriety of establishing a first meridian within the United States, it follows that it ought to be done with the greatest mathematical precision. It is known that the best mode, yet discovered, for establishing the meridian of a place, is by observations on the heavenly bodies; and that to produce the greatest accuracy in the result, such observations should be often repeated, at suitable opportunities, through a series of years, by means of the best instruments. For this purpose an observatory would be of essential utility. It is only in such an institution, to be founded by the public, that all the necessary implements are likely to be collected together, that systematic observations can be made for any great length of time, and that the public can be made secure of the result of the labors of scientific men.[24]

For several years, there the issue remained: issues of scientific accuracy, national need, and political desirability recognized but not reconciled. Lambert again pressed Washington's politicians in November 1818 over the basis to his proposals, stressing that he had not intended for an observatory to be built and that it was of paramount importance to make additional observations "to test the accuracy of the result already obtained." He left it to the wisdom of Congress "to decide whether the object is in itself of sufficient national importance to engage their attention." A report to Congress in 1819 affirmed this. The report additionally stressed how "it would tend to the promotion of science and national credit to fix a first meridian for the United States somewhere within their territories" since any prospect of a single "general meridian"—that is, a universal prime meridian—"would be attended with advantages which the establishment of many meridians would not afford; but this desirable coincidence in opinion and practice is not within our control, and probably never will take place."[25]

In March 1821, three years after Lambert's entreaty over the issue, Congress agreed to a resolution authorizing President Monroe to appoint others to make sufficient observations to determine Washington's longitude accurately. Lambert resigned his clerkship in order to take the further astronomical readings. From November 1821 Lambert was provided with federal support for his efforts, and he presented revisions to his 1809 memorandum based on observations taken throughout the summer of 1821, some of which were hindered by the "unfavorable state of the atmosphere in the nation's capital."[26]

Lambert continued to agitate on the matter until 1824, even distributing two hundred printed circulars of his calculations and arguments to members of the political community. But in the face of a dilatory Congress and before the Washington observatory was completed, no firm decision was reached regarding America's prime meridian. Among America's scientists Greenwich predominated, despite political views to the contrary and differences of opinion. Navigators were of much the same view. Others held strongly to a view asserting the political legitimacy and necessity of an American prime meridian.

America's geographers read their nation differently. In the first quarter of the nineteenth century, America's foremost cartographer was the Philadelphian Henry Schenk Tanner. In the "Geographical Memoir" that prefaces his 1823 *New American Atlas*, Tanner laid out the reasons why, in constructing his work, he adopted the capital of Washington and the U.S. Capitol Building as the base point for his maps:

> The Longitude of Washington from Greenwich has been, in conformity with an act of Congress, recently ascertained by William Lambert, esq., a gentleman well known for his scientific requirements, whose report on the subject is entitled to the highest confidence. By Mr Lambert's observations the Capitol at Washington was found to be 76° 55′ 30″ 54 dec. to the west of the observatory at Greenwich. The exact position of our Capital having been thus determined, with a view, on the part of government, to the establishment of a first meridian for the United States; and as this object will probably ere long be completely effected, by the erection of an observatory, furnished with suitable instruments and apparatus, by which to lessen our dependence on foreign nations for the elements necessary to be used in astronomical calculations; I felt it incumbent on me to contribute my humble aid toward the introduction into use of the proposed meridian, and therefore selected it as the one best suited to an American map, intended for the use of Americans.

Tanner's statement of allegiance to Lambert's work, and thus to a new national prime meridian for America's geographical delineation, was at the same time an admission of the arbitrary nature of that choice for geographical purposes—"it must be admitted that the tracing of first meridians in the construction of maps, is quite arbitrary"—and a call for alternate prime meridians within America to be set aside. "Boston, New York, Philadelphia, and Washington, have each had their advocates in the persons of their resident geographers, who appear to have selected a first meridian for their maps, as their convenience or their fancy dictated." This practice, hoped Tanner, "will be entirely abandoned, since the position of our metropolis has been determined with sufficient exactness for all practical purposes." Tanner admitted that his choice of Washington, justified as it was by Lambert's work, compounded the problems arising from the presence of many different prime meridians that were "already too numerous." Yet it was necessary for America's geographical sense of self and until such time as a universal prime meridian might be established. On that matter "there is no reasonable ground to hope for the accomplishment of such an object, notwithstanding the facilities it might afford to navigation, or however desirable in other points of view."[27]

From the second quarter of the nineteenth century, two related developments lent these questions renewed weight. The first was the 1832 renewal of the United States Coastal Survey, under the direction of the Swiss-born surveyor Ferdinand Hassler. Hassler had first been active in this position from 1816 to 1818; the survey, begun in 1807, was poorly managed and largely inactive until Hassler's second period of tenure. Reporting in 1834 on previous work led by the military, Hassler considered much American coastal mapping "unsafe, and in many instances, useless and pernicious." Like Bowditch, Hassler recognized the importance of terrestrial survey and coastal mapping to America's sense of itself. He stressed the importance of accuracy as the purpose of such work: "In all the applications of exact sciences to practical purposes, the main aim must be to obtain the greatest certainty of accuracy in the results; thence also to obtain the means of proving them by the principles of the science itself: it is even necessary to aim at a much more minute accuracy than might be considered satisfactory, if the degree of accuracy shall be secured which is absolutely requisite in the final results."[28] Hassler used rockets, light signals, and gunpowder explosions to calculate time distance from known longitudinal positions and to measure short geographical distances along America's East Coast, as had the British and the French in the late 1780s in recalculating the longitude of the observatories at Greenwich and Paris (see Chapter 1). He understood too the need for a "celestial signal"—that is, astronomical observations and their resultant calculations—as the technique required for longer distances.[29]

The second development was the decision in July 1849 to establish an American nautical almanac. The intention to produce an American nautical almanac along the lines of Britain's *Nautical Almanac and Astronomical Ephemeris* and France's *Connaissance des temps*—a move that Benjamin Vaughan had foreseen and that Nathaniel Bowditch had effectively prefigured from 1802 in his *New American Practical Navigator*—stemmed from the establishment of the American Nautical Almanac Office within the U.S. Naval Department under the superintendency of Lt. Charles Henry Davis. Davis recognized that Americans customarily used Greenwich for nautical, astronomical, and geographical purposes despite decades of political petitioning for a prime meridian in Washington and the work of Tanner and others in geographical publishing. "It has been so much our general practice to count from this meridian [Greenwich], that it constitutes a part of our familiar thought and knowledge." Yet improvements in terrestrial and topographical survey within America, the result of coordinated triangulation and use of the telegraph to coordinate the measurement of America's interior longitudes, were not being matched by longitudinal coordination beyond America. The effect, argued Davis, was that positioning by Greenwich served Americans less and less well the better they came to know their own nation. He thus supported a different prime meridian for the United States—but not Washington.

A New-Old American Prime Meridian

Davis's proposal in his 1849 *Remarks upon the Establishment of an American Prime Meridian* rested upon his observation that New Orleans was almost 90° west of Greenwich, which would make it both convenient and appropriate as a prime meridian. To Davis's mind, calculations using 90° of difference, a six-hour time difference, were more easily undertaken than any more accurate figure. In ways that have a striking similarity to the earlier French adoption of 20° as the difference between the prime meridians of Ferro and of Paris, Davis was prepared to sacrifice accuracy for expediency: "Round numbers are easy in their use and application." This prime meridian had a further advantage, argued Davis. Placing America's prime meridian in this way would remove the necessity for Americans to calculate the space of the Atlantic in longitudinal calculations; that is, it would remove any necessity of working from Greenwich. Davis omitted detailed reference to the "Washington option." What he was proposing was in effect an initial meridian, which had first emerged in America with the Louisiana Purchase but that in its relative positioning was dependent upon a prime meridian, Greenwich, despite the end in view being a "new" prime meridian for America.

Davis first presented his proposals on July 31, 1849, to William Ballard Preston, secretary to the U.S. Navy. In correspondence between the two, it was suggested that Davis present his ideas to the AAAS at its upcoming meeting in Cambridge, Massachusetts. He did so on August 15, 1849, delivering there a paper titled "Upon the Prime Meridian," which was virtually the same as his proposal to Preston. Davis's intention was not to bypass either Congress or the navy office but rather to bring his views to the attention of America's "principal mathematicians and astronomers." With several others, he recognized the convenience to all civilized nations "if a general meridian were adopted by common consent; if all longitudes were counted in the same manner and from a single origin." He acknowledged that Lambert's proposals with respect to Washington showed "an enlightened apprehension of the benefits that would result from the establishment of a general meridian." Admitting, however, that a single initial meridian was unlikely ever to happen, Davis reiterated the view that an American meridian was vital for America's needs: "The scientific importance of assuming, at present, an American meridian is undoubted."[30]

His choice of meridian, and the principles he advanced to explain and justify it, elicited a mixed response. Davis intended to solicit views on which meridian within the United States would serve most conveniently for the intended nautical almanac. Instead he provoked a renewed debate "on whether a meridian within American borders should be used at all as a prime meridian."[31]

Prompted by Congress, the AAAS acted upon the issue by appointing a specially constituted committee of twenty-two leading scientists, naval officials, and others. The work of this committee as they addressed Davis's proposal clearly exposed the different positions held.[32]

A few on the committee supported Davis's proposal strongly, arguing that an American meridian would advance American astronomical science, that an American meridian fixed at 90° from Greenwich would reduce any uncertainties in longitude as measured from Greenwich, and that an American meridian was in America's commercial best interests. The great weight of opposition came from those with strong maritime interests. A "remonstrance" protesting against Davis's proposals chiefly on the grounds of established practice—that America's sailors have "hitherto computed from Greenwich, which, being thus common to them and Great Britain, forms the basis of the longitude of four-fifths of the commerce of the world"—was signed on behalf of 732 merchants, underwriters, and shipmasters (60 from Nantucket; 58 from New York; 334 from Boston; 33 from Portsmouth, New Hampshire; 60 from Salem, Massachusetts; and 87 and 100, respectively, from Portland and Bath in Maine). The view of John Brune, president of the Baltimore Board of Trade, was typical of the many letters sent to Preston from the Eastern Seaboard communities: "The committee on commerce having examined the subject of the establishment of a prime meridian at a point in the United States, report that they can find no possible advantage likely to arise from such a step; that, on the other hand, were a new prime meridian to be established, the greatest confusion and perplexity would result to the navigator; all his charts would become useless, all his calculations would have to be made by himself, and the tables of reckonings, so important to uniformed shipmasters, would be thrown aside."[33] America's maritime community was not of like mind with her naval administrators.

Some members of the committee noted that international scientific exchange between nations increasingly necessitated a single universal prime meridian. William Smyth of Bowdoin College took the view that the establishment of a new national prime meridian "by so large a maritime power as the United States" would "hasten the compromise by which the long desired object, a universal prime meridian, will be established." Professor Lovering of Harvard University was adamant, however, that such a thing would never happen: "There is no reason to think that one single first meridian can ever be established by the common consent of nations. All desire it; but how many will agree upon the choice?" He and others commented that Davis's scheme from which to reckon longitudes in America was in itself insufficient justification for having an American prime meridian when a suitable prime meridian overseas, Greenwich, was already functioning as one effective baseline for America's national and global measurement. Some argued that if it was important to have a prime meridian in America for astronomical purposes, consideration should be given to separate astronomical tables based on such a meridian and of course to use the newly built National Observatory in Washington. Professor Coaklay of the AAAS Committee said, "I cannot bring my mind to any other conclusion than that such a meridian should pass through the central wire of the best transit instrument in our National Observatory." For most men on this twenty-two-man committee, political expediency was ruled out. America had suffered too long, one critic stated, "from a foolish and futile attempt to reckon American longitude from some unestablished point in the city of Washington ... merely because it was our National Capital."[34]

America was divided over different lines. America's merchant marine opposed Davis's proposal. The country's scientists were divided in their views. Some saw it as helpful in distinguishing American science (even though a few held to the idea that Washington was better than New Orleans). Others— in the minority—were strongly opposed to another defining meridian for America. They saw advantages in a single universal prime meridian. Davis's 1849 proposal and the reactions to it may be seen as one key moment at which, for some, the requirements of science overruled the pressures of politics. Davis himself was less concerned with the political prestige of America's prime meridian and was not of the view that it be located in Washington on those grounds. He was motivated much more by the practical benefits that would accrue for the "three classes of persons to whom this subject is one of special and daily interest—astronomers, geographers (in which two classes I mean to include topographical surveyors,) and navigators—and it is in relation to their several wants and pursuits that I propose to consider this question."[35]

In addition to a lengthy response contained within papers to Congress, Davis spelled out his arguments in a forty-page pamphlet, which to judge from its publication date was printed even as further letters of evidence were being sent to him and to Ballard Preston. Davis reiterated the importance of accuracy to the work of the coast survey already underway and for topographical survey in the American interior, "a work not yet begun." Davis understood the objections of the maritime community. As Bowditch in 1802 had had recourse to correct the works of Robertson, Moore, and Maskelyne, so Davis knew that his proposal demanded adjustment to Bowditch's calculations: seven of the fifty-six tables in the latest edition of Bowditch's *New American Practical Navigator* would require alteration. Among the seafaring communities of New York and Boston, a proposal had even been circulated arguing that there was no need at all for an American nautical almanac: a British nautical almanac adapted to American data should be used instead. The fact that J. Ingersoll Bowditch and George Blunt of Boston led this proposal suggests that this was about managing a perceived threat to their established publishing interests with respect to *The New American Practical Navigator*. Recalibrating to a new American meridian, countered Davis, would in the longer term compensate for any inconveniences in the short term: "If an American meridian comes into use at sea, the charts and books of navigation will be fully adapted to it."[36]

Considered nationally, Davis's plan reflected his authority within the Nautical Almanac Office and the decision to produce an American version of the nautical almanacs compiled and published by Britain and other nations.[37] Looked at transnationally, the debate in America that Davis's 1849 proposal occasioned for a New Orleans prime meridian centered upon the same issues as in Europe: an observed meridian; an ephemeris or nautical almanac; national utility; disagreements among practitioners; difficulties of implementation; and contrasting views over the benefits to astronomy, geography, and navigation.

Davis's opponents also took to print in defense of their views. Joseph Lovering, professor of mathematics and natural philosophy at Harvard and a member of the AAAS Committee, sought support in Bowditch's 1810 and 1811 *Monthly Anthology and Boston Review* papers to bolster his own view against any American prime meridian at all, much less a new one based on New Orleans. For Lovering, "There is no reason to believe that a single first meridian can ever be established by the common consent of nations." Given, in his mind, that global agreement would never be reached, America's long-standing dependence upon Greenwich made absolute sense for its navigators, astronomers, geographers, and topographical surveyors. An American prime meridian for astronomers was "of some convenience to the single observatory through which it might pass." But, reasoned Lovering, since astronomical observations there or anywhere else depended upon accuracy of record and calculation, calibrating American astronomical measurements from Washington was equally possible from Greenwich: "A little more time and care will ensure as much accuracy in the latter case as in the former." In conclusion, noted Lovering, "I must deprecate any attempt to bring about a change in the first meridian used in this country, by a misplaced appeal to our national pride. This is a question, partly of science, but much more of common prudence."[38]

Faced with such contrasting and deeply held views and no easy means to reconcile them, the AAAS Committee resolved to bring its members together. But in this aim, geography defeated them: "In consequence of the remoteness from each other of the persons involved . . . all hope was precluded of a general meeting." Failing to gather its members together, the committee dispatched to them the papers that had been read before the association in August 1849, together with a letter from Davis inviting further reflections. Only twelve of the twenty-two-man AAAS Committee replied.

Of this dozen, "five were in favor of retaining the old standard"—that is, of retaining Greenwich for American navigation and all other measurements. Five were "in favor of an American meridian solely." Two were "in favor of retaining the English meridian for nautical purposes, and of establishing an origin on this continent for geographical and astronomical purposes."[39] These responses were forwarded to Frederick Stanton of Tennessee, chair of the Naval Committee of the House of Representatives. As Stanton noted in his report to the House in May 1850, "There is great conflict of sentiment amongst these distinguished gentlemen [the AAAS Committee]; but it is believed that two positions are so well established as to admit of no dispute. These are, first, that in the present condition of navigation, some inconvenience would be experienced by abandoning the Greenwich, and adopting the American prime meridian; and, second, that the establishment of the latter is indispensable to the accuracy and perfection of all astronomical and geographical operations upon this continent."[40] Accordingly, proposed Stanton, "the Greenwich zero of longitudes should be preserved for the convenience of navigators, and that the meridian of the National Observatory should be adopted by the authority of Congress as its first meridian on this continent, for defining accurately and permanently, territorial limits, and for advancing the science of astronomy in America."[41] On September 28, 1850, Congress took up this proposal and passed an act ordering this dual prime meridian for America—"hereafter the meridian of the observatory at Washington shall be adopted and used as the American meridian for all astronomical purposes . . . Greenwich for all nautical purposes."[42]

America's Twin Prime Meridians

America's formalization in 1850 of two prime meridians was the consequence of opposing interests that could not, or more properly would not, be reconciled. The act enshrined in law what in practice had been the case for nearly half a century. America's scientists could not agree among themselves. Scientists did not see eye to eye with naval and government administrators. The needs of practical navigation were not those of geography. Elements of the question aired earlier in the works of Benjamin Vaughan, William Lambert, and Nathaniel Bowditch were given heightened expression in Davis's 1849 proposal and in the 1850 meeting of the AAAS.

Between these earlier concerns and the mid-century issues, however, certain things had changed that lent weight to the move to two prime meridians. The National Observatory had been built. A nautical almanac had been agreed upon. Coastal mapping had greatly improved. Topographical mapping, more advanced on the Eastern Seaboard, was moving westward. The decision to establish an observatory (which Lambert had advocated and Vaughan and Bowditch supported) endorsed Jefferson's view about the siting of the prime meridian. Contra Jefferson, however, it did so on the grounds of advances in astronomical science, not upon political sentiment. The decision in 1849 to establish a nautical almanac was important for American science and for maritime commerce, but it did not of itself require the adoption of an American prime meridian, and certainly not Davis's New Orleans' proposal, which the maritime community strongly opposed. *The American Ephemeris and Nautical Almanac* had a peculiar bipartite form, one part giving information of use to the maritime community in relation to Greenwich while the other, an ephemeris based on Washington, looked to the needs of America's astronomers and geographers. It was the embodiment, in print, of America's decision to have two prime meridians. In practice, topographic mapping could make use of either prime meridian since, while Washington could be taken as the baseline for charting America's geography and was commonly used in mapping until the 1850 act was repealed in 1912—Washington's position could also be calculated in relation to the prime meridian of Greenwich, and so America could be measured and situated relative to Europe. America changed shape during the nineteenth century as its more accurate measurement was embodied in maps: developments in telegraphy meant that the longitudes of the prime meridians of Washington and of Greenwich were amended as distances between America and Greenwich became calculated "with a certainty hitherto impossible," as one contemporary put it in 1859.[43]

From 1850 America's dual meridian was a divided compromise. It spoke to a lack of political will over fifty years from within Congress—not a failure to recognize the problem but too great an attachment to an American prime meridian (Washington, mostly) as a question of political expediency not rooted in scientific unity. The decision also illuminates America's inward-looking national interests in a period when science increasingly depended upon international exchange, and America sought to establish itself among the world's leading economies. Throughout the debates in Congress and in the AAAS, recurrent appeals to accuracy worked in two ways: cited as a requirement of good science, accuracy, or rather its lack, was seen as an impediment to political action. America's dual prime meridian from 1850 highlights differences within America's scientific communities over the nature and function of an American prime meridian and between the different needs of science and of civil society.

Internationalizing Science and America's Prime Meridians, c. 1850–c. 1884

Between the act of Congress of 1850, which legislated a dual prime meridian for America, and the adoption in November 1884 of the resolutions arising from the International Meridian Conference, American scientific communities were increasingly aware of the advantages of a single universal prime meridian and the adoption of universal time (as several of Davis's opponents in 1849 and 1850 had been). At the same time, sharply held differences were expressed, in America and elsewhere, not just over the prime meridians and which might be the world's first but also over the relative advantages of the metric system and of the imperial system for the world's metrology. These issues are discussed more fully in Chapters 3 and 4, but examining them here is helpful in understanding the prime meridian in America after 1850 and as background to the fact that in 1884 at the Washington meeting the American prime meridian of Washington was not on the list of "prime" candidates for the world's initial meridian.

In the United States, the problem of the prime meridian assumed further significance in the second half of the nineteenth century following developments in railway timekeeping and in the standardization of time. Timekeeping became a major feature of the Greenwich Observatory from mid-century: in Britain, railway time was standardized around Greenwich mean time in 1847. In America similar developments occurred in 1870 following initial proposals from Charles Dowd for four railway time zones and later from Cleveland Abbe, who in 1883 established standard railway time in North America on the basis of Greenwich mean time. The adoption of standard railway time helped regulate timekeeping within American society. For Abbe—one of the United States' delegates at the 1884 Washington conference—it meant that U.S. scientists and politicians were already disposed to the idea of universal time set from the prime meridian of Greenwich before it emerged as a recommendation of that meeting.[44]

By the early 1880s, one of the features distinguishing debate in America over the world's prime meridian was exactly what Benjamin Vaughan had argued in 1810: the advantages of a universal prime meridian identified on cosmopolitan principles. Sandford Fleming, one of the British delegates to the 1884 Washington meeting, spoke to just this matter: "The establishment of an initial or prime meridian as the recognized starting point of time-reckoning by all nations, affects the whole area of civilisation, and conflicting opinions may arise concerning its position. Its consideration must therefore be approached in a broad cosmopolitan spirit, so as to avoid offence to national feeling and prejudice."[45] For Fleming, this meant that it was inappropriate "to have it passing through London or Washington, or Paris, or St Petersburg, or indeed through the heart of any populous or even inhabited country." For him, as for many others, "a meridian in close proximity to Behring's Strait suggests itself as the most eligible."[46] Fleming was here echoing the views of M. Bouthillier de Beaumont, president of the Geographical Society of Geneva, who in 1875 had proposed a line in the Bering Strait.[47] Charles P. Daly, president of the American Geographical Society, endorsed this idea in America on the grounds of its intrinsic international benefits, although he noted that Greenwich was adhered to in general use in America and that Americans "would probably be quite willing, as a nation, to unite in its adoption."[48]

As we shall see, concerns over the prime meridian in the nineteenth century were part of wider international debates over which metrological system, metric or imperial, to use, and America was no exception. In America these debates centered upon the Cleveland-based International Institute for Preserving Weights and Measures, which scientists anxious to preserve imperial units in America established in 1879. That body was vociferous in its support of those such as Scottish astronomer Charles Piazzi Smyth, who advocated the continued use of the imperial system and who argued for the Great Pyramid of Giza in Egypt to be the world's prime meridian. Between 1883 and 1886, the institute published a journal, the *International Standard,* devoted "to the discussion and dissemination of the wisdom contained in the Great Pyramid of Jeezah." The *International Standard* became a vehicle for Americans campaigning for the retention of imperial units to air their views and for the institute to solicit funds. Charles Latimer, its president, even wrote to Britain's prime minister, William Ewart Gladstone, as to whether governments or scientists should regulate with respect to the prime meridian. Gladstone's reply was to the effect that in Britain "the government rarely interferes directly in matters connected with scientific observation; and therefore he does not see how he could himself further the object which you have in view [adoption of the Great Pyramid as the world's prime meridian]."[49] In June 1884 the institute's Committee on Kosmic Time and the Prime Meridian published its findings in a report titled *What Shall Be the Prime Meridian for the World?*[50]

To Frederick A. P. Barnard, the views of the imperial lobby and of the pyramidologists (as he termed those pressing for the Great Pyramid to be the world's prime meridian) were utterly without foundation: "The Pyramidical system has never been in use and never will be."[51] Barnard was the staunchly prometric principal of Columbia College in New York and president of the American Metrological Society. In 1849 he had been one of the twenty-two men on the AAAS Committee appointed to deliberate upon the prime meridian in America. There was in his view no scientific basis to the claims of those, in Cleveland and elsewhere, who took Giza to be a divinely ordained metrological center of global calculation and the best place for the world's prime meridian. In his preface to *What Shall Be the Prime Meridian for the World?*, Charles Latimer intimated that their purpose was to remedy the "confusion and inconvenience arising from the adoption of diverse meridians in reckoning longitude."[52] But as his own report four months before delegates gathered in Washington showed, scientists in America were far from in accord over the world's prime meridian, over Washington's validity as a legitimate candidate (despite Congress' ruling in 1850), or even over the units to be employed in science as a whole. In this, and in the confusion that different prime meridians caused, they were not alone.

o   o   o

From the later eighteenth century and in the wake of political independence from Britain, the prime meridian in America was seen as a line in space and a point in time against which the nation could distance itself from Britain and accurately align its new territories. Americans' geographical confusion over the fact of numerous prime meridians at work mirrored that of Europeans'. Yet America's geographical experience before and after 1850 was also distinctive for the confusion resulting from different prime meridians within that nation's bounds—and for the political and scientific discussion this prompted.

Those in the United States who held to Greenwich as America's prime meridian did so by defending arguments in terms of established practice and future utility, appeals to accuracy, and the wider benefits—expressed in terms of the "universal good" or the "cosmopolitan end"—to the global-scientific and mercantile-user communities. Those who opposed these arguments saw in a national prime meridian for America a distinctive symbol of political independence and of geographical identity. Debates over the geography of the prime meridian in America were shaped in private in Philadelphia, in Vaughan's unpublished manuscript concerned for a transnational science; in public in Boston's newspapers in intemperate exchanges between Nathaniel Bowditch and William Lambert in 1810 and 1811; and again in public in criticism from Eastern Seaboard mercantile communities over Davis's 1849 proposal for a new first meridian through New Orleans. When Prof. William Bartlett of West Point wrote to Lieutenant Davis in October 1849 to oppose the latter's proposal for a New Orleans prime meridian for America, his arguments centered less upon the validity of that baseline and focused more upon the collective scientific benefits. "Why not, then," Bartlett appealed to Davis, "rather unite for the purpose of diminishing the number of prime meridians, and, setting aside all national preferences, supply, by common agreement, the little that nature has in this respect left unprovided?"[53]

In 1884 international politics would provide the means to overcome what Bartlett considered nature's deficiencies and what we have seen to be national differences. Before then, politics in America provided a means to inquire into the problem of America's prime meridians but not to solve it. Between 1805 and 1825, in William Lambert's words in particular and in Congress, groups within the nation articulated the problem of America's prime meridian as a political matter. Before the completion of the National Observatory in Washington in 1849, politics overrode the requirements of accurately determining the capital's longitude even though legislators recognized the value of accurate longitudinal measurements of the nation's coast and interior. By the mid-nineteenth century, America had ruled itself anew. The act of 1850 was designed, on the one hand, to separate America from the rest of the world (it now had its own prime meridian for certain purposes) and, on the other, to affirm its connectedness (it would continue to use another prime meridian for different purposes). American navigators had long based their longitudinal positioning upon Greenwich and would continue to do so, a fact reflected in the American nautical almanac. What in 1810 William Lambert had termed a "degrading dependence" upon Greenwich was in 1850 enshrined in law.

There is then, from Lambert, Vaughan, and Monroe in the early nineteenth century to Davis's 1849 proposal, a strong narrative connecting many Americans in their views on the value of a universal prime meridian. What there was not was any shared sense as to how that might be achieved. By the early 1880s, the differences expressed in America and elsewhere over which prime meridian to use and why were bound together with concerns over regulating time and which metrological system to use in measuring the world.

PART II

GLOBAL UNITY?

3

International Standards?

Metrology and the Regulation of Space and Time, 1787–1884

THE PRIME MERIDIAN is a particular expression of a basic human practice—namely, the measurement of things. Measurement varies greatly in nomenclature and in significance. Touring France on the eve of the Revolution, the English agricultural commentator Arthur Young raged against "the infinite perplexity of the measures" encountered. No place was like another in its practices and units of linear, areal, or volumetric measurement: "They differ not only in every province, but in every district and almost in every town." France was not unique in this respect. Across Europe, differences in measurement were apparent everywhere. Shared terms were in common use, but there was little agreement over the standards of such measurements. And such standards as did exist were poorly enforced. National systems of governance were everywhere confronted by significant local variations in practices of measurement, whether in everyday monetary transactions; in their excise systems; or, with respect to the geographies of the prime meridian, in the units used to measure space and time. Metrology, the science of measurement, varied everywhere by geography.[1]

To the French mathematician and astronomer Pierre-Simon Laplace, advances in science, politics, and daily life depended upon removing this diversity in order to achieve the practical benefits of both uniformity and universality. His anxieties expressed in 1800 over the diversity of terrestrial measurement—"The prodigious numbers of measures in use, not only among different people, but in the same nation; their whimsical divisions, inconvenient for calculation, and the difficulty of knowing and comparing them"—extended also to the regulation of time. "It is much to be wished," he said, "that all nations would adopt one common æra, not depending on moral revolutions, but determined by astronomical phenomena alone." The measurement of longitude, and from that the authority of one prime meridian over others as a metrological baseline, was also singled out: "It is desirable that all the nations of Europe, in place of arranging geographical longitude from their own observatories, should agree to compute it from the same Meridian, one observed by nature herself, in order to determine it for all time to come. Such an arrangement would introduce into the science of geography the same uniformity which is already enjoyed in the calendar and the arithmetic, and, extended to the numerous objects of their mutual relations, would make of the diverse peoples one family only." For Laplace, the Paris meridian should be France's and Europe's baseline: "This labour [that is, working from the observed meridian of Paris], the most useful to geography which has yet been performed, is a model which every enlightened nation will no doubt, hasten to imitate."[2]

Laplace was not the only Frenchman looking anew at time in this revolutionary age. Late in 1797 Jean-Alexandre Carney, a schoolteacher from the Hérault in the South of France, presented propositions on space, time, and measurement to his fellow citizen members of the Societé libre des Sciences et Belles-Lettres in Montpellier. The date of his address used the terms of the French Revolutionary Calendar: *le 16 nivose an VI* (the sixteenth of December in year six—that is, 1797). That system, which divided the year not into weeks but into months, each with three *décades* of ten days, had been formally introduced on November 23, 1793 (and would be abandoned, by order of Napoleon, from January 1, 1806).

Carney's paper, less revolutionary than Laplace's perhaps, shared his contemporary's concerns. Carney stressed the importance of astronomy to questions of space and of time in posing his key question: "Are we to presume that within a fairly modest number of years there could be a universal era and a universal prime or initial meridian?" Answering himself in the affirmative, Carney justified his response by noting that the "instruments and methods" that would allow such things were being "honed every day" because, for all eras, "at least to those which begin with a solstice or an equinox, one can attach an Initial Meridian." Carney was less concerned with the idea of a geographical or an astronomical prime meridian for universal use. He was aware of Paris's role with respect to France and dismissed Cap Ferro's earlier positioning with respect to Paris as a "Meridién Fictif."

What motivated him was the possibility that the French Revolution might usher in a new revolution in time, a new universal age whose dates would be set in relation to astronomical and geographical measurements. Although one contemporary praised his "very ingenious idea," Carney's December 16, 1797, paper was not acted upon: as short-lived as the French Revolutionary Calendar, it has not been cited by later commentators until now.[3]

This chapter addresses the connections between metrology and modernity in the century or so from 1787 as they help explain the geographies of the prime meridian. Behind Laplace's appeal to "moral revolutions" and possible changes in timekeeping lay two metrological consequences of France's political revolution between 1789 and 1799: the invention and uptake of the metric system, centered on the Paris meridian, and the adoption across France, albeit short lived, of a new "standard" "Revolutionary Time," based on decimal divisions of time. Between its invention in 1790 and the signing in Paris in May 1875 of the International Metre Convention, debates over the meter and its geodetic utility were crucial in shaping the prime meridian question. This is illustrated here by the example of that particular "candidate" in the 1860s for the world's prime meridian, the Great Pyramid of Giza.

If metrology is a key theme in what follows, so are two of the most important inventions of the nineteenth century: telegraphy and the steam locomotive. In everyday life, telegraphy facilitated the growth of global communication. For astronomers and geodesists, it provided a new way to determine longitude and measure space. But even as it helped advance geodesy by aiding in more accurately fixing the several prime meridians then in use, it also exposed the increased need for a single global prime meridian. At the same time, the growth of the railway system in Britain, across Europe, and throughout the United States necessitated the use of a standard time to govern schedules. People associated with these technologies pressed vigorously for a single prime meridian from the 1870s, as we shall see (Chapter 4). But the groundwork for these twin influences was laid earlier, not least by the facts of Franco-British collaboration in the geographies of triangulation, which I addressed in Chapter 1 and return to next.

International Regulation c. 1790–c. 1837: Fixing Prime Meridians by Triangulation

Conjoint British and French work aimed at fixing the Paris and Greenwich meridians effectively ceased between the French Revolution in 1789 and the fall of Napoleon in 1815. Within each nation, however, metrological work continued. For the British this centered from 1791 upon the activities of Ordnance Survey under the leadership of William Roy and William Mudge and involved the triangulation of Britain, beginning with the one-inch-to-one-mile mapping of Kent in 1801. In France between 1792 and 1798, Mechain and Jean Baptiste Joseph Delambre undertook the remeasurement of a meridian arc based on the Paris Observatoire, from Dunkirk in the North to Barcelona in the South. In doing so they used that new and revolutionary unit, the meter, as part of a completely distinct metrological system for calculating national space, the earth's dimensions, and the prime meridian. By 1817, when the French resumed plans to map their nation and in 1821 as the British and the French resumed collaborative inquiries about the Greenwich-Paris measurements, collaborative metrological concerns had again moved to the fore on both sides of the English Channel.[4]

Responsibility on the British side for the resumption of "the operations for connecting the meridians of Paris and Greenwich" rested with Capt. Henry Kater and Capt. Thomas Colby (a later director of Ordnance Survey). Henry Kater's capacities as a metrologist and surveyor were honed in India in the late eighteenth and early nineteenth centuries during his assistantship to William Lambton and his Great Trigonometrical Survey of the Indian subcontinent and in Britain from 1814 by work for Parliament and the Royal Society concerning Britain's weights and measures. Colby and his "steady men," accompanied by the French mathematician and astronomer François Arago, crossed the Channel between Folkestone and Calais several times in the autumn of 1821. One of the difficulties they encountered was relocating the exact spots in England that Roy had employed in the later 1780s: a mill used by Roy had since been demolished, and one of the guns marking the ends of his baseline on Hounslow Heath had been removed and lost. The summer of 1822 was spent in establishing triangulation points and sight lines, including "the temporary meridian mark erected near Chingford for the Royal Observatory [Greenwich]" (Figure 3.1). This station was chosen "in order that a side of one of our triangles might coincide with the meridian of Greenwich, and that the azimuths of the different stations, with respect to that meridian, might thence be deduced with greater accuracy than might have resulted from observations of the pole star": in other words, geographical observation and calculation based on careful positioning was thought more likely to produce accurate results than astronomical reckoning alone.

Despite such claims, accuracy was compromised in several ways: by "the unsteadiness of the building" at Hangar Hill Tower; by the fact that "the

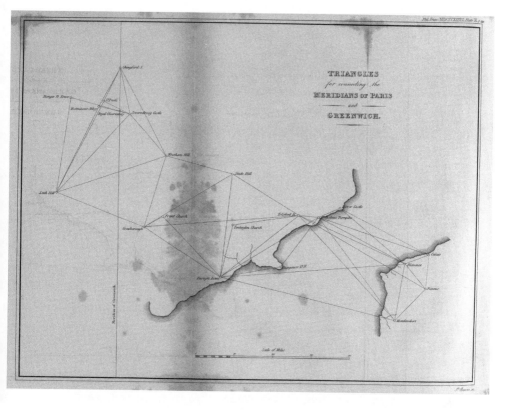

FIGURE 3.1 Henry Kater's map of 1828 shows clearly the triangulation scheme pursued in his and Thomas Colby's joint work with Francois Arago between 1821 and 1823. However, perhaps because of doubts over William Roy's metrology, Kater disguises the scheme's dependence, on the British side, upon Roy's earlier work (see Figure 1.11). *Source:* Henry Kater, "An Account of Trigonometrical Operations in the Years 1821, 1822 and 1823, for Determining the Difference of Longitude between the Royal Observatories of Paris and Greenwich," *Philosophical Transactions of the Royal Society of London* 118 (1828): 153–239, plate 11, opposite p. 199. Reproduced by permission from the National Library of Scotland.

intervening smoke of London" meant that only once was the signal erected upon that tower actually seen; and because by mid-November 1822 the men moving and establishing the theodolite in these several stations were suffering the effects of prolonged periods of outdoor work "upon a wet and clayey soil." Not the least of the difficulties, reported Kater, was the fact that Roy had used different linear measurements from those now in use: "By the comparison of various British standards of linear measure, published in the Phil. Trans. for 1821, it appears that the standard employed by General Roy for the measurement of the base upon Hounslow Heath differed from the Imperial standard yard; and in consequence it becomes necessary to multiply General Roy's distance by .0000691 to obtain 5.82, the correction to be added to such distance in order to convert the feet of his survey into Imperial feet." Roy's use of a different yard extended to his British mapping work: "The sides of the Trigonometrical Survey of Great Britain are, I believe, derived from bases measured by General Roy's standard, and they will therefore require the same correction as that employed above, should it be necessary to convert them into Imperial feet."[5]

There is a certain symmetry in this calculative and thematic narrative. In 1790 Roy had cautioned the French that their measurements had miscalculated the extent of France. In the 1820s Kater counseled likewise that Roy's metrology, and from that the mapping of Britain, similarly required adjustment to take account of different standards. Fixing the prime meridian depends vitally upon the use of a standard metrology. The reverse is also true: in order to produce and represent geographical space in map form, the prime meridian or other baseline must be accurately known. Little things can make a big difference. On the morning of October 2, 1821, for example, the weather was so bad at Blancnez on the French Coast (see Figure 3.1) that Kater made no use of the readings taken that day: "It blew so violently, that from this, or from some other cause which I cannot discover, these observations, though agreeing well among themselves, differ so widely from those made on the evening of the 3rd, under more favourable circumstances, that I have declined employing them." As he stressed, "The truth of the preceding work wholly depends upon the degree of reliance that may be placed upon the base on Hounslow Heath; and as the accuracy of this is in some measure questionable, it is certainly desirable that a new base should be measured." In closing his account of the labors shared with Arago and others, Kater made reference to others' work on the Paris-Greenwich problem:

It is to be regretted that our excellent associate M. Arago had not yet published the results of his operations in France; and I must therefore, in the absence of higher authority, take the longitude of Calais, as given in the Connaissance des Tems, to be 0° 28′ 59″ west of Paris. Adding this to 1° 51′ 18″.73 the east longitude of Calais from Greenwich, given by the present work, we obtain 2° 30′ 17″.73 for the difference of longitude between Paris and Greenwich. This converted into time is 9^m $21^s.18$, differing only $0^s.28$ in defect from the admirable results obtained by the operations with fire signals, reported in the Philosophical Transactions for 1826, by Mr. Herschel.[6]

Astronomer John Herschel was employed on the Paris-Greenwich problem in the summer of 1825 on behalf of the Board of Longitude and the Royal Society, assisted by Capt. Edward Sabine and partnered by Colonel Bonne and Lieutenant Largeteau of the French corps of geographical engineers. Their instrument of choice, in addition to four chronometers lent by the Admiralty and a detachment of artillery lent by Wellington, was the rocket. A series of rockets was fired from the observation points established by Colby on the English side and by Arago on the French side—the scheme thus bringing to mind Whiston and Ditton's 1714 plan to solve the longitude problem. Triangulation and calculation followed from these sightings. In contrast to Kater's experiences, bad weather was no impediment: "The observations were continued during 12 nights, 10 signals being made at each rocket station every night. The weather throughout the whole of this time was magnificent, and such as is not very likely to occur again for some years; a circumstance of the last importance in operations of this nature." At each station, teams of observers worked to coordinate things: eyes to the rockets' glare, rocketry to chronometry, and chronometry to trigonometry, all with a view to calculating longitude and the difference between the meridians of Paris and Greenwich. Some results were rejected due to poor observing conditions. Herschel's paper, a testament to rigorous methods uniformly applied, presented longitude differences on a day-by-day basis for most of July 1825 before concluding with the mean of the observations taken: "On the whole then, 9^m $21^s.6$ may be assumed as a result not very likely to be altered a whole tenth of a second, and very unlikely to be altered to twice that extent, by future determinations."[7] Herschel's statement is at once a declaration of precision and a judgment of tolerance concerning instrumental and mathematical error.

But Herschel had erred or, rather, others had on his behalf. Herschel's figures from Greenwich were derived from calculations made there "and officially communicated to him from the Astronomer Royal." In the 1820s this was John Pond, whose skills extended to the translation in 1809 of *System of the World* by Laplace, whose views on the prime meridian we have seen. The man who pointed out Herschel's mistake was Scottish astronomer Thomas Henderson. Pond's seemingly small error—of one second, in one table, for one day—had significant consequences. Specifically, the effect was, as Henderson wrote in 1827, "to redeem the result of the observations of 21st July from the suspicion which attached to it [the 21st was one of the days whose results were discarded owing to the weather conditions]." Additional errors "in the comparison of the chronometers employed for the observation of signals at Greenwich and Paris with the transit clocks, are also to be apprehended." This was regrettable since, as Henderson wrote, "In this important national operation, the utmost accuracy is desirable, and it has therefore been thought proper to subject the whole observations to a new computation." The effect of such recalculation, expressed in terms whose tolerance regarding error parallels that of Herschel; Kater; and earlier, Roy, was "that 9^m $21^s.46$, or to the nearest tenth of a second, 9^m $21^s.5$ is the most probable value of the difference of meridians in question; that it is likely that this determination is within two tenths of a second of the truth; and that additional observations, even to a very considerable number, would not materially diminish the small uncertainty that still exists."[8]

The problem of accurately measuring one nation from its prime meridian or in fixing different prime meridians relative to one another lay in a combination of things: the metrologies employed in the baselines used, the period of observation, the number of measurements taken, the instruments employed, the capacity for calculative error, the level of trust in one's fellow operatives, the visibility of the sighting stations, and the weather. What sustained the personnel involved in fixing the positions of the Paris and Greenwich prime meridians after 1821 was the idea of accuracy in an absolute sense, just as it had for Roy in earlier years and as it had in America for Nathaniel Bowditch in criticizing William Lambert's celestial observations and adjusting Maskelyne's astronomical calculations. As men like Méchain, Roy, Kater, Arago, and Herschel came to realize, the practical accomplishment of accuracy was always a relative achievement: a consequence of the instruments chosen; of the operators' tolerance; and, often, of very local circumstances.

Edward Dent, the eminent maker of watches and clocks, knew this from firsthand experience. Dent, who the Royal Observatory had employed to repair its chronometers, was commissioned in the late 1830s by George Biddell Airy, Astronomer Royal, and François Arago, director of the Paris Observatoire, "to ascertain the meridian difference" between Paris and Greenwich by the transit of chronometers. Distinguishing between what he termed the "travelling rate"—the amount each chronometer gained or lost during the journey to Paris and back—and the "stationary rate," taken from the chronometers' timekeeping in Paris and in Greenwich, Dent claimed that his results "coincide to fourteen hundreds of a second, making the longitude of Paris Observatory east of Greenwich, 9^m 21^s, 28 by stationary rates, 9^m 21^s, 14 travelling rate." Reporting on what he termed the "official errors" received from the respective observatories—those derived from astronomical calculations—Dent reminded his readers of a more direct cause for error: "It should be remarked, that by passing through the paved towns, both in England and in France, the chronometers were exposed to severe and continual concussion."[9]

Study of the calculations of the Paris and Greenwich prime meridians in the half century from 1787 reveals that their locations were only ever proximate. Imprecision and confusion in geography and in time was, it turned out, an almost unavoidable consequence of the numerous methods and metrologies used to determine different points and lines in space.

The Politics and Geography of Metrology, 1790–1878

Metrology is the science of measurement. It helps us to understand the dimensions of the world. It reflects that world in what we usually assume to be a neutral fashion. But metrology also produces the world. The act of measurement is in fact laden with value and is often an expression of power: in central governmental standards laboratories, for example, or when a government measures a territory as part of its claim on that territory. Sometimes power is a by-product of measurement in the way that authority might accrue to a central government agency indirectly as it sets particular standards and makes certain choices about units and methods that others are obliged to follow—in the development of telegraphy, for example, or of standard units of measurement in physics. As Simon Schaffer notes, "The history of metrology demonstrates that its institutional regulation is also, and precisely, a value system. Metrology's apparently contradictory demands for institutional insulation and ever wider spatial integration stem from and embody the political and economic conflicts of the modern social order."[10]

The complex connections between metrology and modernity that underlie the geographies of the prime meridian are especially apparent in Britain in debates between 1814 and 1878 over units of measurement and in France between 1791 and 1875 following the "invention" and adoption of the meter in an age of revolution and its hoped-for ratification at the International Metre Convention.

Imperial Measures of Difference

In Britain the history of weights and measures between the mid-eighteenth century and the third quarter of the nineteenth had four main phases: one of grudging diversity, one of legislative activity (not altogether successful) to reduce diversity, a period of argument between advocates of metrication and decimalization on the one hand and adherents of imperial units on the other, and a period in which imperial standards were slowly adopted.

A great variety of weights and measures were in use in Britain before 1814. What Arthur Young encountered in France he would have met with anywhere in Britain. In some places in late eighteenth-century England, the acre in common use was 75 percent larger than the so-called statutory acre; a stone of London beef was half the weight of a stone of beef in Scotland; bushels, bolls, yards, and ells varied within and between counties and between towns and country. Everywhere, differences in units of length, capacity, and weight were commonplace, and everywhere the differences were reckoned damaging to the nation's economy. In the 1740s the Royal Society made note of the diverse standards used throughout London and compared French and British linear units without offering a judgment on their relative merits. Attempts in the later 1750s and 1760s following the 1758 Carysfort Commission's exposure of Britain's metrological diversity and the inadequacies of existing legislation failed for want of parliamentary time. In May 1790 further parliamentary debate followed in the wake of an invitation from Louis XVI of France to George III, via France's Constituent Assembly, for Britain and France to address the problem of different weights and measures to their common good. This came to nothing, and the onset of war with the French brought further discussion to an end.[11]

Contemporaries knew the issue of different weights and measures to be important but could not agree upon a solution. In his 1788 pamphlet, Thomas Williams proposed London as the base point for a recalculation of the earth's dimensions in order that "an universal comparison of foreign measures with our foot being made and recorded, all those measures might by justly obtained." A remeasured world and a reconsidered English foot might then be the basis to a new metric and decimal standard for all: "Let the 100.000th part of the 52nd meridional degree after having been accurately found, be called a Geographical yard; and the 10th of each yard, or the millionth part of the degree, be called a Geographical foot: and let these measures be an universal standard of comparison for all known measures." Sir George Shuckburgh reported at length to the Royal Society in 1798 on the problems of different standard lengths. For the political economist Sir James Steuart, the problem demanded the rejection of all established units: "I conclude that the best scheme of any to be adopted, is to depart entirely from every measure whatsoever now known in this country." To Steuart, who favored the introduction of decimalization, the question of which unit should be used "is a matter of absolute indifference in regulating a standard." What most vexed him was the lack of legislative intent and any determination to disseminate a standard within Britain and farther afield: "If the British Parliament shall come to a sensible, determined, and animated resolution, to establish their measures upon solid principles; why not, for the glory of their prince, and honour of their country, make a small step farther for the good of mankind?"[12]

As if belatedly heedful of Steuart, early nineteenth-century Britain was distinguished by successive parliamentary commissions determined to resolve these issues. In 1814 a House of Commons select committee reported, echoing Laplace, that "the great causes of the inaccuracies which have prevailed, are the want of a fixed standard in nature." The 1815 bill "for establishing and preserving an uniformity of weights and measures" failed after its second reading. A Royal Commission "to consider the Subject of Weights and Measures" reported three times between 1819 and 1821 without significant effect. Acts of 1822 and 1823 failed; the first in the Commons, the second in the Lords. Only in 1824 was an act passed in Britain to introduce what became the imperial system and to render null and void all previous legislation on variant weights and measures. The act came into force on May 1, 1825. Local and customary measures were not finally abolished by law until the acts of 1834 and 1835. At the heart of this metrological reform—as a member of the 1814 select committee and of the Royal Commission between 1819 and 1821—was Henry Kater.

Kater's credibility for the task of establishing a standard metrology in Britain rested on several things. He was a leading figure in the Royal Society. He had paid careful attention to others' earlier attempts to develop standards, notably, those of Roy and Shuckburgh from 1795 and 1798. And he had undertaken considerable experimental work, motivated by the requirements of regulation, of instruments, and of the state. In 1818, for example, Kater developed the reversible pendulum, a refinement of the seconds pendulum. This instrument was used for accurate timekeeping, and because the pendulum swung equally to either side of its vertical position, the pendulum's swing in time could be used to measure space. That same year he experimented with comparing French meters with English units using microscopes and temperature differences. From May 1, 1825 and following the act of 1824, the imperial standard yard was decreed to be based on a unit of length that had been in use since 1743. Provision was made that should it be lost or damaged, the now-standard yard—of 36 to 39.1393 inches (with additional length at either end to hold the brass yard itself)—was to be recovered by reference to the length of Kater's seconds pendulum in a vacuum and to the latitude of London at sea level.

Kater produced numerous papers detailing his metrological work. Each evidences an almost obsessive attention to precision in the methods and instruments used. In one he discussed the comparative thickness of the knife blades he used to repair experimental equipment. Kater tested others' earlier metrology—the Royal Society's work of the 1740s, Roy's, Shuckburgh's, and France's—in order to offer corrections to it and to site and cite his own measurements with respect to the "Trigonometrical Survey" of Britain, the Ordnance Survey. So when, as noted, Kater commented reprovingly in 1821 upon Roy's use in the 1790s of the "Ordnance Yard," he did so from a position of personal authority, political patronage, and expertise gained through years of experimentation designed to perfect the philosophical and practical pursuit of precision. More than anyone, Kater helped create, measure, and manage Britain's imperial standards, whose linear expression the "Imperial Yard" would replace variant metrologies; improve Britain's mapping; and, in liaison with the French, position the prime meridians of Greenwich and of Paris more accurately.[13]

Or it might have, had not everything gone up in smoke. In an inferno witnessed, among thousands of others, by the painter J. M. W. Turner, the Houses of Parliament were destroyed by fire on October 16, 1834 (Figure 3.2). Kater's imperial yard, the official yardstick of Britain's linear standards, housed there for safekeeping, was damaged beyond repair, as were all the other standard weights and measures. Steps to reconstruct the yard from copies distributed elsewhere were not begun until 1838 (Kater died in 1835). Among the commissioners appointed to restore the imperial measures were the Astronomer Royal, George Biddell Airy, and John Herschel. Airy, Herschel, and their fellow commissioners did not feel constrained by the act of 1824. In their 1841 preliminary report, they recommended among other things that elements of Kater's work be redone in order to reestablish the standard yard. Crucially, although the commission decided that "no change be made in the value of the primary units of the weights and measures of this kingdom," it favored the decimalization of Britain's metrology (including the coinage) and did so on the grounds of economic convenience

FIGURE 3.2 *The Burning of the Houses of Lords and Commons, October 16, 1834,* by Joseph Mallord William Turner. Oil on canvas, 1834–1835. 36 ¼ × 48 ½ in. (92.1 × 123.2 cm). *Source:* Philadelphia Museum of Art: John Howard McFadden Collection, 1928.

and general ease of use. The commission formally began its work in 1843. Its recommendations were enshrined in law in an act of 1855. The yard measure was retained, with numerous type specimens prepared. The commissioner with final responsibility for the reestablishment of the standard yard was George Biddell Airy. Airy's account of his work and that of his fellow commissioners recommended that "the genuine Standard of that measure of length called a yard" be deposited at London's Exchequer Office, with four copies available "in case of loss" to be held at the Royal Mint, the Royal Observatory, Westminster, and the Royal Society. Multiple copies— "accessible representatives," as they were called—were distributed to observatories in Europe and foreign governments. Further small refinements were made to the 1855 act in the Weights and Measures Act of 1878, the last major legislation on metrology in Victorian Britain.[14]

This is a summary of a complex history and geography. Between 1814 and the act of 1878, Britain's weights and measures were the subject of debate, report, bill, and act in Westminster on 119 separate occasions. What Laplace had observed of universality in 1800, John Herschel stressed of uniformity in 1849: "Uniformity in nomenclature and modes of reckoning in all matters relating to time, space, weight, measure, &c., is of such vast and paramount importance in every relation of life as to outweigh every consideration of technical convenience or customs." What distinguished Britain between the appeals of these men and for longer periods was neither universality nor uniformity in its measures but diversity. The imperial yard was formalized only in 1824, made law in 1826, and made law again in 1855 after being reestablished in 1843.

The philosophical and practical pursuit of accuracy and precision that underlay the schemes of national mapping and transnational plans to fix different prime meridians might be undone at any moment by the circumstances of daily life, experimental error, house fire, customary usage, or legislative indecision. At the same time, the standardization of imperial units in Britain was paralleled by a recognition of the metric system. What Steuart had proposed in 1805 and commissioners in 1843 had likewise advocated was returned to in 1864 in the Metric (Weights and Measures) Act. This permitted the use of metric weights and measures in Britain. Further acts in 1867, 1868, 1871, and 1873 reinforced the meter's place in certain quarters of British life. In effect, Britain recognized the metric system but did not formalize it in law. Rather than standardize and make uniform one metrological system, Britain from 1864 permitted two.[15]

The Metric System and Terrestrial Measurement

In France things were more complicated. The nation's metrology in the eighteenth century was, as Arthur Young experienced and Pierre-Simon Laplace testified, "utterly chaotic." The significance of the meter and its imposition after 1791 can be simply described: an arbitrary measure based upon the dimensions of the earth itself and imposed by revolutionary political authority became the basis to a new metrology. Others have discussed the actually far-from-simple rise to dominance of the metric system as a revolutionary "invention" designed to replace myriad seigneurial measures. This being so, the principal features are recounted here only as they relate to the prime meridian.[16]

Proposals for a metric system in France were initiated in May 1791 in a report presented to the Académie des Sciences by Pierre-Simon Laplace; Joseph-Louis Lagrange; Jean-Charles de Borda; Marie-Jean-Antoine,

Caritat de Condorcet; and Joseph-Jérôme Lalande. They recommended that, rather than use the seconds pendulum as a device of measurement, nature itself should be employed. The meter was to be one ten-millionth part of the arc of the earth's quadrant—a quarter of the globe—measured from the equator to the North Pole and based on the Paris prime meridian. For this to be effective, however, several things had to be implemented. These included remeasuring old baselines, notably, those used by Cassini II in his 1739–1740 survey and that underlay his 1744 map of France (see Chapter 1 and Figure 1.4). In this regard the arc of the meridian in question needed to be extended north and south, from Dunkirk to Barcelona and if possible farther still, and new observations should be made on the basis of triangulation to verify this scheme. The proposed new unit of length also had to be calibrated in relation to existing linear standards. Rather than take the many existing toise in customary usage, the toise de l'Academie, better known as the toise de Perou, was adopted, this being the standard that La Condamine, Godin, and Bouguer used in their 1735 survey of a meridian arc in Peru to determine the length of the seconds pendulum and the meridian degree at the equator.[17]

That is why Delambre and Méchain set to work between 1792 and 1798 in implementing a new means to measure France—and the world—based on the Paris meridian and using a revolutionary new unit, the meter. Delambre was in charge of the northern measurements (Cassini IV declined to undertake the fieldwork); Méchain those of the south, toward Spain. Méchain, after all, had experience: he had been part of the Anglo-French team calculating the difference of longitude between Greenwich and Paris in 1787.

Geography, politics, and other Frenchmen, however, all combined to thwart the metric remeasurement of France. Just as Kater in the early 1820s could not locate the measuring points used by Roy in his triangulation schemes, so Delambre and Méchain in the early 1790s could not locate, or if found use, many of the survey points used by Cassini II in 1739–1740. Conflict with the British meant they were unaware until later of Roy's reservations regarding the work of Cassini III. Carrying royal warrants to secure safe passage and explain the task—resurveying France from its capital's prime meridian—was no guarantee of safety in a time of revolutionary politics. Some citizens were alarmed by military engineers using devices that could see and site places over distance. On the day planned for laying out initial measurements from the Paris Observatoire to Montmartre, parts of Paris were in flames. Delambre was arrested several times and his work, intermittent at best, was regarded with suspicion. Méchain had no easier a time, being treated as a spy, falling ill, and producing inconsistent figures from his observations. In June 1795 the observatory in Paris was placed under the authority of a new state-controlled body, the Bureau des Longitudes. Weeks before, France's National Convention had formally ratified the metric system in France, leaving the definition of the meter pending until completion of the geodetic work. This was completed in 1798. The metric system was legalized in France on December 10, 1799. It was slow to be adopted, however, and not until January 1, 1840, was the meter made obligatory in France.

Delambre and Méchain's metrological odyssey and the later formalization of the metric system matters for the prime meridian, both because of its result and its implications. Their findings as applied to the meter calculated that the arc from Dunkirk to Barcelona amounted to $9°40'45''$ and measured 551,584.72 toise and that the length of the meridian quadrant was 5,130,740 toise, assuming a particular ellipticity of the earth and that the new unit of length in question, the meter, was, at 10^{-7} of the meridian quadrant, equivalent to 0.5130740740 toise. Put simply, their work was remarkably accurate. In its implications, however, it was problematic. The results of the expeditions to Lapland and Peru in the 1730s had shown that the earth was not a uniform sphere but an oblate ellipsoid—that is, was flattened at the poles. But the earth is not uniformly elliptical. What the measurements of France from the Dunkirk to Barcelona arc showed—and by inference the rest of the meridian quadrant—was that the flattening of the earth was different in different segments. What had been born in an age of political revolution as an authoritative, even a democratic, Earth-commensurate unit was in fact not so: "The metre of 1799, and indeed the metre of today, is in fact about one fifth of a millimetre too short if it is to represent truly the actual 10^{-7} of the quadrant."[18] The earth's geography confounded a uniform metrology. Politics complicated the possibilities of a universal metrology. Not all meridians were the same.

Geodesy, Metrology, Telegraphy, and Railway Time, c. 1837–1883

The advent and wider use of the electric telegraph—often in association with the growth of the railways—transformed the speed and the nature of communications during the nineteenth century. As the speed with which messages could be sent in time greatly reduced the influence of geographic space upon the nature of human interaction, so space and time were "compressed," even "annihilated." Geographical space gives way, so to say, to time space; linear distance gives way to time distance. Yet geographical space does not reduce to simple Euclidean dimensions. Space in vital respects is a human construct. Even if it exists as an objective "thing in itself," the meaning we give it is at once social, temporal, experiential, and relational—it is a way of conceiving of the relationships between objects. Rather than "annihilating space and time," as some have put it, new technologies such as the electric telegraph and the railway created different notions of space, new forms of connectedness, and new social spaces.[19] No less than the meter and the yard, time as we experience and measure it is a social construction that varies geographically.

Wiring the World

Herschel hinted at the potential of telegraphy in discussing the Paris-Greenwich problem in the summer of 1825.[20] It came into its own, however, following work on commercial telegraphy by William Cooke and Charles Wheatstone. Drawing on this work, from the 1840s geodesists and astronomers increasingly came together as an international network deploying telegraphy from particular local sites, such as national observatories. Equipped with these global connections, geodesists and others could use telegraphy to address questions about the geographical size of their nations. Answers would be determined by the relationships between the accurate baseline positioning of individual points of observation and recording and the units of measurement used. The geographical consequences of such "world wiring" were threefold: Greenwich became the telegraphic base point from which geodetic readings were taken with a view to correcting Britain's shape; Europe beyond Paris grew more and more connected by a telegraphic network that had Greenwich as its central node; and Europe was better linked to North America through Greenwich, particularly to the eastern United States.

In June 1852, as he was working to restore Britain's imperial measures and as part of his *Report to the Board of Visitors of the Royal Observatory Greenwich*, George Airy declared his faith in the possibilities arising from these "galvanic connexions." Yet he was under no illusion concerning their implications: "With the improvement of our instrumental means, our practical results have undoubtedly improved in accuracy. But our speculative difficulties have by no means diminished. This, however, is what happens in the advance of every science of observation. That the questions which now trouble us will soon receive a satisfactory answer I do not doubt: but I anticipate that they will be followed by others, perhaps more perplexing, which have not yet presented themselves."[21] Numerous telegraphic connections were by then underway or planned. The "determination of the difference of longitude with the Paris Observatory had long been contemplated as one of the important uses of our galvanic connexions," wrote Airy, and he spent much time liaising with Arago about this before the latter's death in 1853. Procedures followed practices established in the triangulation schemes of preceding decades: exchanging staff between the two observatories to ensure operational consistency; carrying out sequences of observations and measurements; and finally calibrating with a view toward accurate determination of the two principal meridians. In a tone of almost casual confidence, Airy reported in 1854 on the work of the year past. The result, he wrote, "claims a degree of accuracy to which no preceding determination of longitude could ever pretend. I apprehend that the probable error in the difference of time corresponds to not more than one or two yards upon the earth's surface." Similar declarations of accuracy a year later sought additional legitimacy by reference to others' earlier attempts: "The number of days considered available for longitude, in consequence of Transits of Stars having been observed at both Observatories, was twelve; and the number of Signals was 1,703. Very great care was taken on both sides, for the adjustments of the instruments. The resulting difference of longitude, 9m. 20s. 63, is probably very accurate. It is less by nearly 1s [second] of time than that determined in 1825 by Rocket-signals, under the superintendence of Sir John Herschel and Colonel Sabine."[22]

New lines of communication were opened. Following the establishment of a telegraphic connection between the Greenwich and Edinburgh observatories in 1853, Airy sought to use the latter point for "the determination of the longitude of a more distant point"—Lerwick, in Shetland—as part of "the azimuthal accuracy of our great National Survey." Greenwich and the observatory at Pulkova, Russia, under the direction of Russia's imperial astronomer Otto Struve, were soon linked in similar fashion. Struve was keen that Greenwich and Pulkova should be more closely connected so Britain would be wired into Europe's geodetic circuits. In 1857 Struve called for the repeat calculation of some earlier British, French, and Belgian triangulation in order to better connect Britain to Eastern Europe so that "a new determination of the longitude of Valentia [in Ireland] by the galvanic telegraph might be recommended." And given the different metrologies that were in use within and between nations, "a new comparison of the units of measure employed on the different base lines might be made."[23]

Measurement of the world by the mid-1850s might, in Airy's view, have been accurate to within "one or two yards" of the world's surface. But not everyone used yards or meters. Since in Britain the yard was only then re-gaining stability, global terrestrial accuracy might at the very least be compromised and at worst, unachievable. Airy knew that what was commonplace within Britain and across Europe was difference, not uniformity. Struve's cautionary remarks about needing to calibrate different units of linear measurement in order to effect accurate transnational metrological comparison were timely. The principal triangulation of the United Kingdom had been completed in 1851. The longitudes of the stations employed in putting Britain to shape had been calculated by the use of astronomical measurements from the Royal Observatory at Greenwich, by the countrywide transmission of chronometers in the way Dent had done between Greenwich and Paris, and by triangulation between these fixed points. The result was a lattice-like web of geodetically and telegraphically determined recording points (Figure 3.3). Lt. Col. Henry James of Britain's Ordnance Survey produced a map showing the position of all the measured arcs of meridians that had been undertaken by then, including Lambton and Everest's in India, Airy's in Britain, and Delambre and Méchain's for France and parts of the British Isles (Figure 3.4). Further plans to connect the triangulation of Britain with that of Europe needed to know the units used at the different baselines: "However accurately the trigonometrical observations might be performed, it is obvious that without a knowledge of the exact relative lengths of the Standards used as the units of measure in the triangulation of the several countries, it would be impossible accurately to express the arc of parallel in terms of any one of the Standards."

The solution arrived at was "that a comparison of the Standards of Length should be made." Between 1862 and 1864, using "a building and apparatus expressly erected for the purpose of comparing Standards," James and his deputy, Capt. Alexander Clarke of the Royal Engineers, oversaw the measurement "with the greatest accuracy" using four different European linear units. Each linear unit was its country's standard unit of length as used by the Ordnance Survey or employed in Britain's colonies in Australia, India, and the Cape of Good Hope. The purpose was not to propose a single universal standard or to determine one unit as being "better" than another. Rather it was to determine the differences between the units to allow for that difference when extending Greenwich-based networks of triangulation beyond Britain. James reckoned the task globally important: "Before the connection of the Triangulation of the several countries into one great network of triangles extending across the entire breadth of Europe, and before the discovery of the Electric Telegraph, and its extension from Valentia to

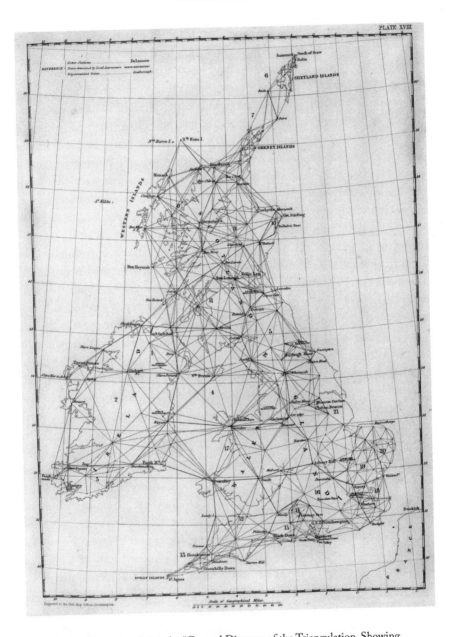

FIGURE 3.3 This map of 1858, the "General Diagram of the Triangulation, Showing the Connection of the Different Figures in the Preceding Plates," clearly reveals the results of the completed triangulation of the United Kingdom by the mid-nineteenth century. *Source:* [Ordnance Survey], *Account of the Principal Triangulations* (London: Eyre and Spottiswoode, 1858), plate 18. Reproduced by permission from the National Library of Scotland.

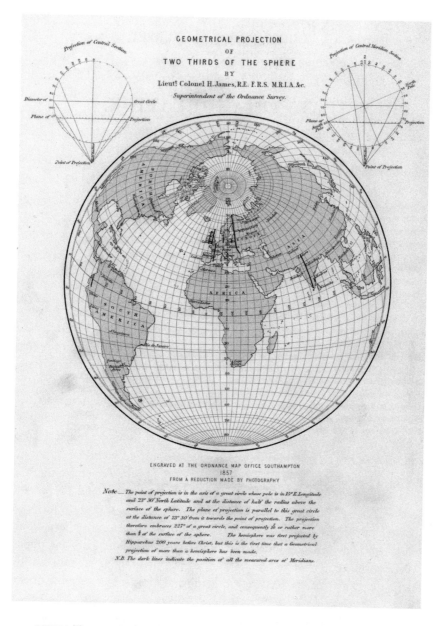

FIGURE 3.4 The several meridian arcs in existence by the mid-nineteenth century are clear in this map produced as part of the Ordnance Survey's comparative examination of the metrologies of different nations. *Source:* [Ordnance Survey], *Account of the Principal Triangulations* (London: Eyre and Spottiswoode, 1858), facing title page. Reproduced by permission from the National Library of Scotland.

the Ural Mountains, it was not possible to execute so vast an undertaking as that which is now in progress. It is in fact a work which could not possibly have been executed at any earlier period in the history of the world."[24] Struve's and James's reference to Valentia was to Valentia Island in southwest Ireland. From the mid-1860s, this was the western link of Europe's geodetic and telegraphic network stretching from Kerry to the Urals. In 1866 Valentia became Europe's principal point of telegraphic contact with North America. What James in 1866 saw as world defining, Airy reported upon more guardedly in June 1867: "A most important determination of longitude has been made."[25]

Conceptions of global regulation and standardization were one thing; implementation quite another. The world was being measured by different people using different units from different observational baselines. The world's geography had no agreed single base point. Europe's nations continued to employ various metrologies in differing national base points even as America worked with two separate prime meridians.

For these several reasons, telegraphy is an important element in the geographies of the prime meridian: it was a means for calculating space by time; for facilitating international exchange; for remeasuring national and vital international dimensions; and for aiding in the regulation of time. This was especially so for the United States. As Stachurski points out in discussing telegraphy, longitude, and map making in North America, "the explosion of gunpowder, the launch of a skyrocket, or the extinguishing of a bright light" may all have worked over shorter distances and for those countries with the resources to measure a continental coastline "one tiny increment at a time." For longer distances—such as the American interior or the trans-Atlantic distances between the United States and Europe—a "celestial signal" (telegraphy) was needed. Beginning in 1847 the American Coastal Survey used the Harvard Observatory in Boston as the nation's reference point for longitudes measured in America and later for trans-Atlantic connections routed through Heart's Content in Newfoundland. To Alexander Dallas Bache, successor to Ferdinand Hassler and superintendent of the United States Coastal Survey between 1843 and 1867, telegraphy spoke to new possibilities altogether—to know one's nation, to fix longitude, and to advance science: "Preliminary determinations of longitude from Europe have been made by the best methods known to science. The telegraphic method of longitudes, perfected in connexion with the survey, has enabled us to connect the distant points—Calais, in Maine, and New Orleans, in

Louisiana—with a certainty hitherto impossible. I am satisfied that a few signals by telegraph from America to Europe will enable us to determine the difference of longitude with a degree of accuracy which neither long-continued astronomical observations nor the transportation of chronometers have yet reached, or can ever reach."

Bache's words have a familiar ring. Like Roy, Kater, Herschel, Arago, and Airy, the authority of accuracy lay in revising or dismissing extant work, technical advances, and promises for the future. Numerous telegraphic measurements were made from the later 1860s under the direction of Bache's successor, Benjamin Althorp Gould. Gould worked with Airy to determine the longitude from the cliffs at Foilhummerun on Valentia to Greenwich following his arrival in London in 1867, after having "secured the friendly assistance of the Directors of the Atlantic Cable." By the early 1870s, "galvanic communication" linked the United States (via the observatory at Harvard) with continental Europe through Paris, Brest, and Saint Pierre; Munich and Paris were used to help determine the longitudes of Vienna and Berlin. Lisbon was being linked "as part of a chain of longitudes extending from South America." Russia was linked to the United States. All used Greenwich as their connecting node.[26] Telegraphy brought the world closer together. But the continued use of various prime meridians as the baselines for national and global geodetic calculation meant that it remained a world of difference.

Railways and the Standardization of Railway Time

Historians, sociologists, geographers, and metrologists of time have each shown how in the early modern world notions of time and the routines of daily life were governed by what we might term "natural time"—hours of daylight, the rhythms of the seasons as they affected manual labor, and so on. With the advent of industrial capitalism and developments in communication systems such as telegraphy, both the regulation of time and social mechanisms for timekeeping, such as public clocks, needed to work to commonly held standards. The use of solar time, for instance—taking noon to be the moment when the sun is directly above the point of observation—was inadequate as a basis to the more accurate measurement of time. It was also profoundly confusing since, as the sun orbits the earth westward, noon is later, locally, the farther west one travels from a fixed point of observation or record. Modernity required that natural time, local time, and the customs of recording time be replaced by new standards of chronometrics.[27]

At the cutting-edge of modernity in this task was the railway. The principal purpose behind the standardization of railway time was to avoid the confusion that followed the railway's use of local times. In 1840 the Great Western Railway became the first railway company in Britain to standardize time. By 1855 time signals from Greenwich, using Greenwich mean time, were regulating time in nearly all of Britain's cities and railways. Not everyone in Britain acceded graciously to having to dispense with local times and to being measured by Greenwich. As an anonymous commentator to *Blackwood's Edinburgh Magazine* put it upon the implementation there of Greenwich mean time in 1848, "What, in the name of whitebait, have we to do with Greenwich more than with Timbuctoo, or Moscow, or Boston, or Astracan, or the capital of the Cannibal Islands?" The answer was uniformity of time— "Simply, that there may be a uniformity of time established by the railway clocks," as the Edinburgh writer conceded. Greenwich grew more and more the center for the national regulation of time. Greenwich mean time became the standard time throughout Britain in 1880, and Greenwich was the coordinating center of Britain's railway time by mid-century, as it was for networks of triangulation and national mapping and as it would become in 1867 for transnational networks of telegraphy. The report of the Royal Observatory in 1874 exemplifies this central calculative function: "The Westminster clock records its correctness and errors at Greenwich as does also the clock at the Lombard Street Post Office [the local referent for the Bank of England]. The 10.00 am current is most extensively used for the Provinces. . . . Many of these offices re-distribute the time-signal to the offices radiating from them, so that practically from the 10.a.m. current from Greenwich most of the post office and railway clocks in the Kingdom are regulated."[28]

In Europe, most countries' railways held to national time—Germany to Berlin time from 1874, Sweden to standard time (one hour different from Greenwich) from 1879—until sometime after 1884. In America, as Bartky has documented, the multiplicity of local civic times and of different railway times was a source of national concern. The solution arrived at in 1870 by the educationalist Charles F. Dowd was to divide the United States and Canada into four separate "time belts," each 15° of longitude wide. In Dowd's scheme the uniform time within the respective belts would be determined by their relationship with four lines of longitude whose respective positions would be determined by multiples of 15°—that is, by one hour's difference in time in relation to Greenwich—75°, 90°, 105°, and 120° west of Greenwich, respectively. Dowd's plan was not immediately implemented, but under the direction of William F. Allen, secretary to the General Time Convention established in 1874 by American railway company owners, four uniform standards of time, or "time zones"—eastern, central, mountain, and Pacific standard time—were established across America on November 18, 1883. As with comparable metrological regulation—the meter, the imperial yard, or Greenwich mean time—standard railway time in the United States was not adopted everywhere and at once. From 1883, however, Greenwich set the time standard for America's railways.[29]

The World's Baseline?: Competing Metrologies and the Great Pyramid of Giza, 1859–1884

From the later 1850s, these questions of metrology, precision, accuracy, and standards in the earth's measurement came together in what to us now is the surprising consideration of the Great Pyramid of Giza as a possible site for the world's prime meridian. They did so because of increasing archaeological, geographical, and colonial interest in the Near East and a profound public disquiet over the possible replacement of imperial standards in Britain and in America. In early July 1863, the *Times* (London) reported upon the matter in portentous tones: "A very great trial is impending over this free and happy country. It is not the loss of our cotton trade, or our colonies, of our prestige, or our maritime supremacy. It is not the exhaustion of our coal fields, the deterioration of our racehorses, or the downfall of the Established Church. It is a change that would strike far deeper and wider than any of these; for there is not a household it would not fill with perplexity, confusion, and shame." The threat was metrication: "From a division in the House of Commons yesterday, it appears we are seriously threatened with a complete assimilation of all our weights and measures to the French system." In Britain the 1864 Metric (Weights and Measures) Act had legalized the metric system, confining it to scientific purposes and prohibiting its use in business and trade. But there and in America, where in 1866 the legislature had also sanctioned the use of the meter in specific contexts, the perception of metrication was enough for those who considered the meter "foreign"—a unit born of revolutionary politics now being adopted by other nations.[30]

A key figure in articulating these concerns was the London publisher John Taylor—who included among his output Herschel's astronomy. Taylor not only viewed Britain's imperial units to be ancient in their origin and use but also believed they were enshrined in the Pyramids of Egypt. Taylor presented his views in several publications: *The Great Pyramid? When Was It Built, and Who Built It?* (1859); a pamphlet of 1863 titled *The Standards of*

Length, Capacity, and Weight, Established in Egypt Four Thousand Years Ago, and Still Preserved in the Measures and Weights of Great Britain; and, in 1864, *The Battle of the Standards: The Ancient, of Four Thousand Years Ago, against the Modern, of the Last Fifty Years—the Less Perfect of the Two.* Taylor's belief in a combination of numerology, Egyptology, and metrology had deep roots. Among his Victorian contemporaries in addition to Herschel, Taylor's ideas influenced Charles Piazzi Smyth, Regius Chair in Astronomy at the University of Edinburgh and, from 1846, Scotland's Astronomer Royal.

Piazzi Smyth had conducted astronomical observations based on Tenerife in 1856 before turning to archaeological and metrological investigations in Egypt. His biographers date this interest in pyramidology—specifically, his belief that Egyptians used a measure equivalent to British imperial units to construct the Great Pyramid of Giza—to 1863 and to the writings of John Taylor.[31] They may predate this. In a lecture in April 1859 to Edinburgh's Chamber of Commerce, Piazzi Smyth had declared himself much concerned with longitude, chronometry, telegraphy, and steamship navigation—"all equally expressive of the general movement that is characterising the mind of the nineteenth century."[32] Piazzi Smyth was in favor of Britain's imperial system and against the adoption of the French metric system then being debated in Parliament. In this he mirrored Herschel, who had resigned from the government's Standards Commission when it came out in favor of the metric system in 1864, albeit only in prescribed usage. Piazzi Smyth published his views on the Great Pyramid in his *Our Inheritance in the Great Pyramid* (1864), which went into five editions by 1890, and in his *Life and Work at the Great Pyramid* (1867). He claimed not only that the monument was based on an equivalent of British imperial units but also that it should serve as the world's prime meridian because of its position and its embodiment of metrological authority. Piazzi Smyth's views on imperial metrology were warmly acclaimed—notably, by robust imperialists and patriots already persuaded that the Great Pyramid was a monument to God; that its base units were essentially imperial; and that imperial metrology was ordained: "The whole Christian and scientific world [is] under a deep and lasting obligation," wrote William Cooke of London to Piazzi Smyth. W. S. Chauncey, an engineer from New South Wales, wrote to Piazzi Smyth: "I greatly admire the resolute way in which you uphold all that is really excellent in our English standards."[33]

Why should Piazzi Smyth think this? Why does this matter in terms of the geographies of the prime meridian? In April 1860 Herschel had contributed to the debates on metrology in letters to the *Athenaeum*, later published as a pamphlet "On a British Modular Standard of Length." Herschel there drew attention "to a simple numerical relation between our actual parliamentary standard of length and the dimensions of the earth." This, he continued, "puts us in easy possession of a 'modular system,' which might be decimalized, and which, abstractedly considered, is more scientific in its origin, and, numerically, very far more accurate than the boasted metric system of our French neighbours. It is simply this,—if the British Imperial standard inch were increased by one thousandth part, it would be, with all but mathematical precision, one five-hundred-millionth part of the earth's axis of rotation." For this fact, which he corroborated by reference to other European astronomers, Herschel acknowledged that he was indebted to John Taylor, who claimed in *The Great Pyramid* that the diameter of the earth at the Pyramids was exactly 500 million "English inches." Herschel was proposing a new metrological system—the "British modular system"—one easy to use and which, he argued, required no legislation to make official: "It is so easy to convert 'Imperial standard' lengths, of whatever denomination, into 'British modular' lengths of the same denomination by subtracting (or modular into imperial by adding) one thousandth, that it is not worth while to legislate on the subject." Herschel did not agree with Taylor in the latter's view that the diameter of the earth at Giza was 500 million English inches. He believed only that it was very near that and that with adjustment—"all but mathematical precision"—such a unit could be used to delimit the earth. Herschel had earlier objected to the metric system on the grounds that the formal definition of the length of the meter—that is, one ten-millionth part of a quadrant of the earth—was unsatisfactory. He was dissatisfied because the definition depended on the length of a particular meridian, the Paris meridian, and upon the measurement of an earth that was not uniformly elliptical. Herschel was right about the imprecision of the meter. Herschel's 1860 modular system proposals were no more precise—but his was an imprecision of a different sort—slightly less than 500 million British inches for the length of the polar axis: *if* the length of the British inch were to be increased by a thousandth part, or the thickness of a human hair, the earth's polar axis would be exactly 500 million British "geometrical inches," and the geometrical inch would become a scientific, Earth-commensurable unit of length.[34]

Herschel did not adopt Taylor's specific claims, but Piazzi Smyth placed much weight upon Herschel's engagement with Taylor's work, and from his own reading of Taylor and the measurements of the Great Pyramid, he drew together a case for its metrological significance. Taylor had argued that the Great Pyramid enshrined British imperial units (give or take the thickness of a human hair, or "a spider's line," as one critic later put it). Because of its position, reasoned Taylor, very nearly on the thirtieth parallel of north latitude, the Great Pyramid's interior temperature corresponded to the mean temperature of the entire inhabited world. Further, there was more land and less sea on an arc of the meridian north and south of the Great Pyramid than for any other point on the earth's surface. The pyramid had been built, it was finally claimed, by a great number of people migrating eastward specifically to construct the monument, using what were now British units, as testimony to the glory of God. In short, for metrologists disposed to dismiss the revolutionary, fallible, and Earth-incommensurate meter and to see authority in imperial units, the Great Pyramid was a global point of origin—a "Metrological monument," as Piazzi Smyth had it—embodying the biblical ordination of British imperial standards.[35] It helped, of course, that Egypt was also increasingly inside Britain's imperial orbit of power.

At the same time, advocates of metrication were making their metrology move. In addition to debates in Westminster and in Congress, the meter was promoted through international scientific meetings; notably through the International Geodetic Association, which first met in Berlin in 1864 and again in 1867, and by the International Metric Commission in 1870. At its Newcastle meeting in 1863, the British Association for the Advancement of Science established a twenty-four-man committee, which included George Biddell Airy, with a brief to report upon the "best means of providing for a uniformity of weights and measures, with reference to the interests of science." The committee strongly advocated the use of the metric system for scientific purposes, pointing out the advantages of the metric system in scientific research and teaching and urging its formal adoption in all walks of British life. This view was a significant prompt to adoption of the metric system in scientific measurement the following year. To the American metrician Charles Davies writing in 1871, "The mind of the civilized world, brought into sympathy and close connection by the Wires of the Telegraph, is now earnestly directed to the question of uniformity in the language of business relations; and the Metric System of France is presented as a means of effecting such uniformity." To the editor of *Nature* in 1870, the "battle of the standards" was already over: "The Metre has gained the victory." When the International Metric Commission oversaw the Metric Treaty in 1875—"a landmark in the history of measurement, international cooperation, and globalization," as one commentator put it—what many nations, with the exception of Britain, were accepting was easily usable and certainly shared, but it was not a natural Earth-commensurate standard.[36]

Piazzi Smyth returned to the Great Pyramid and to the question of the world's prime meridian in 1883 as a member of the Committee on Kosmic Time and the Prime Meridian and in an intellectual and political climate energized by debates over metrology. This committee was established in 1879 by a group of scientists anxious to preserve imperial units in America, the Cleveland-based International Institute for Preserving and Perfecting Weights and Measures.[37] The committee's report, titled *What Shall Be the Prime Meridian for the World?*, was published in Cleveland in June 1884, as we have seen (Chapter 2). In the report Piazzi Smyth commented upon the four sites then being debated for the world's single first meridian—"a prime meridian for all mankind from this time henceforward," as he put it: Alaska (the Bering Strait), Washington, Greenwich, and the Great Pyramid of Giza. The Bering Strait option he dismissed because it did not permit measurement. Too much of it passed over the ocean "where nothing can be fixed, nothing exactly known, nothing certainly referred to. . . . How these negative qualities can be considered an advantage for enabling the exact science of modern times to trace out a definite, easily distinguishable, and permanent 'prime meridian line'—from, and in terms of, which every existing government in the whole world is to measure its difference of longitude with surpassing accuracy—is beyond my comprehension." Washington he dismissed as reflecting the interests of America. Greenwich was too British and too occidental to serve "either the two hundred millions of souls of fellow subjects governed by the British in India, or of the many hundred millions ruled over by Russia, China and Japan." In contrast, the Great Pyramid of Giza was eminently suitable:

The meridian of the Great Pyramid passes over solid, habitable, and for ages inhabited, land through nearly the whole of its course from north to south. Its line is capable therefore of being laid out along almost all that distance by trigonometrical measurement, and marked by masonried, station signals; and that is the only unquestionably accurate, permanent and sufficiently visible method of setting forth the one base for longitude measuring in the future before all varieties of men. . . . The Great Pyramid itself, has occupied its position, indexed too remarkably by Nature, through more than 4,000 years; and after all that lapse of time and growth of science therein, is acknowledged to be still, above its other high fulfillings, the grandest as well as best built surveying station-mark and monument that has ever been erected the whole world over.[38]

To the International Institute for Preserving Weights and Measures, Piazzi Smyth's arguments established—in their minds beyond doubt—that those who had built the Great Pyramid "so planned and shaped it as to make it a combination of geometrical, geodesical, and stellar truths, worthy to be the grand standard of a world-system of weights and measures."[39] In his preface to *What Shall Be the Prime Meridian for the World?*, Charles Latimer had made it clear that the purpose was remedying the "confusion and inconvenience arising from the adoption of diverse meridians." As his report showed and as Chapter 2 disclosed, America's scientists and politicians were far from in accord over the world's prime meridian, over Washington's validity as a legitimate candidate (despite Congress having determined it so in one respect in 1850), and over the units to be employed in science.

Seen against the seemingly rational choice of Greenwich for the prime meridian from October 1884, these discussions over the Great Pyramid, imperial standards, and the "pyramid inch" strike the modern reader as singularly odd; a mix of biblical zeal, mysticism, numerological delusion, and unjustified historical scholarship. So interpreted, the case of Giza is a distraction from the "proper" history of the prime meridian in Paris, Greenwich, and Washington. That is to miss the point. John Taylor's and Piazzi Smyth's claims over pyramidical units, the possibility for the Great Pyramid as the world's prime meridian, and Herschel's views over standards were all part of a formative context over competing metrologies. As Schaffer has shown, the Great Pyramid and Egypt as a whole were of cartographic interest to Henry James and Ordnance Survey from 1866 in the wake of their Southampton-based metrological calibration. The Near East and metrology were also at the heart of what was becoming British archaeological science at this time. The Great Pyramid was of importance as a sight line to Airy in his astronomical observations over the transit of Venus in 1874. With Roy, Kater, Arago, Delambre, Méchain, and others, the views of Piazzi Smyth and his contemporaries over that possible site were also part of long-running narratives over different standards, precision, uniformity, and accuracy: the very "stuff" out of which science emerged in this period.[40]

Precision and the units selected for its measurement are not innate. Authority over accuracy lies in the claims made about measurement, not in the units themselves. Nature does not provide the means to its own revelation:

meters, yards, railway time—and prime meridians—are social constructions invented and mobilized for particular purposes.

o　　o　　o

The evidence presented here with reference to the prime meridian in the period from 1790 to 1884 affirms the importance of the themes revealed in earlier periods. Fixing the location of different prime meridians—notably, in this period, Paris and Greenwich—continued to be a matter of importance to astronomers, geographers, mapmakers, and politicians. Fixing the position of different prime meridians depended upon accuracy and precision. The first term encompasses the values arrived at in determining, say, an observatory's position astronomically or geodetically relative to a presumed true value. The second term means the degree to which the results of measurement, by different people in different places and perhaps even using different means, show the same result. The prime meridians of Paris and of Greenwich continued as a focus for particular attention because the work of Roy, Kater, Arago, Herschel, and others was at once tolerably accurate and persistently imprecise.

This chapter has also revealed how the several prime meridians in use during the nineteenth century were measured. It has shown that their delimitation was in various ways part of the gradual, geographically uneven, and socially contingent emergence of science itself. Triangulation was made to work with gunpowder and rockets. It worked too with better timekeepers as corroborating devices. Even in the hands of skilled observers, improved instruments might offer different readings despite the operator's skill and protestations of precision in the published results. Because of differences in metrology and in timekeeping, uniformity was more sought after than realized, whatever the form of measurement used. Universality, the consistent application everywhere of a standard unit established by a community of practitioners, remained only an end in view.

Like triangulation, trigonometry, and rocketry, telegraphy promised much. Airy's galvanic connections were important in establishing a new means to delimit national space and to connect practitioners across international space. But in association with the expansion of the railways, the advent of telegraphy revealed more starkly than before the persistence of different times—and of different measuring units—within various national contexts. This was a world of metrological inconstancy.

4

Globalizing Space and Time

Getting to Greenwich, c. 1870–1883

THIS CHAPTER EXAMINES the debate over the prime meridian and universal time between Otto Wilhelm Struve's influential paper on these topics in Saint Petersburg in 1870 and delegates' deliberation upon them at the International Geodetic Association (IGA) meeting in Rome in 1883. It discloses how the linked questions of a single global prime meridian and of universal time were articulated and argued over in print, in speech, and in scientific settings—notably, at the International Geographical Congresses (IGCs) between that of Antwerp in 1871 and that of Venice in 1881—to become the subject of international consensus not in Washington in 1884 but in Rome in 1883. What follows draws upon the cases made for different prime meridians and the gathering sense, expressed in various settings during the nineteenth century, that while a single global prime meridian might be desirable there seemed no easy way of achieving it.

Several American scientists voiced this view as they debated Charles Davis's mid-century proposal to use New Orleans. In France the prime meridian was several times the subject of review in the Paris Geographical Society from the mid-1840s. In an address in 1844, geographer and ethnographer Jean Baptiste Gaspard Roux de Rochelle, three times the society's president, stressed the preeminence of Cap Ferro even as he identified other prime meridians then in use. He called for a single line to serve that "republic of letters and of science which embraced the whole world," but no action was taken. In March 1851 through the efforts of orientalist and historian of science Louis-Piérre-Eugène-Amélie Sédillot and geographers Roux de Rochelle, Antoine d'Abbadie, and Edmé-François Jomard, the society called upon the governments of the "Principal States of Europe and in America" to adopt a common prime meridian in order to settle, once and for all, the use by different countries of diverse longitudinal base points. Their solution, an imaginary line "au milieu de l'Ocean," was effectively a later version of the early modern agonic prime meridians associated with the Azores (see Chapter 1). Later reports in the society's bulletin show a rather episodic enthusiasm for a single prime meridian. By then the issue was the subject of international attention.[1]

In Britain, Henry James of Ordnance Survey, whose work in the early 1860s on comparative metrology we have noted, was by 1868 unequivocal on the matter, perhaps as a result of the difficulties he and his colleagues encountered as they worked with varying units of measurement: "It is greatly to be desired that a first meridian and a uniform system for maps should be adopted for all nations." James stressed the advantages of cartographic uniformity—"the series of maps made in each country would then exactly correspond." His views on the world's prime meridian left little room for doubt (although his reasons were not strongly scientific): "As Greenwich is nearly in the centre of the habitable portion of the globe . . . the meridian at Greenwich should be adopted as the first meridian for all nations."[2]

What distinguishes discussion about the prime meridian and universal time in Saint Petersburg, Paris, London, Venice, Rome, and in other places between 1870 and 1883 from earlier debate is its overtly international, even global, tenor.[3] The change can be explained in part by the rising international profile of the specific issues of metrology and the standardization of time and of geodetic measurement already discussed. It may be explained, too, by the ways in which science more generally was increasingly internationalized during the late nineteenth century. For Elisabeth Crawford, "the universe of international science" in the half century from 1880 was distinguished by three features: cognitive homogeneity (the sharing of common problems and methods within and among discrete disciplines), standardized communication (with international associations and journals increasingly crossing national and disciplinary borders), and new agreements over technical standards (as with the metric system, for example, but also with shared methods in the sciences more generally and in an emerging consensus over the conduct of scientific research in laboratories and in universities).[4]

These issues are explored under three headings. The first examines how and why and where the prime meridian became a topic of debate in international geographical meetings between 1871 and 1881. The second examines proposals made from the 1870s over universal time, with particular reference to the work of Sandford Fleming and his emphasis upon a single prime meridian as the first stage of cosmopolitan timekeeping. The third explores how these issues came together at two meetings, the 1881 IGC in Venice and the 1883 IGA meeting in Rome, and examines how these meetings helped shape the Washington conference of 1884.

Internationalizing Geography and the Prime Meridian, 1870–1881

A key moment in shaping the geography of the prime meridian as an international issue came in early 1870 in Russia. Much as geographer Roux de Rochelle had done in Paris in 1844, astronomer Otto Struve began his talk to the Imperial Russian Geographical Society in Saint Petersburg in early February 1870 by summarizing the prime meridians in use. For Struve, three prime meridians were predominant, then and in earlier periods. These were the (in origin) Ptolemaic line known and named variously as Ferro, Cap Ferro, or Hierro; the Paris meridian, as taken from the Paris Observatoire; and the Greenwich meridian, taken from that observatory. Struve took it as understood that Ferro was exactly 20° west of the Paris meridian. If he knew of those early eighteenth-century debates over the arbitrariness of Cap Ferro's "accurate" positioning vis-à-vis the Paris Observatoire—a "Meridién Fictif" as Carney had put it in 1797—he made no mention of them. Nor did he mention attempts ongoing since the later eighteenth century to position the prime meridians of Paris and of Greenwich relative to one another. Unlike Roux de Rochelle and his Parisian compatriots, who favored a single prime meridian based on the Azores, Struve declared that, of the three, the best choice for the world's prime meridian was Greenwich. As a leading European astronomer and as someone who had undertaken significant metrological and survey work in Russia and Eastern Europe, Struve's view in 1870 concerning Greenwich had enduring influence. To appreciate why, we need to consider both his reasons for advancing that prime meridian over others and the means by which his arguments traveled to be debated elsewhere.

Struve stressed the requirements of science. Science, he said, should take precedence over national interests and customary usage regarding different prime meridians: "The question of the unification of meridians does not depend on any consideration of political economy, it concerns the scientific world alone." Struve endorsed moves toward metrological uniformity, by which he meant the advantages of the metric system and its wider take-up through telegraphy, in geodesy, in astronomy, and for topographical surveys. He used evidence from education and cartography to support his case. Although Ferro was the most commonly used prime meridian in elementary school geography books and atlases, particularly in Germany and in Eastern Europe (French geography books and atlases used either Ferro or Paris, sometimes both), Greenwich was the most widely employed on maps and charts "of a scientific nature" (and in British geography and astronomy texts). He particularly endorsed the utility—because of its accuracy—of Britain's *Nautical Almanac and Astronomical Ephemeris.* In Struve's view *The American Ephemeris and Nautical Almanac,* first published in 1855, compared well to the British text in terms of accuracy. But because of America's congressional decision in 1850 to adopt two prime meridians for two different purposes, *The American Ephemeris and Nautical Almanac* incorporated tables based on the Washington prime meridian employed by the U.S. Naval Observatory as well as tables based on the prime meridian of Greenwich. If America's *American Ephemeris and Nautical Almanac* was a hybrid, France's *Connaissance des temps* he reckoned no longer accurate. Russia, Struve reminded his audience, had stopped producing its own ephemeris in 1853 in favor of the British text.

Struve's views on Greenwich and his justification for it was an endorsement of the universal utility of science. But Struve cautioned that the possible future adoption of Greenwich would not be without problems. Longitude was not used in standard ways, either on maps or at sea. Because a prime meridian based on Greenwich bisected Europe and Africa, its adoption as the world's initial meridian would mean that longitude would have to be signaled as either positive or negative in the different "halves" of those two continents. Should Ferro be used, no such system would be necessary since for Europe and for Africa all longitudes would share the same sign. Ferro split no continents. Struve explained that other prime meridian lines that did likewise—that is, which did not divide populated land masses and which were based either on exact hours from Greenwich or multiples of 15° (one hour in time)—were also possible. One lay within the Atlantic, 30° west of Greenwich. The other, at 180° from Greenwich and so effectively its "antimeridian," passed through only part of that almost entirely uninhabited peninsula of Asia in the Bering Strait and corresponded more or less with that area of the Pacific Ocean where navigators added or subtracted a day in their reckoning of time. This possibility for the world's prime meridian—what we may later think of as the "Bering Strait option"—made good sense if its adoption as a prime meridian could be linked to the adoption of universal time and a standard civil day.[5]

Struve's 1870 paper summarized the key arguments that had to be invoked in choosing a single prime meridian on internationalist principles: scientific utility, universal good, practicability, association with the regulation of time, and the repeal of national interest and established practice. It was, notes Bartky, "the first salvo in what turned out to be a half century of skirmishes aimed at having the world adopt Greenwich as the common meridian for longitudes."[6] Struve's paper became widely known, in part because his central arguments were summarized in a French geographical digest. Vivien Saint-Martin, vice president of the Paris Geographical Society and the periodical's editor, made clear Struve's reasons for favoring Greenwich in summarizing the Russian's talk—and added his own thoughts on the imperatives of the metric system.[7]

Struve's paper also featured in discussions on the prime meridian at the world's first international geographical meeting, the Antwerp IGC on August 14–22, 1871. Delegates to Antwerp came as professional geographers, cartographers, or members of regional societies or scientific bodies, not as official representatives of their governments. Geography was not then the established, professionalized, and institutionalized subject it would later become. That is not to say it was not widely practiced throughout Europe and in North America. The subject was routinely taught in schools, in universities, and in military academies. There was a diverse and well-established publishing industry devoted to geographical schoolbooks, gazetteers, and atlases. A range of practical activities under the collective label "exploration" was apparent in new and revised maps. Exploration commanded enormous public interest and extended the geopolitical and commercial reach of European nations in particular. National geographical societies appeared during the nineteenth century to direct and report upon these activities. Twenty-two such bodies were founded before 1871: the Paris Geographical Society (1821); the Berlin Geographical Society (1828); the Royal Geographical Society in London (1830); the Imperial Russian Geographical Society (1845); and the Geographical Society of New York (1852), to name only a few. In the ten years from 1871 to 1880, thirty-nine geographical societies were established. Across Europe and North America, these and other civic institutions helped shape geography as a sternly empirical science of empire and of commerce.

Yet professional geographers were few and far between and for the most part produced narratives of exploration or topographical maps, sometimes both. Geography departments in universities appeared only late in the

nineteenth century. For all its presence and practices, geography was not everywhere an agreed-upon science nor yet international in its languages and
epistemological reach. For these reasons Antwerp and later geographical
meetings should not be seen as the fully formed associational outcomes of
a subject whose intellectual definition was certain and whose practitioners
shared common problems in standard ways. Discussions about a single prime
meridian and its global utility were not a consequence of geography's established disciplinary identity and shared cognitive content; the relationship
might almost be understood in reverse. That is, the debate over a single global
prime meridian in late nineteenth-century international meetings did not
reflect geography's emergent status as an international science so much as
help produce it.[8]

Initial Discussion: Antwerp 1871

The Antwerp meeting was a new beginning for the sciences of the earth. The
vice president of the Organising Committee, Charles d'Hane-Steenhuyse,
intended the gathering to "arouse the geo-cosmographical sciences from
their long sleep." One of the features of the first IGC was the emergence of
structures that helped to imagine and manage what such a congress should
be. For the purposes of academic discussion, parallel sessions or "groups" were
organized to focus upon delegates' areas of common interest: later, delegates
met in a general session to vote on agreed-upon specific questions. This general session became the executive forum or congress of the IGC.

In Antwerp the prime meridian was the focus of attention in groups
devoted to cosmography, navigation, and international commerce. The chair
of the cosmography group was the British naval officer and Arctic explorer
Vice Adm. Erasmus Ommaney. Ommaney was active within the British
Association for the Advancement of Science and was a council member of
the Royal Geographical Society in London. Uniformity in cartography in
mapping scales and symbolization and the crucial question of a single initial
meridian were the main topics addressed by Ommaney's group. The issue of
a common first meridian came before delegates six times in group and collective discussions during the Antwerp meeting. Because the substance of
Struve's 1870 Saint Petersburg lecture had been outlined in an early general
session, delegates soon understood what we can think of as the parameters
of the prime meridian question as an international matter: Should there
be one at all? Which of several candidates should it be? On what grounds
should it be chosen?[9]

Ommaney presented his fellow delegates with a single direct question: "Could we not agree to adopt the same prime meridian?" For some delegates the answer lay in the advantages to science (their responses emphasized issues of accuracy and error). Others stressed matters of daily practicality (common usage and convenience or inconvenience) or the need for cartographic and metrological standardization. Delegates understood that having so many prime meridians was awkward given the constant need to calculate the longitudinal and astronomical differences among them. The use of different prime meridians heightened the risk of accidents at sea from having incommensurate base points in navigation. Cartographers stressed the desirability of having a standard geographical o° base point, keen as they were to effect greater standardization in map production. Most delegates considered Britain's *Nautical Almanac and Astronomical Ephemeris* to be the most accurate ephemeris. Later, reporting on the Antwerp meeting to colleagues in the Royal Geographical Society, Ommaney drew attention to the fact that "several questions of international importance" were dealt with there, including "the possibility of adopting the same first meridian by all nations," but neither he nor the society's council took the matter any further.[10]

In Antwerp the proposal was made that Greenwich be adopted as the world's initial meridian, as Struve had suggested. Delegates also understood, however, that the Paris prime meridian was widely used in scientific studies and had a claim on primacy. As a result of these contrasting views, the proposal finally arrived at was limited in reach and due to its wording likely to always be limited in effect: "The Congress expresses the opinion that for maritime routing charts [that is, pilot or navigational charts], a first meridian be adopted, that of the Greenwich Observatory; and that after a period of time, say in ten to fifteen years, this initial position is to be made absolutely obligatory of all charts of this nature."[11]

This resolution, voted upon and accepted in a general session, is the first expression of a conjoint international resolution to name a single initial prime meridian for the world. It was, however, weak, being concerned with only one sort of map employed by one user community—the navigational. As a result it was never likely to be formally ratified outside this geographical gathering. The Antwerp IGC had no authority to insist that governments accept it, and not everyone thought it sound. The proposal left geographers and topographical surveyors using different o° start points for their maps. Astronomers continued to use the observed prime meridian relative to their observatory. The status quo ante was maintained: differences in practice existed among various users and adherents to different prime meridians. For French hydrographer Adrien Germain, secretary to group I in Antwerp, the obvious candidate for global adoption was Paris, not Greenwich, given the Paris Observatoire's historical importance and because the French employed *le meridién/ la meridiénne* as an observed prime meridian on maritime charts and in the nation's topographical mapping and geographical teaching. To Émile Levasseur, however, Germain's French colleague in Antwerp, the precedence of Paris was undeniable, but the choice—on practical grounds—had to be Greenwich, since the great majority of nautical charts in existence used Greenwich.

Antwerp, in short, was important in marking the first international resolution toward a global prime meridian formulated through collective discussion. But the Antwerp IGC itself achieved almost nothing.[12]

Continuing Debate: Paris 1875

Otto Struve's 1870 paper and his recommendation of Greenwich was the subject of an address given to the Paris Geographical Society in May 1874 by naval officer Guidoboni Visconti. This was translated and published in the society's bulletin early in 1875. Visconti's attention to Struve's work had a local significance in Paris of which he may not have been aware.

Paris and the geographical society there played host to the second IGC in August 1875. The local organizer, Adrien Germain, had spoken out in Antwerp on behalf of the Paris prime meridian and against the claims of Greenwich and Struve's recommendation. As plans for the Paris IGC developed, Germain used his position to limit the opportunities for collective discussion of the prime meridian lest delegates reach agreement over Greenwich. To Germain, no country should have to adopt a neutral shared meridian, France least of all. Germain stressed the status of the Paris prime meridian and denied Struve's claim that the French *Connaissance des temps* was an inferior national ephemeris. While this might have been true in 1870, over the years it had subsequently been "much improved in its accuracy" and was now every bit the equal of Britain's *Nautical Almanac and Astronomical Ephemeris.* In acting thus, Germain set a tone for the Paris meeting that at least initially kept the Paris prime meridian in delegates' minds and sought to limit collective debate over a single prime meridian in case discussion strayed in unwelcome directions.[13]

Participants in the specialist groups, however, which were the key academic debating spaces in the IGC, did discuss the question. Group II— hydrography and maritime geography—returned to the issues of cartographic uniformity, Greenwich's predominance on the maritime charts, and the Antwerp proposal, but only to reiterate the need to adopt a common first meridian. The delegates of group I—which embraced mathematical geography, geodesy, and topography—took the issue further. For them, the choice of a shared initial meridian could not meaningfully be separated from questions of a uniform metrology, universal time, and the common good of a shared baseline for the world's measurement. Alexandre-Émile Béguyer de Chancourtois, the French mineralogist and superintendent of the mines, favored a prime meridian that ran through the Azores. He argued his case on the grounds that it more or less neatly separated the Old World from the New and that its position in the mid-Atlantic made it a good baseline against which to change the civil day. Others argued for a prime meridian in the Pacific between the Asian and American continents at 180° from either Paris or Greenwich.

British geographical representatives were less involved in Paris in 1875 than Ommaney had been in Antwerp in 1871. They were so in part because of the Royal Geographical Society's rather haughty attitude toward the Paris IGC as it was being planned. Late in 1874 the president of the Royal Geographical Society had requested that Britain's foreign secretary, Lord Arthur Russell, should write to his French counterpart to advise him "that at present the English Government show no inclination to take any official part in the Geographical Congress & Exhibition as not being held by desire or under the Authority of the French Government, but as the action of a private Society." Whatever justification there may have been for this prickliness over protocol, the differences in attitude thus exposed over the prime meridian as a governmental problem would be more starkly revealed in later meetings—and not just by the British.[14]

In Paris a representative of the Geographical Society of Geneva argued that the world's prime meridian should center upon Jerusalem. It was "neutral territory"; no country could object to siting the world's baseline there. Opponents argued that the initial meridian could not be sited there because there was no observatory against which to position an observed meridian. This view was simply countered: the line should be determined first as a statement of universal or transnational principles, with the astronomical observatory built afterward to affirm this collective enterprise. Others argued for Ferro. Otto Struve revised his 1870 recommendation concerning Greenwich, arguing now on the grounds of the greatest neutrality among scientific nations. On this basis, argued Struve, the world's prime meridian should be the antithesis of Greenwich, 180° from the observatory there.

To judge from this evidence, Germain's efforts in Paris in 1874 and 1875 to limit discussion over the world's prime meridian were only partially successful. Antwerp began the internationalization of the prime meridian in formal meetings yet ended in a weak and limited proposal. Debate within Paris highlighted persistent national differences, even, as with the Jerusalem proposal, adding to the confusion. Yet it is possible to see in the discussions in Paris further shared recognition of the virtues of a single meridian, although the reasons advanced reflected the interests of one national or scientific community over another. Broadly, astronomers were content with different national observed prime meridians and with the calculations this demanded. Some, such as French astronomer and historian of navigation Antoine François Joseph Yvon-Villarceau, saw merit in the universal acceptance of a common initial meridian, provided related amendments were made to the reckoning of time. Metrologists and cartographers emphasized the virtues of a common initial meridian for topographical maps. Maritime communities looked commonly to Greenwich, as they had done in Antwerp, but Greenwich was not strongly endorsed by others. The few British in Paris remained silent, perhaps because they had no intention of using any prime meridian other than Greenwich.

At Paris, participants in group I were sufficiently convinced of the merits of a common initial meridian to vote upon the issue, although strictly with reference to its use in world maps and atlases. The prime meridian they chose, by a large majority, was Ferro. This was taken to be 20° west of Paris. Delegates proposed that this point should be described by all atlas publishers as "the meridian of origin." The IGC forum did not accept this recommendation (records are silent as to why). It was perhaps because of Germain's influence, because French delegates outnumbered those from other countries, because British voices were relatively silent, or because delegates recognized that the Paris prime meridian had considerable legitimacy as a world-leading prime meridian.[15]

International Differences: The Prime Meridian
and Geographical Societies, 1875–1881

The issue of a common initial meridian continued to be debated within geographical societies and elsewhere precisely because of perceived shortcomings in the IGC. Reporting to colleagues in Geneva, Swiss delegates to Paris identified three key stumbling blocks to agreement: multiplicity (the existence of many prime meridians), confusion (caused by adherence to multiple initial meridians for differing purposes), and inertia (lack of agreement over which meridian to use for a specific purpose or for all purposes). In their view the proposals agreed to in Antwerp in 1871 regarding maritime charts and the recommendation over topographical mapping in Paris in 1875 were both failures. But what those meetings did do was put the prime meridian on the agenda of national geographical bodies and in the public's mind.[16]

In 1876 in a paper to the Italian Geographical Society in Rome, the president of the Geographical Society of Geneva, Henri Bouthillier de Beaumont, lent weight to Struve's amended view over the initial meridian by proposing that it run through the Bering Strait, from which position longitude would be counted east and west of this line. De Beaumont had been present at the Paris IGC and following that meeting had deliberated for a year upon the advantages of multiple prime meridians. By 1876 he declared himself in favor of a prime meridian that ran through the Bering Strait, 150° west of Ferro. This was not the same solution Struve had advanced in his 1875 revision: that the prime meridian be 180° from Greenwich. The extension of his meridian over both poles would, de Beaumont argued, form a meridian arc passing through numerous European countries and the African continent—but not through any nation's capital or an astronomical observatory. This positioning for what he called the *médiateur*—perhaps in reference to the equator (Fr. *equateur*) or because his proposition diplomatically mediated between competing options—was, he continued, the best solution for a global prime meridian that was everyone's and no one country's.[17]

As with Struve's 1870 Saint Petersburg paper, de Beaumont's 1876 Rome paper was influential beyond its local context. Copies of it were distributed first in original and later in expanded form. His idea was picked up by geographical societies, at international geographical meetings, and by the popular press. The prime meridian was not, however, debated at the 1876 Brussels international geographical meeting, which focused on the division of sub-Saharan Africa by Europe's imperial powers. It was next raised at the International Congress of Commercial Geography in 1878 in Paris, where a resolution was passed drawing attention to the inconvenience of having numerous prime meridians, but no solution was offered. In the 1879 meeting of this body in Brussels, delegates again recognized the collective benefits of a common shared meridian (many of them proposing Greenwich), but after a day's debate, they proposed only what was in effect the same limited resolution that had been advanced in Antwerp. The French journal *L'exploration* summarized de Beaumont's work in early January 1879, as did the *Times*

(London) and *Nature* the following month. The June 1879 edition of *Popular Science Monthly* lauded the term "mediator" in its summary of the prime meridian problem, the author declaring himself in favor of de Beaumont's "project," seeing in its use advantages to many nations—"the common property of all the civilized nations . . . neutral ground, a position independent of all political power, and under guarantee of all the states of the civilized world."[18]

From an American perspective, Chief Justice Charles P. Daly, mayor of New York and president of the city's geographical society, considered de Beaumont's plan a possible solution to continuing national differences. In 1879 in his annual review of geographical matters, Daly reported that "several attempts have been made during the past few years to get the nations of the world to agree upon a common meridian, instead of having each adhering to its own, like the meridians of Greenwich, Paris, Washington, &c." Since "the meridian of Greenwich is the one most extensively found on maps and charts," Daly observed how Americans have "generally been disposed to adhere to it, and would probably be quite willing, as a nation, to unite in its adoption." But, he continued, when the question of doing so had arisen at the IGC at Antwerp, "the disposition of the French members was to adhere to the meridian of Paris, and as it seems to be difficult to get one nation to adopt the meridian of any other, the object might be effected by adopting a common meridian solely upon geographical grounds."

Daly reckoned de Beaumont's proposal for the prime meridian to be sited in the Bering Strait 150° from Ferro as sensible for several reasons: It could "be very easily connected" (in astronomical and mathematical calculations) with other "principal meridians," and it would divide Europe into east and west, "thus giving a division which has been tacitly recognized for ages." Passing as it would through many nation-states, "it would become really an international meridian, as each nation might establish a station or observatory on the line of it." In essence, Daly judged de Beaumont's proposal expedient:

As this would be, for the reasons above suggested, a very desirable first meridian, and as there appears to be no other way of getting over the disposition of nations to adhere to their own and of avoiding the confusion of having so many, I fully concur in M. de Beaumont's suggestion, and hope, as a practical relief from an existing difficulty, that it may hereafter be generally adopted. There is in my opinion so much good sense in the suggestion, that I think, in the course of time, the merit of it will become so fully recognized, that the force of public opinion, in each nation, will ultimately lead to the adoption of a line in which all can agree, without interfering in any national preference.[19]

Daly's hope that the solution would be arrived at from "the force of public opinion" was misplaced. Editorial comments in the *Times* and other outlets on de Beaumont's proposal were generally favorable, but they were few. There is no evidence that they prompted more general debate upon the matter. Differences thus continued by nation, in terms of customary usage, within the scientific world, and among the geographical community. By 1879 and even by early 1880, little had changed in either Europe or North America to alter the centuries of geographical confusion, with the inconsequential exception of that international congress of European meteorologists that recommended the worldwide adoption of Greenwich in the production of synoptic maps of climate and weather.[20] This state of affairs, however, soon came under further scrutiny.

Toward Resolution: Venice 1881

The third IGC was organized by the Italian Geographical Society and held in Venice in September 1881. There, the prime meridian was group I's principal business. Unlike the meetings in Antwerp and Paris, the Venice IGC was distinguished by the presence of delegates from North America representing organizations that had not hitherto been formally involved in IGC discussions. Their interests centered upon metrology, measurement, and mapping, and they included the American Metrological Society, the U.S. Army Corps of Engineers, and the American Geographical Society, among others. This involvement would prove to be important, not least ensuring the renewed interest of the U.S. Congress regarding the prime meridian.

Discussing the confused state of affairs over the world's prime meridians that delegates in Venice sought to remedy, George Wheeler of the U.S. Army Corps of Engineers drew attention to the extent of the problem (citing both Pierre-Simon Laplace and Henry James of Britain's Ordnance Survey in doing so):

Upon examining specimens of the extended general topographical map series of Europe fourteen separate and independent meridians of reference are found: (1) Greenwich, for the United Kingdom and India; (2) Paris, for France, Algeria, and Switzerland; (3) Lisbon, for

Portugal; (4) Rome, for Italy; (5) Amsterdam, for Holland; (6) Isle of Ferro, westernmost of the Canaries, for Prussia, Saxony, Wurtemberg, and Austria; (7) Ferro and Christiana [Oslo], for Norway; (8) Copenhagen, for Denmark; (9) Madrid, for Spain; (10) Stockholm and Ferro, for Sweden; (11 and 12) Ferro, Pulkowa, Warsaw, and Paris, for Russia; (13) Brussels, for Belgium; and (14) Munich, for Bavaria.

Both Greenwich and Washington, principally the former, have been used for maps of land areas in the United States.

In addition to these topographical cartographic prime meridians, Wheeler identified those in use in hydrographic mapping: "The meridian of Greenwich is used on the Government marine charts of England, and India, Prussia, Austria, Russia, Holland, Sweden and Norway, Denmark, and the United States; while France employs Paris; Spain, Cadiz; Portugal, Lisbon, and Naples is found on some Italian as well as likewise Pulkowa on certain Russian hydrographic charts." Further, noted Wheeler, "As nautical and astronomical tables came into more general use a number of meridians of reference were established, as at Toledo, Cracow, Uranibourg, Copenhagen, Goes, Pisa, Nuremberg, Augsburg, London, Paris, Rome, Greenwich, Washington, Vienna, Ulm, Berlin, Tubingen, Venice, Bologna, Rouen, Dantzig, Stockholm, St. Petersburg, &c."[21]

The debate on these national and thematic differences in Venice centered upon three papers. Paper "A" was given by Charles Daly of the American Geographical Society on behalf of his fellow American Frederick A. P. Barnard (Barnard being deaf). In 1881 Barnard was president of Columbia University and, as we have seen, of the American Metrological Society. It was from this latter position that in 1883 he inveighed against Piazzi Smyth and his views over imperial metrology, "pyramid units," and the suitability of the Great Pyramid at Giza as the world's initial meridian (Chapter 3). The Barnard-Daly paper argued that the prime meridian to be used was "the meridian situated in longitude one hundred and eighty degrees, or twelve hours distant from the meridian of Greenwich ... which meridian passes near Behrings Straits and lies almost wholly on the ocean." This view was shared by Gen. William B. Hazen, also of the American Metrological Society, whose paper "B" was given on his behalf by George Wheeler. Paper "C" was given by Sandford Fleming, the Scots-born railway engineer and chancellor of Queen's University in Kingston, Ontario, who spoke on behalf of the Canadian Institute of Science in Toronto and the American Metrological Society.

Fleming's paper was to the point: it argued for "the establishment of a Prime Meridian and Time-zero, to be common to all nations." Recognizing the need for "scrupulously avoiding offence to local prejudice or national vanity," Fleming's argument revolved not around the prime meridians then in use but upon "the relations of time and longitude and the rapidly growing necessity in this age for reform in time-keeping." Because time was in effect kept only locally and not measured worldwide in any standard or universal way, "there can be no absolute certainty with regard to time unless the precise geographical position be specified as an important element of the date." Fleming's paper gave notice that the coordinating systems for the world's measurement of space and time were at odds with one another, even as telegraphy and the railway were binding the world together. Prime meridians at different 0° and systems of timekeeping were "irregular . . . inconvenient . . . irksome." Fleming stressed the connections between a global prime meridian and universal time: "It is obvious," emphasized Fleming, "that the world's time-zero should coincide with the prime meridian to be used in common by all nations for reckoning terrestrial longitudes." The standardization of what he called "cosmopolitan time" was his and others' end in view, but "the first step towards its introduction is the selection of an initial meridian for the world."[22]

Fleming concluded his paper with seven resolutions, arrived at from discussion among the North American delegates (Table 4.1).[23] After further debate the delegates urged that the proposed "International Commission" should be made up of "scientific men such as Geodicians, Geographers and men who represent the interests of commerce, etc" and that each nation might be allowed three such representatives.[24]

These papers and resolutions in Venice focused on two proposed initial meridians, either the Bering Strait mediateur (150° from Ferro) or the Greenwich antimeridian (180° from the Greenwich Observatory), and upon the prime meridian's connections with universal time. Resolutions were advanced over the advantages to all nations of a single prime meridian. Washington is mentioned as a venue for the proposed meeting of an international commission. For these reasons it is tempting to regard the 1881 Venice proposals as formative precursors of the recommendations arrived at in Washington in 1884. More than any other delegate at Venice in 1881, Fleming knew that the views of scientists counted for little unless they subsequently made formal representations to their own governments with a view to legislation. There is certainly a sense among the scientists involved that Venice was important. Otto Struve even anticipated its significance. Fleming had

Table 4.1 Resolutions toward "the Unification of Initial Meridians" from Sandford
Fleming's paper to the 1881 International Geographical Congress

Resolution number	Resolution
1	That the unification of initial meridians of reference for computing longitude is of great importance in the interests of geography and navigation
2	That the selection of a zero-meridian for the world would greatly promote the cause of general uniformity and exactness in time-reckoning
3	That in the interests of all mankind it is eminently desirable that civilized nations should come to an agreement with respect to the determination of a common prime meridian, and a system of universal time reckoning
4	That the Governments of different countries be appealed to immediately after the close of Congress, with the view of ascertaining if they would be disposed to assist in the matter by nominating persons to confer with each other and endeavour to reach a conclusion which they would recommend their respective governments to adopt
5	That in view of the representations which have come to this Congress from America, it is suggested that a Conference of Delegates who may be appointed by the different governments be held in the city of Washington, and that the Conference open on the first Monday in May, 1882
6	That the gentlemen whose names follow be an Executive Committee to make arrangements for the Meeting of Delegates, and to take such steps as may seem expedient in furtherance of the objects of these Resolutions.[a] And that all communications in respect thereof be transmitted to General W. B. Hazen, Meteorological Bureau, War Department, Washington
7	That the Italian Government be respectfully requested to communicate these resolutions to the Governments of all other countries

Source: Sandford Fleming, *The Adoption of a Prime Meridian to Be Common to All Nations. The Establishment of Standard Meridians for the Regulation of Time, Read before the International Geographical Congress at Venice, September, 1881* (London: Waterlow and Sons, 1881), 13–15.
[a]The men listed were, in order given, Dr. F. A. P. Barnard, president, American Metrological Society, New York; Capt. George M. Wheeler, Corps of Engineers, Washington, DC; Chief Justice Charles Daly, president, American Geographical Society, New York; Justice Field, U.S. Supreme Court, Washington, DC; Gen. G. W. Cullum, vice president, American Geographical Society, New York; Gen. W. B. Hazen, director, Meteorological Bureau, Washington, DC; Judge Peabody, American Geographical Society, New York; Prof. Cleveland Abbe, Signal Office, Washington, DC; David Dudley Field, American Geographical Society, New York; James B. Francis, president, American Society of Civil Engineers, Boston; Dr. Daniel Wilson, president, Toronto University, Toronto; John Langton, president, Canadian Institute, Toronto; Sandford Fleming, chancellor, Queen's University of Canada, Ottawa.

written to him in late 1880 and had asked him to distribute copies of Fleming's pamphlets on time reckoning to Russian scientists and scientific societies. In reply, Struve wrote in January 1881: "You will know that the question of the first meridian was already discussed at the geographical congress at Anvers [Antwerp 1871], but there the French influence was too predominant. I expect the chances for carrying your propositions will be considerably more favourable at Venice. I should like to assist at the meeting, but I am not sure that I shall be able to do it."[25]

To appreciate the significance of the Venice 1881 papers, we have to understand how the IGC worked. Fleming's resolutions were proposals from only one set of delegates within group I, not the unanimous views of the group. Others in group I held fast to Ferro. One delegate proposed a variant to Ferro, setting his scheme for universal time and the prime meridian on Ferro's antimeridian, Kamchatka in the Russian Far East (Figure 4.1). Béguyer de Chancourtois, who in Paris in 1875 had proposed the Azores, offered several prime meridians and urged universal acceptance of the metric system as he did so. At the same time, he promoted his own scheme for a new system of geographical instruction that coincidentally combined different initial meridians and the metric system (Figure 4.2). Henri Bouthillier de Beaumont's appeal for a Bering Strait prime meridian would have placed London eleven hours and twenty minutes in advance of that mean midnight (Figure 4.3).

Given these and other differences, group I brought forward only a limited proposal: "Group I expresses the hope that within a year Governments appoint an international commission for the purpose of considering the subject of an initial meridian, taking into account the question of longitude, but especially that of hours and dates.... The President of the Italian Geographical Society is requested to undertake the steps necessary to realize this view via his Government and the foreign geographical societies." This was never endorsed by the Venice IGC as a whole: after Venice only the limited resolutions adopted in Antwerp and in Paris remained in force.[26]

Attention to the "afterlife" of conferences—a theme taken up in detail in Chapter 6—is also instructive here. Several months after the Venice meeting, the formal printed record still had not appeared: as George Wheeler noted to Fleming in March 1882, "Very little has apparently been done in pushing forward the results of the Congress & as you know Italians are proverbially slow." Wheeler had been engaged in checking the text of the proces-verbal and had observed that both de Beaumont and Chancourtois had significantly amplified their papers postdelivery. The official record

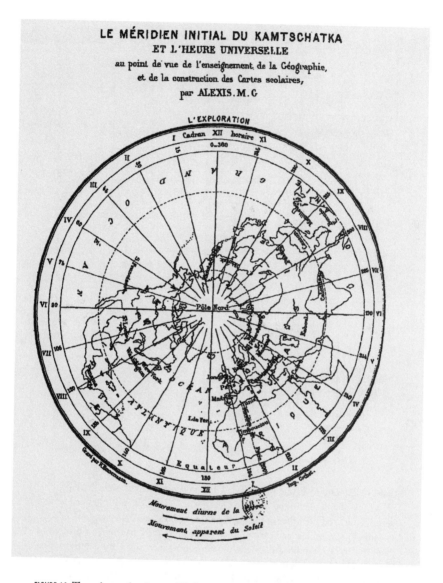

FIGURE 4.1 The scheme for the world's "initial meridian" and universal time to be centered on Kamchatka was presented as a paper to the 1881 Venice International Geographical Congress by M. G. Alexis of the Belgian Geographical Society. Like Alexandre-Émile Béguyer de Chantcourtois's proposals (see Figure 4.2), this was part of a plan to simplify the teaching of longitude and time in geographical education. *Source:* M. G. Alexis, "Le meridien initial du Kamtschatka et l'heure universelle," *L'exploration,* August 18, 1881, 518–522.

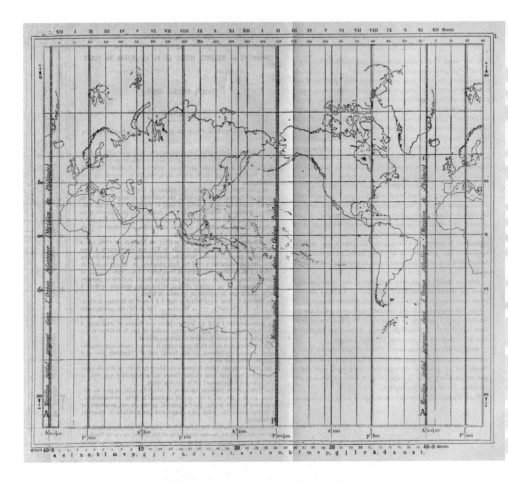

FIGURE 4.2 Alexandre-Émile Béguyer de Chancourtois's map of two different prime meridians, one in the Atlantic Ocean (the "Meridien de Ptolomée," as he described it) and the other in the Pacific Ocean, was part of his plan for a new system of geographical and cartographic instruction. De Chancourtois first outlined this in 1874, modified his views in 1881 in his paper to the Third International Geographical Congress in Venice, and further promoted his scheme in pamphlets in January and May 1883. *Source:* A.-E. Béguyer de Chancourtois, "Etude de la question de l'unification du meridien initial et de la mesure du temps pour suivie au point de vue de l'adoption du système decimal complet," in *Programme d'un système de géographie,* 1874, figure facing p. 14. Reproduced by permission from the Bibliothèque de l'Observatoire de Paris.

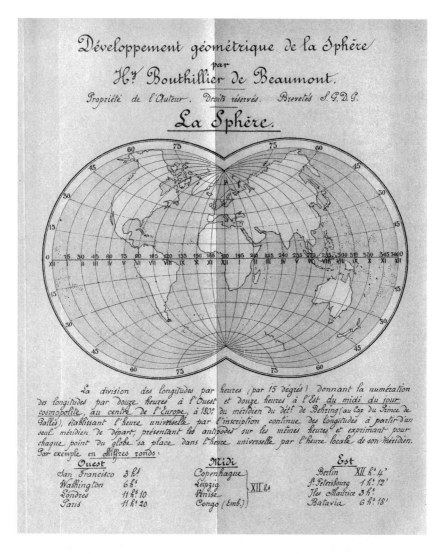

FIGURE 4.3 Swiss geographer Henri Bouthillier de Beaumont's proposal in 1875 for a prime meridian was for a line 150° west of Ferro, the classical origin point favored by Ptolemy. Had the time altered at mean midnight at such a prime meridian, midday would have been at a line in Central Europe close to the cities of Copenhagen, Leipzig, and Vienna. *Source:* Carte de H. Bouthillier de Beaumont, "De la projection en cartographie et présentation d'une nouvelle présentation de la sphère entière comme planisphère" (Geneva, 1875). Reproduced by permission from the Bibliothèque de l'Observatoire de Paris.

of the meeting, the *compte-rendus*, was not ready due to the Italian chair, Professor Dalla Vedova, being much delayed. Clearly, there had been a plan within the Venice IGC not to support the views of group I. "I have now learned to a certainty," Wheeler confided, "only what could be suspected at Venice, that the result of the action of the group was intended to be weakened if not nullified by the manner of expediency 'le voice' & in not calling for approval in General Session." Quite how this was done is not clear, but several Italian officials spoke against the proposal, and as Wheeler observed, "There is a variety of feeling for and against the subject partly personal & professional & partly national."[27]

The Venice IGC was important nevertheless. It brought together leading members of influential institutions from North America and demonstrated their shared commitment to a solution. More so than in previous IGCs, delegates at Venice stressed the related importance of the prime meridian and of universal time for reasons of science, accuracy in measurement, and daily commercial transactions. Wheeler stressed this in summarizing "the principal propositions" discussed, and through his connections with Prince Teano, the chair of the Italian Geographical Society, that society issued invitations to others calling for a meeting devoted solely to the question of establishing a single prime meridian. For Wheeler—and for many others but by no means all—this necessitated "an initial meridian passing through Greenwich, measuring directly therefrom, or from a point 180° distant in longitude." The view of an Australian participant that a future "International Commission" should be appointed by all governments "in order to fix with one accord a standard meridian" echoed the views of Fleming and others over the need for official representation and governmental legislation.

The 1881 Venice meeting provides the clearest articulation of views at the early IGCs regarding a shared first meridian, the connections between the prime meridian and universal time, and the implications of both to the modern world. Fleming's paper in particular made clear the links between modernity, a single prime meridian, time reckoning, and the need for a formal meeting to decide upon those issues. The Venice IGC helped shape these issues but it did not—because it could not—bind legislation upon them.[28]

Time Reckoning and "Cosmopolitan Time," 1870–1883

An unregulated railway network's failure to run on time could be fatal. When two trains of the Providence and Worcester Railway Company collided near Pawtucket, Rhode Island, on August 12, 1853, leaving thirteen dead and scores injured, the cause was attributed to one train being "behind time"—a consequence of operating on different local timetables. Although this and similar incidents exposed citizens to the hazards of various time zones, legislators were slow to reach a consensus on the subject. In the United States, the railways ran on a regulated time system only from 1870 following the work of Charles F. Dowd in advancing "national time," which divided the country into one-hour time zones. From 1883 this system was based on standard railway time measured in relation to Greenwich (Figure 4.4). This was one practical demonstration of the use of separate prime meridians, America's (Washington, DC) and Britain's (Greenwich), for various civic and scientific purposes in one country. Dowd was the first "to suggest a workable system of uniform timekeeping that could span a continent."[29] From the mid-1870s, however, the principal figure involved in developing systems of uniform timekeeping to span the globe was Sandford Fleming.

Sandford Fleming and Cosmopolitan Time

Sandford Fleming was an archetypal Victorian. An emigré Scot who had built a new life in Canada as a railway engineer and sometime chancellor of Queen's University in Kingston, Ontario, Fleming had been concerned throughout his life with the imperatives and consequences of technical change, including the standardization of time. Fleming's first publication on the question of uniform timekeeping for the world was the 1878 pamphlet *Terrestrial Time*. His twin concerns were the irregularity and inconvenience of local time systems throughout the world and the attendant advantages of a universal system of global timekeeping. Fleming's specific scheme was for a set of terrestrial-time meridians based on a mean solar day divided into twenty-four equal regional time zones of one hour. In innovative fashion, he proposed the use not of numbers but of letters from A to Y (he did not use J and Z), each letter marking, respectively, equally spaced meridians on the surface of the globe. In proposing the use of watches and clocks that would use two dials, he advocated the retention of local time alongside the adoption of a single terrestrial or global universal time: one dial to show terrestrial time (based on Greenwich time), designated as *G*, and a second dial to be rotated to one of twenty-four positions to mark local time. *Terrestrial Time* makes only passing reference, however, to the prime meridian: "It is suggested that the initial meridian be established through or near Behring's Straits, passing from pole to pole through the Pacific Ocean, so as to avoid all Continents and Islands." Fleming makes no reference in *Terrestrial Time*

4

EXPLANATION.—Central Time is based upon that of the
one hour faster than Central Time, or four minutes slower than
Mountain Time is based upon the 105 meridian, and is one hour s[?]
colors upon the above map represent the localities governed by the

INTER-COLONIAL STANDARD TIME			
At Charlottetown, Prince Edward			
Island ....is 10 min. faster than Solar Time.			
" Halifax, N. S	5	" faster	" "
" Moncton, N. B...	20	" faster	" "
" St. Johns, N. B...	24	" faster	" "
" St. Stephens, N. B	29	" faster	

EASTERN STANDARD TIME			
At Albany, N. Y.....is	5	min. slower than Solar Time	
" Baltimore, Md....	6	" faster	" "
" Bangor, Me.......	25	" slower	" "
" Boston, Mass.....	16	" slower	" "
" Buffalo, N. Y.....	16	" faster	" "
" Cambridge, Mass.	16	" slower	" "
" Charleston, S. C..	20	" faster	" "
" Columbia, S. C...	24	" faster	" "
" Danville, Va......	18	" faster	" "
" Hamilton, Ont....	19	" faster	" "
" Hartford, Ct......	9	" slower	" "
" Montreal, Que....	6	" slower	" "
" New Haven, Ct...	4	" slower	" "
" New London, Ct..	11	" slower	" "
" New York, N. Y..	4	" slower	" "
" Ottawa, Out......	3	" faster	

At Philadelphia, Pa .
" Pittsburgh, Pa....
" Port Hope, Ont...
" Portland, Me.....
" Portsmouth, Va..
" Providence, R. I...
" Quebec, Que.....
" Richmond, Va....
" Toronto, Can ....
" Washington, D. C.
" Wilmington, N. C.

CENTRA

At Atchison, Kan. ..
" Atlanta, Ga.......
" Augusta, Ga .....
" Burlington, Ia....
" Chicago, Ill.......
" Cincinnati, Ohio .
" Cleveland, Ohio..
" Columbus, Ohio..
" Detroit, Mich.....
" Dubuque, Ia......
" Galveston, Tex ..
" Hannibal, Mo.....

MAP SHOWIN

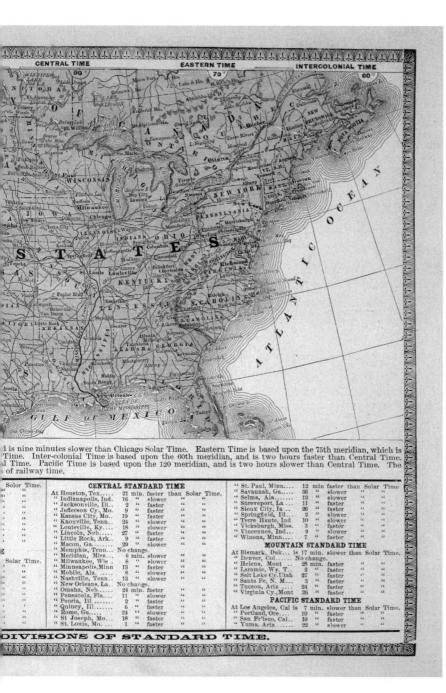

CENTRAL TIME   90   EASTERN TIME   75   INTERCOLONIAL TIME   60

l is nine minutes slower than Chicago Solar Time. Eastern Time is based upon the 75th meridian, which is
Time. Inter-colonial Time is based upon the 60th meridian, and is two hours faster than Central Time.
l Time. Pacific Time is based upon the 120 meridian, and is two hours slower than Central Time. The
of railway time.

CENTRAL STANDARD TIME

At Houston, Tex....	21 min. faster than	Solar Time.	
" Indianapolis, Ind.	16 " slower	"	"
" Jacksonville, Ill..	2 " faster	"	"
" Jefferson Cy., Mo.	9 " faster	"	"
" Kansas City, Mo..	19 " faster	"	"
" Knoxville, Tenn..	24 " slower	"	"
" Louisville, Ky....	18 " slower	"	"
" Lincoln, Neb.....	27 " faster	"	"
" Little Rock, Ark..	9 " faster	"	"
" Macon, Ga.......	29 " slower	"	"
" Memphis, Tenn...	No change.		
" Meridian, Miss...	6 min. slower	"	"
" Milwaukee, Wis..	8 " slower	"	"
" Minneapolis,Minn	13 " faster	"	"
" Mobile, Ala.	8 " slower	"	"
" Nashville, Tenn..	13 " slower	"	"
" New Orleans, La..	No change.		
" Omaha, Neb.....	24 min. faster	"	"
" Pensacola, Fla...	11 " slower	"	"
" Peoria, Ill	2 " faster	"	"
" Quincy, Ill	6 " faster	"	"
" Rome, Ga.....	24 " slower	"	"
" St Joseph, Mo...	18 " faster	"	"
" St. Louis, Mo. ...	1 " faster	"	"

" St. Paul, Minn....	12 min faster than	Solar Time	
" Savannah, Ga.....	36 " slower	"	"
" Selma, Ala.....	12 " slower	"	"
" Shreveport, La ...	11 " faster	"	"
" Sioux City, Ia ..	26 " faster	"	"
" Springfield, Ill...	2 " slower	"	"
" Terre Haute, Ind	10 " slower	"	"
" Vicksburgh, Miss.	2 " faster	"	"
" Vincennes, Ind...	9 " slower	"	"
" Winona, Minn....	7 " faster	"	"

MOUNTAIN STANDARD TIME

At Bismark, Dak...	is 17 min. slower than Solar Time.		
" Denver, Col......	No change.		
" Helena, Mont ..	28 min. faster	"	"
" Laramie, Wy. T..	2 " faster	"	"
" Salt Lake Cy.Utah	27 " faster	"	"
" Santa Fe, N. M...	3 " faster	"	"
" Tucson, Ariz	24 " faster	"	"
" Virginia Cy.,Mont	26 " faster	"	"

PACIFIC STANDARD TIME

At Los Angeles, Cal	is 7 min. slower than Solar Time.		
" Portland, Ore....	10 " faster	"	"
" San Fr'isco, Cal..	10 " faster	"	"
" Yuma, Ariz......	22 " slower	"	"

DIVISIONS OF STANDARD TIME.

FIGURE 4.4 The regulation of America's railways and from 1883 the standardization of American standard time in relation to Greenwich is clear from this "Map Showing the Divisions of Standard Time." *Source:* George Franklin Cram, *Cram's Unrivaled Family Atlas of the World* (Des Moines: O. C. Haskell, 1883), 4. Photo courtesy of the Newberry Library, Chicago (Baskes_g109_g463_1863).

to Struve's initial suggestion of Greenwich, to the Russian's modified recommendation, or to Henri Bouthillier de Beaumont's idea for a line in the
Bering Strait fixed in relation to Ferro as positions for the world's initial
meridian (cf. Figure 4.3).[30]

By May 1879, however, Fleming had become aware of Struve's paper
of 1870 and de Beaumont's of 1876. Fleming's later publications, *Time-
Reckoning and the Selection of a Prime Meridian to Be Common to All Nations*
and the shorter *Longitude and Time-Reckoning,* were published in Toronto
in the *Proceedings* of the Canadian Institute, the city's leading scientific
organization (cofounded by Fleming in 1849). They were initially given as separate spoken addresses to that body. It is not clear why the pamphlets were
published separately rather than as a single work, as their central concerns
are clearly related. *Time-Reckoning* in particular mirrors Fleming's *Terrestrial
Time* in content and argument; each refers to the other.[31]

Much of *Time-Reckoning* discusses the history of systems of timekeeping
as essentially sets of arbitrary measuring schemes codified locally through
customary practice. The requirements of timekeeping in the modern world,
stressed Fleming, were different given the demands of science and the effects
of telegraphy and the railway: "These extraordinary sister agencies having
revolutionized the relations of distance and time, having bridged space, and
drawn into closer affinity portions of the earth's surface previously separated
by long, and in some cases, inaccessible distances." The modern age required modern time modeled, in his scheme, on a twenty-four-hour system
marked by letters, as in *Terrestrial Time.* But what to call this scheme? "Either
of the designations, 'common,' 'universal,' 'non-local,' 'uniform,' 'absolute,' 'all
world,' 'terrestrial,' or 'cosmopolitan,'" Fleming wrote, "might be employed.
For the present it may be convenient to use the latter term."

If he was uncertain what to call his scheme, Fleming was in no doubt
as to its purpose, seeing "the establishment of a common prime meridian
as the first important step, and as the key to any cosmopolitan scheme of
reckoning." These issues were a matter of national and local difference, but
their solution had to transcend nations' interests:

Under the system of cosmopolitan time, the meridian which corresponds with zero would practically become the initial or prime
meridian of the globe. The establishment of this meridian must necessarily be arbitrary. It affects all countries, more especially maritime
countries, and in consequence of prejudice and national sentiment, it

is possible that delicacy and tact and judgment may have to be exercised in the consideration of the subject. There ought not, however, to be much difficulty in dealing with the question. Matters of scientific concern are not and should not be made subservient to national jealousy. Science is cosmopolitan, and no question can be more thoroughly so than that which we are attempting to investigate.[32]

In *Longitude and Time-Reckoning*, Fleming focused more centrally on the need for a single prime meridian. He began by summarizing the history of various prime meridians (their classical positioning, France's 1634 directive, and Piazzi Smyth's advocacy of the Great Pyramid are each noted) before making several key points about a universal global baseline. The first concerned the impracticality of linking cosmopolitan time and concomitant changes in days to any established center of population. Because his scheme for cosmopolitan time meant that days and dates would have to change in line with the selected initial meridian no matter its location, it was, as Fleming put it, "inexpedient to have it passing through London or Washington, or Paris, or St. Petersburg, or indeed through the heart of any populous or even inhabited country." Rather, he wrote, "We should look for a meridian, if possible, to pass through no great extent of habitable land, so that hereafter the whole population of the world would follow a common time-reckoning; and simultaneously human events would be chronicled by concurrent dates." An examination of the terrestrial globe revealed two possibilities: One was a meridian "drawn through the Atlantic Ocean, so as to pass Africa on the one side and South America on the other without touching any portion of either continent, avoiding all islands except a portion of eastern Greenland." The other was "a meridian being similarly drawn in the opposite hemisphere so as to pass through Behring's Strait, and through the whole extent of the Pacific Ocean without touching dry land." His initial possibility clearly reflected classical and early modern debates over the positioning of the prime meridian; the latter endorsed the revised arguments of Otto Struve. Either positioning would work, noted Fleming, "but a meridian in close proximity to Behring's Strait suggests itself as the most eligible."

To make a common prime meridian effective on cosmopolitan principles, reasoned Fleming, it would be advantageous to align it with one of the several systems of longitude then in use. Further, it would be "a still greater advantage if the new initial meridian could harmonize with the longitudinal

Table 4.2   The use, by country, of different prime meridians and their associated
volume of maritime commerce in the late nineteenth century

Country	Ships of all sorts		First meridian used
	Number	Tonnage	
Great Britain and the British Colonies	20,938	8,696,532	Greenwich
United States	6,935	2,739,348	Greenwich
Norway	4,257	1,391,877	Oslo, Greenwich
Italy	4,526	1,430,895	Naples, Greenwich
Germany	3,380	1,142,640	Ferro, Greenwich, Paris
France	3,625	1,118,145	Paris
Spain	2,968	666,643	Cadiz
Russia	1,976	577,282	Pulkova, Greenwich, Ferro
Sweden	2,151	462,541	Stockholm, Greenwich, Paris
Holland	1,385	476,193	Greenwich
Greece	2,036	424,418	(Not given in original)
Austria	740	363,622	Greenwich, Ferro
Denmark	1,306	245,664	Copenhagen, Paris, Greenwich
Portugal	491	164,050	Lisbon
Turkey	348	140,130	(Not given in original)
Brazil and S. America	507	194,091	Rio de Janeiro, Greenwich
Belgium	50	38,631	Greenwich
Japan, &c., Asia	78	39,931	Greenwich
Totals	57,697	20,312,093	

Source: Sandford Fleming, *Longitude and Time-Reckoning* (Toronto: Copp, Clark, 1879), 56.

divisions most in use in the navigation of the high seas." The words "most in use" are important. To determine this, Fleming identified what he called "the *anti* or *nether* meridians" of those European capital cities that were in common use as prime meridians, observing that some of their antimeridians passed near the Bering Strait, one of his two "neutral" options: "The addition or subtraction of 180° would, in any one case, be a ready means of harmonizing the proposed new zero with the old reckoning of longitude." On this criterion six European capital prime meridians were identified: Oslo, Copenhagen, Greenwich, Naples, Paris, and Stockholm. Based on what he obscurely termed "the best authorities within reach," Fleming presented tables to show the number and tonnage of steamers and sailing ships belonging to the world's leading maritime nations and reveal how those nations, and with what volumes of commerce, used the several prime meridians in question. This was his means of demonstrating which initial meridian was "most in use," at least in global commercial terms (Tables 4.2 and 4.3).[33]

Table 4.3   The volume of global shipping in relation to different prime meridians in the late nineteenth century

Location of prime meridian	Ships of all kinds		Overall totals (%)	
	Number	Tonnage	Ships	Tonnage
Greenwich	37,663	14,600,972	65	72
Paris	5,914	1,735,083	10	8
Cadiz	2,468	666,602	5	3
Naples	2,263	715,448	4	4
Oslo	2,128	695,988	4	3
Ferro	1,497	567,682	2	3
Pulkova	987	298,641	1.5	1.5
Stockholm	717	154,180	1.5	1
Lisbon	491	164,000	1	1
Copenhagen	435	81,888	1	0.5
Rio de Janeiro	253	97,040	0.5	0.5
Miscellaneous	2,881	534,569	4.5	2.5

Source: Sandford Fleming, *Longitude and Time-Reckoning* (Toronto: Copp, Clark, 1879), 56.

Of the twenty countries listed, twelve prime meridians were in use (Table 4.2). The most commonly used was that of Greenwich, which was employed alone or in combination with other initial meridians by twelve countries. Table 4.3 reveals that for the world as a whole 65 percent of all ships and 72 percent of total tonnage based their longitude upon Greenwich. The Greenwich prime meridian was by some distance the world's leading baseline for longitude so far as marine commerce was concerned. If the world's single initial meridian should be the longitudinal antithesis of a European prime meridian, Fleming reasoned, it would clearly be most convenient if it were the "meridian drawn 180° east and west of Greenwich.... By the adoption of this as a common prime meridian, there would be no disarrangement in the charts, the nautical tables, or the descriptive nomenclature of nearly three-fourths of the ships navigating the high seas." Fleming illustrated his joint scheme for the use of cosmopolitan time and local time, with clocks and watches whose lettered and numbered dials would allow the measurement of both, and clearly marked his "Proposed Common Prime Meridian" (Figure 4.5). This illustration is the first diagrammatic integration of Fleming's time and longitude proposals. If it were to be adopted, continued Fleming, "there would be no favoured nation, no gratification of any geographical vanity. A new prime meridian so established would be

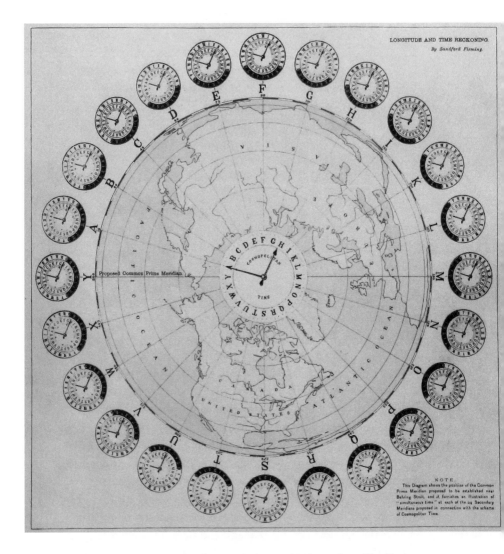

FIGURE 4.5 In *Longitude and Time-Reckoning*, as this figure shows, Sandford
Fleming brought together his scheme for the universal use of cosmopolitan time
and the use of local time on each of twenty-four hourly meridians, together with
the "Proposed Common Prime Meridian" positioned "near Behring Strait", 180°
from Greenwich. *Source:* Sandford Fleming, *Longitude and Time-Reckoning*
(Toronto: Copp, Clark, 1879), opposite p. 60.

essentially cosmopolitan, and would tend towards the general benefit of humanity."[34]

Sandford Fleming's Proposals and the Geographies of International Debate

Fleming did not suppose that his scheme as outlined in *Longitude and Time-Reckoning* would be taken up at once. He wanted to show, however, that it would be practical to do so provided certain key assumptions were made and to demonstrate the principles by which countries might jointly agree on a universal time and a single initial meridian. His overall aim was to show "that there is no impediment to the establishment of a prime meridian for the world unmarked by national pre-eminence, a meridian in itself admirably adapted for the important purposes referred to in connection with the notation of time, and the accurate reckoning of chronological dates in every country on the surface of the earth." That is why Fleming, working in Canada and corresponding with Frederick Barnard, Cleveland Abbe, and others in the American Metrological Society, was so exercised when he learned of the work of "eminent geographers in Europe" then addressing the question and was so attentive to de Beaumont's mediator in particular (Fleming dismissed Germain's 1875 scheme as "national and non-cosmopolitan").

De Beaumont's and Fleming's proposals were similar but not identical. De Beaumont had proposed that the common meridian should run 150° west of Ferro or what was nearly 180° from a meridian passing "through or at no great distance from Copenhagen, Leipsic, Venice and Rome" (see Figure 4.3). Fleming's line was 180° from Greenwich. The only question then remaining, as Fleming termed it, "is which of the two would least interfere with present practices; least disarrange charts, tables and nautical nomenclature; which would most accommodate and best satisfy the greatest number of those who use and are governed by the maps and forms and astronomical almanacs now in use." Fleming knew the answer to his own question given the weight of evidence (Table 4.3) that underlay his Bering Strait option, Greenwich's antimeridian (Figure 4.5).

Sandford Fleming's 1879 pamphlets represent both an evolution and an integration of contemporaries' thinking about the problem of standardizing time and of establishing a single initial meridian. His intention that cosmopolitan time should serve citizens everywhere and depend upon a single prime meridian is unequivocal. From early May 1879, two further developments gave Fleming's work greater geographical reach and significance and further connected the European and the North American communities engaged with the issues.

The first was work undertaken by the American Metrological Society. In its "Report on Standard Time," available from May 1879 but not printed for public use until March 1880, the society's Committee on Standard Time under the chairmanship of Cleveland Abbe recommended that all American railways adopt the time defined by one of the regional meridians that temporally divided America and was based upon Greenwich. Variations upon local time in the United States should be dropped, the committee argued, in favor of local time determined by the time zones set in relation to Greenwich (see Figure 4.2).[35]

The second was the decision in May 1879 by Canada's governor general, the Marquis of Lorne, to send Fleming's two 1879 papers to Sir Michael Hicks-Beach, Britain's secretary of state for the colonies, accompanied by a memorial from Daniel Wilson, president of the Canadian Institute in Toronto. This memorial stressed the importance of Fleming's proposals to Canada; the country's sheer size making the issues of universal time and a standard common meridian as pressing as Dowd, Allen, and Abbe had shown it to be for the United States. In his memorial Wilson urged the governor general to see this as the moment in which Canada might help Britain advance toward "an acceptable solution of a problem of international importance." The office of the governor general "could exert influence upon official and scientific authorities in Great Britain, and those of foreign Governments, to it." Wilson's memorial ended by stressing the simplicity of Fleming's Bering Strait proposal, noting that it was "free from the sources of international jealousy which have hitherto neutralized the efforts of scientific men to remedy practical evils which are universally recognized."[36]

The prime meridian's currency as a scientific and a political issue by the 1870s is evident in how this material was dealt with in British governmental circles. In London, Fleming's papers and the accompanying memorial were referred by Hicks-Beach to George Biddell Airy, Astronomer Royal and director of the Royal Observatory at Greenwich. Airy was asked to judge the scientific bases to Fleming's papers and the civic consequences of the solutions proposed. Airy took Fleming's arguments and proposals—the inconvenience of local time, the importance and nature of cosmopolitan time and the scheme for its operation, and the option of the Bering Strait as the world's first meridian—and dismissed them all, one by one. Fleming's claims over the inconvenience of local time were denied, Airy citing the example of the railway station in the city of Basle, where clocks showing French and

German railway times and Basle local time were all used without difficulty. Airy rejected out of hand Fleming's proposal for a letter-based twenty-four-hour cosmopolitan time: "I see not the slightest value on the remarks extending through the early parts of Mr. Fleming's paper." On Fleming's suggestion over the prime meridian, he remarked:

If a Prime Meridian were to be adopted, it must be Greenwich, for the navigation of almost the whole world depends on calculations founded on that of Greenwich. Nearly all navigation is based on the Nautical Almanac, which is based on Greenwich observations and referred to Greenwich Meridian. . . . But I, as Superintendent of the Greenwich Observatory, entirely repudiate the idea of founding any claim on this; let Greenwich do her best to maintain her high position in administering to the longitude of the world, and Nautical Almanacs do their best, and we will unite our efforts with special claim to the fictitious honour of a Prime Meridian.

Airy was here endorsing the importance of astronomical measurements (and by implication his own authority) and the role of *The Nautical Almanac and Astronomical Ephemeris,* not Fleming's figures of the world's marine tonnage used in association with different prime meridians (Tables 4.2 and 4.3): he made no remark on that score.

Airy was also asked for advice on the governor general's request that the British government help make officials and scientific bodies more aware of the pressing nature of the issues discussed in Fleming's papers. Airy's advice was twofold: first, he distinguished between political agency, civic need, and scientific evidence, and second, he suggested that the memorialists themselves should contact scientific societies to this end:

That it has been the custom of her Majesty's Government to abstain from interfering to introduce novelties in any question of social usage, until the spontaneous rise of such novelties has become so extensive as to make it desirable that regulations should be sanctioned by superior authority. That it does not yet appear that such extensive spontaneous call in reference to the subjects of the Memorial, has yet arisen. That it appears desirable that the question should be extensively ventilated by the memorialists, and should be submitted by them to the principal Geographical and Hydrographical bodies, including (perhaps with others) the Royal Geographical Society, and the Dock Trustees or other commercial bodies, in London, Liverpool and Glasgow.[37]

In response, the Canadian Institute dispatched 150 copies of Fleming's two 1879 papers together with the initial memorial, asking that they be distributed via the Colonial Office to scientific societies and the representatives of foreign governments in London and to scientific bodies in numerous countries. In Britain the papers were sent to seven institutions. Through these official channels, the Canadian Institute would be "glad to learn how far the solutions to the difficulties referred to may be generally acceptable."[38]

British institutions and individuals were in general unsupportive. The Admiralty dismissed Fleming's claims on a lack of evidence of public demand, rather loftily assuming that it mattered only whether the British government thought it expedient to alter the world's timekeeping practices: "It does not appear to their Lordships that there is a sufficient demand by the public to justify Her Majesty's Government in attempting to change the existing practice, and that before seriously considering the question, they would be glad to learn that it had been more extensively discussed among the geographical and nautical bodies who are more interested in it."

Charles Piazzi Smyth was scathing about Fleming's choice of prime meridian, it being "in a part of the world where there are either no inhabitants at all, or, if a few do reside near one end of the line, they are a miserable driblet of wretched Kamschatkan savages, prowling with difficulty for food over snowy wildernesses under the doubtful rule of Russia!" He scorned Fleming's universalist principles, as if "the grand object now of advanced civilization is to consult in everything the utmost development of internationality." More predictably, Piazzi Smyth called for the prime meridian to be not west of Greenwich but east of it to reflect both the millions of people in Asia and his own choice of the Great Pyramid of Giza, which had been "found by scientific examination, and weights and measures" to be the world's metrological (and imperial) baseline.

The Council of the Royal Astronomical Society took no action. It believed that the proposals for cosmopolitan time were more likely to be adopted than the idea of a shared initial meridian, given that such a thing involved "the susceptibilities of individual nations." No scheme for a single prime meridian would succeed, they reasoned, until such time as there was "a general readiness on the part of civilised nations seriously to entertain the question."

The response of the Royal Geographical Society, via its president Gen. Sir John Henry Lefroy, was in one sense direct and dismissive: "It appears to me that there is nothing to be said against the proposal, except its impracticability, which is such that no Scientific body is likely to urge it seriously." In another sense Lefroy recognized Fleming's proposals as symptomatic of wider contemporary concerns. This did not extend to support for Fleming: Lefroy knew that the worldwide adoption of such proposals "requires the convenience of legislatures, and a vast advance in the education of the world." But there was, he considered, "a measure trending in the same direction which this society could do much to promote." Lefroy was alluding not to the growing internationalization of the prime meridian problem but to a limited proposition that geographers and cartographers had pressed for in the Antwerp and Paris IGCs: "I mean the general addition on all Maps and Charts of the Time equivalent of the Longitude on every meridian inserted. I believe that the practical convenience and the Educational value of this trifling and Easy addition to our maps and charts would in a few years found to be very great."[39]

This varied but largely dismissive reception afforded Fleming's proposals did not daunt Daniel Wilson at the Canadian Institute. In a further memorial of April 5, 1880, Wilson prefaced the institute's second distribution of Fleming's 1879 papers, this time targeting geographical and scientific bodies throughout Europe.[40] Wilson reiterated the importance of universal time and a single prime meridian to the world's "practical economy" and stressed the collaborative involvement of the American Metrological Society (whose "Report on Standard Time" was distributed with Fleming's pamphlets). As Wilson cautioned, Fleming's proposals in isolation made little sense; each depended upon the other: "The establishment of Cosmopolitan Time involves the primary determination of an initial Meridian, as a zero for computing the revolutions of the globe on its axis; and it is only by common consent that such Prime Meridian can be determined."[41]

Of the eleven European organizations to whom Wilson distributed the Abbe-Fleming papers in May 1880, only four replied. From Saint Petersburg, Otto Struve reflected upon his 1870 recommendation and his 1875 amendment to it. Struve's recommendation of Greenwich in his 1870 lecture had been "the simplest solution" given that observatory's preeminence in advancing mathematical geography and navigation and because "the great majority of charts now in use upon all the seas are made according to this Meridian and about 90 per cent of the navigators of long-standing are accustomed to take their longitudes from this Meridian." He had revised this viewpoint at the 1875 Paris IGC for two reasons. If Greenwich were the world's prime meridian, cartographers, navigators, and educators would face the disadvantage of having to use longitudes of different signs in different "halves" of the globe because longitude's notation, as + or − degrees east or west from the first meridian, was not agreed upon. Further, given its long usage, it was possible that "French geographers and navigators of other nations" would still employ the Paris meridian as their baseline "from custom, from a spirit of contradiction or from national rivalry." It was far better then "to choose as Prime Meridian another Meridian situated at an integral number of hours east or west of Greenwich, and among the Meridians meeting this condition I have indicated in the first place the Meridian proposed to-day by scientific Americans as that which would combine the most favourable conditions for its adoption."

In thus supporting Fleming's proposal on the Bering Strait's prime meridian, Struve stated his reasons clearly:

1. It does not cross any continent but the eastern extremity of the North of Asia, inhabited by people very few in number, and little civilized, called Tschouktschies.

2. It coincides exactly with that, where, after the custom introduced by an historical succession of maritime discoveries, the navigator makes a change of one unit in the date; a difference which is made near a number of small islands in the Pacific Ocean, discovered during the voyages made to the east and west. Thus the commencement of a new date would be identical with that of the hours of Cosmopolitan time.

3. It makes no change to the great majority of navigators and hydrographers except the very simple addition of 12 hours, or of 180° to all longitudes.

4. It does not involve any change in the calculations of the ephemerides most in use among navigators, viz., the English Nautical Almanac, except turning mid-day into mid-night and *vice versa*. In the American Nautical Almanac there would be no other change to introduce. With a cosmopolitan spirit, and in the just appreciation of a general want, the

excellent ephemerides, published at Washington, record all data useful to navigators, calculated from the Meridian of Greenwich.

Struve was less anxious about cosmopolitan time. He recognized the advantages of standardized time to civil society and as an astronomer knew that "generally in all questions requiring an exact determination of time, the adoption universally of one Time would be a valuable advantage and might be easily effected." He was in no doubt as to the choice of prime meridian and advised his Saint Petersburg colleagues accordingly: "I would therefore recommend the Academy to pronounce without hesitation in favour of the universal adoption of the meridian situated 180° from Greenwich as the Prime Meridian of the globe."[42]

From Berlin in July 1881, the German geodesist Dr. G. V. Boguslawski recommended to the Berlin Geographical Society that it should endorse Struve's view. The Belgian Geographical Society discussed Wilson's memorial and Fleming's proposal (as well as de Beaumont's 1876 paper) in July 1880, but there is no record of that body's reply to the Canadian Institute. In March 1882, a response was received from Don Juan Pastorin, hydrographer to the Spanish Navy, who had discussed the issue in April 1881. Pastorin supported Fleming's proposal for the prime meridian. He pointed out that Spain had at one time or another employed fifteen prime meridians and that in 1869 its government had appointed a commission to regulate the defining base initial meridian for Spain's map and charts. However, the commission had ceased its labors without agreement (as Pastorin observed was true of the Antwerp and Paris IGCs).[43]

Fleming's biographers have concentrated on his advocacy of universal time rather than the need for a single prime meridian.[44] But the evidence examined here shows that his plans for universal time were inextricable from his ambition to establish a single prime meridian against which the world could set its clocks. This was, to repeat his words, "the first important step . . . the key to any cosmopolitan scheme of reckoning." In his 1878 and 1879 pamphlets, Fleming outlined proposals for the establishment of cosmopolitan time on universalist—"common," "all world"—principles and as a means to accommodate local time should his other recommendations be taken up.

Fleming is significant principally because of the substance of his ideas. But he was influential too because of his timing and his collaborative work with men like Cleveland Abbe and with the American Metrological Society, the American Society of Civil Engineers, and the American Association for the Advancement of Science (AAAS) to focus the shared interests of various national scientific bodies on what was understood to be a global problem. His solution—Greenwich's antimeridian—was not the only one being advanced. The weight of opinion in London and Paris was most obviously against him. But he had support in Saint Petersburg and in Berlin, as well as in Toronto and in Washington and elsewhere in the United States. In the early 1880s, it clearly remained far from inevitable that Greenwich would become the world's prime meridian.

Toward Global Accord: Rome 1883

In the wake of the Venice IGC, the question of the prime meridian had what we might think of as a "double geography"; that is, it was debated by national organizations with both individual nation's interests and those of the whole world in mind.[45] As Fleming wrote to the AAAS in July 1882, "The subject has attracted considerable attention on both sides of the Atlantic. In Europe, within the last two or three years, it has been considered by scientific societies in Russia, Prussia, Italy, Switzerland, Spain, France and England. On this continent it has been discussed in the United States and Canada: at the American Metrological Society, the American Society of Civil Engineers, at the Royal Society of Canada, and the Canadian Institute, Toronto." At the IGA meeting in Rome in October 1883, persons from all across the world found a setting in which to articulate a global solution to these questions.[46]

After Venice, Fleming continued to work with Cleveland Abbe and the bodies in America then campaigning for standard time, particularly the Special Committee on Standard Time of the American Society of Civil Engineers, which had been established only weeks before the 1881 Venice IGC. At its meeting in Washington, DC, in May 1882, this committee brought forward a proposal to the U.S. Congress, urging it to take the steps necessary to establish a shared first meridian for longitude. It cautioned that should such an agreement fail, "the people of the Western Continent should determine a zero meridian for their own use and guidance, with a view of establishing, as speedily as possible, a suitable time system for the United States, Canada and Mexico."

It is difficult to imagine such an agreement over a single prime meridian ever materializing given the diverse uses of the 0° baselines of Washington and of Greenwich in the United States; Fleming and the Canadian Institute's preference for Greenwich's antimeridian; and the varied use of Washington and of local prime meridians in Mexico. Yet the fact that such a statement was made at all indicates the contemporaries' strength of feeling, especially over standard time. Reviewing Fleming's cosmopolitan scheme for regulating time, Simon Newcomb, superintendent of the American almanac office, rejected it on the grounds of American exceptionalism: "A capital plan for use during the millennium. Too perfect for the present state of humanity. See no more reason for considering Europe in the matter than for considering the inhabitants of the planet Mars. No; we don't care for other nations, can't help them, and they can't help us." Writing to the AAAS, Sandford Fleming reiterated his commitment to solving time reckoning throughout North America, calling for a convention of scientific bodies, chambers of commerce, and government bodies to be established. At the same time, Cleveland Abbe worked to make the U.S. Signal Office the institutional center for the keeping of uniform time in the United States.

In the early 1880s, with all this international and domestic debate, the time seemed right in the United States for action. In June 1882 Congress announced that "international agreement upon this subject is demanded more imperatively every day, both by science and by trade." Following a recommendation to this effect from its Committee on Foreign Affairs, Congress agreed to authorize the president "to call an International Congress to fix and recommend a common prime meridian." Legislation to this end was agreed upon on July 31, 1882, and in October 1882 President Chester Arthur made inquiries to foreign governments about the possibility of a future international conference directed at a common prime meridian and universal time. At much the same time, a small group of men within the American Geographical Society, including Charles Daly and General Hazen, director of the U.S. Signal Office, recommended to the society that it hold discussions on the benefits of a single global prime meridian. In Congress and in its leading geographical bodies, the United States understood the benefits of global uniformity and was prepared to act. This atmosphere of mounting impatience explains why General Hazen, Wheeler, and Fleming were in Venice in 1881.[47]

Even in Europe, where there was little consensus within and among nations and no coordination among scientific bodies over questions of the prime meridian or of universal time, momentum picked up. Because the Italians had hosted the most recent IGC, in Venice in 1881, and recommendations there were later directed to their politicians (see Table 4.1, Resolution 7), the Italian government was charged with bringing Fleming's resolutions to the attention of other governments. In July 1882 the president of the

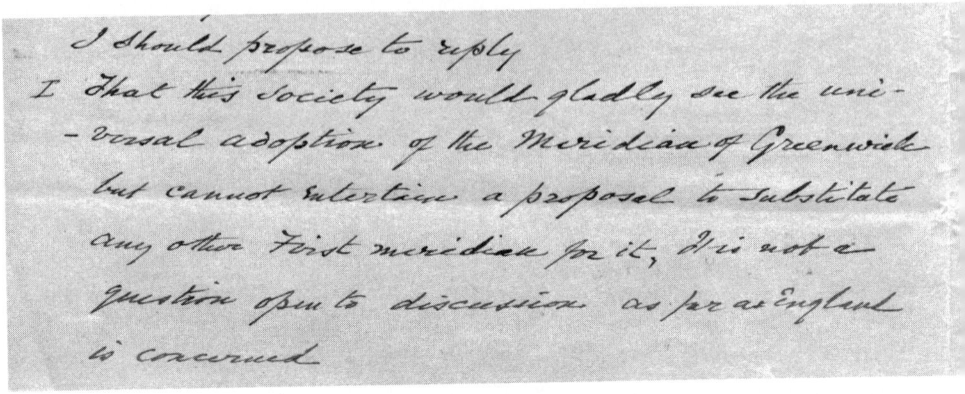

FIGURE 4.6 The Royal Geographical Society's refusal to consider any prime meridian other than Greenwich in any proposed "International Convocation" is evident in Gen. John Henry Lefroy's recommendation to his colleagues Clements Markham and Douglas Freshfield on November 19, 1882: "That this society would gladly see the universal adoption of the Meridian of Greenwich but cannot entertain a proposal to substitute any other First meridian for it, It is not a question open to discussion as far as England is concerned." *Source:* Royal Geographical Society (with the Institute of British Geographers) Archives, MSS JMS / 21/49, November 15, 1882. Reproduced by permission from the Royal Geographical Society (with the Institute of British Geographers).

Italian Geographical Society, Professor Giuseppe Dalla Vedova, wrote to the societies that had attended the Venice IGC over the desirability of an international commission. Most replies favored such a proposal. Britain's Royal Geographical Society did not.

General Lefroy reported to the society's council in November 1882 that an invitation had been received to "take part in an International Convocation proposed to be held next year. To be composed of scientific persons, Surveyors, Geographers and representations of Commerce. . . . Three from each state. To come to an understanding as to the adoption of a universal First Meridian." He reminded his fellow geographers that "the subject was before the Council 2 or 3 years ago." Lefroy was more bullish in 1882 about Britain's position regarding Greenwich as the world's prime meridian than he had been in 1879. In his view the society should not for a moment consider an alternative to Greenwich (Figure 4.6). As to a universal scheme for time reckoning, which in turn could involve adopting the twenty-four-hour system, "the Society might express a readiness to take part in such discussions."

Lefroy's recommendation regarding Greenwich as the world's prime meridian prompted no debate within Britain's leading geographical society, no doubt, because his report reflected existing opinion on the question. His remarks upon timekeeping hint at only a partial rapprochement on that issue. Having read the replies sent to the American Society of Civil Engineers from the 137 authorities they had consulted over standard time, Lefroy had conceded that in science the use of the same time system would help—provided it was based on Greenwich: "It would conduce to simplicity, if scientific men could agree on such occasions to use Greenwich Time." In concluding his recommendation to his colleagues, Lefroy reiterated his personal preference—"the educational effect of expressing Longitudes in Time rather than in degrees"—which, he noted, had been sanctioned by the Royal Geographical Society Council in 1879 but never introduced. This would, noted Lefroy, "inevitably follow the adoption of a universal First meridian."[48]

The Italian Geographical Society's invitation letter of July 1882 thus reflected group I's resolutions at the Venice IGC. But it was not the will of the 1881 IGC as a whole. In Europe, nevertheless, the actions of the Italian Geographical Society gave that proposal renewed credibility. In the United States, the decision by Congress to direct the president to hold a conference on a common prime meridian and a scheme of global timekeeping reflected the practical concerns of organizations focused on the issue of standard time. In October 1883 in Rome, these concerns came together.

The International Geodetic Association Meeting, Rome 1883

The IGA was founded in 1862 to enable geodesists across the world to work together to determine the earth's shape and eliminate the metrological differences in "standard" units of length that Henry James and Alexander Clarke of Britain's Ordnance Survey had encountered as they sought to extend topographical mapping and telegraphy beyond Britain in the early 1860s. The questions of a single prime meridian and of universal time were on the agenda at the 1883 IGA meeting in Rome for the same reasons they had been on the table at IGCs and in more general debates over cosmopolitan time: geodesy in general and European topographical survey in particular, delegates heard, would benefit from uniformity; geography and cartography (marine and terrestrial) required simplification around a single 0°; and the railways and telegraphy demanded standard systems of timekeeping.

The key figure at the 1883 IGA meeting was Adolphe Hirsch, director of the Neuchatel Observatory in Switzerland and secretary to the International Committee for Weights and Measures, which had been established after the passing of the Metre Convention in Paris in 1875. As Bartky has detailed, Hirsch's skills in Rome were diplomatic, scientific, and practical. They were so in part because Hirsch had come to his conclusions beforehand.

In January 1883 Hirsch had written to the engineer Carl Siemens for advice, intimating that "the old question of a universal first meridian, the solution of which has been so many times unsuccessfully attempted, has been revived and apparently with a better prospect of a final solution, as, in addition to the requirements of navigation a universal intimation of time is necessary for the home service of the telegraph and railways." What is clear from this correspondence is that Hirsch had already come to a view over the prime meridian and had done so in conversation with colleagues in Switzerland and with members of the geodetic association: "For the position of the first meridian which from a purely scientific view is of no great consequence as long as it is defined by one of the principal observatories, we advocated Greenwich as having the first claim and the best chance of ultimate adoption."

What Hirsch wanted was not to alert Siemens to these propositions but to ask his help in securing the involvement of the British at the forthcoming Rome meeting. As a member of the International Committee on Weights and Measures, Hirsch knew that Britain was not a signatory to the 1875 Metre Convention. Britain had to date evinced "scant sympathy with similar international objects" (referring to the Metre Convention and the 1862 German-led scheme for European topographical triangulation using the meter). Yet it was vital that Britain should take part in Rome given the importance of longitude and the issue of a single prime meridian. Hirsch wanted Siemens's help in "uniting the two questions" (the adoption of the meter and the adoption of Greenwich): "The choice of a first meridian lies after all between London, Paris, and Washington, and much if not all would be gained if the French could be induced to favour Greenwich. This would be accomplished in my opinion if England accepted as a sort of counter-concession the metric system, and joined the Meter Convention. Do you think it possible to find adherents to this scientific barter in the official and scientific circles of England?"[49] Although no evidence suggests that Siemens acted as go-between in this way, the issue as a quid pro quo to Greenwich's adoption was recognized as a concern by the British.

Britain's delegates in Rome were William Christie, the Astronomer Royal (George Biddell Airy had retired in 1881), and Alexander Clarke, one-time deputy to Henry James, the director of Ordnance Survey. Richard Strachey, the Indian administrator and surveyor, was involved but did not travel to Rome. Their appointment was coordinated by Col. John Donnelly, secretary to the Treasury, on behalf of the government's Science and Art Department. Briefing Christie in August 1883, Donnelly made the British government's position clear: "It will be seen that the special question announced for discussion is that of a common first meridian, treated from a scientific point of view. It is presumed that any decisions of the Congress will be 'ad referendum.' But it is advisable that you should make it clear that the Government, although willing to consider any opinions that may be expressed, does not undertake to be bound in any way by the decision of the majority of the Congress if this should be to alter the first meridian at present in use in this Country." The related meter question was to be dealt with as a matter of science and of protocol: "It has been suggested that the question of the adoption of the metrical system of weights and measures may be brought before the Congress. Should this occur, I am to point out that as the question does not arise in the invitation, you have received no instructions with regard to it, and that you can hold out no hopes that the Government will take any measures toward the general adoption of the metrical system in this country. Any part therefore that you may take in the discussion should simply be regarding it as a scientific question."[50] For the British at least, participation in Rome was on prescribed terms: like Hirsch, minds had been made up before setting off.

As the rapporteur in Rome, Hirsch stressed that the scientists present had the authority to discuss the issues on behalf of their government. Although any resolutions arrived at could not be binding, future meetings—such as that proposed by the United States—might take the issues forward. Hirsch coordinated discussions through his "Report on the Unification of Longitudes by the Adoption of a Unique Initial Meridian, and on the Introduction of a Universal Time." Hirsch's paper had four main points: the unification of longitude; the necessity for the world's single prime meridian to be based on an observed prime meridian—that is, to be based on an astronomical observatory; the relative advantages of the prime meridians that fulfilled this criterion; and a uniform time system for the world that depended upon the unification of the civil day and the astronomical day. As delegates also understood, behind these concerns lay matters of metrological unification.

Hirsch recognized that printed astronomical ephemerides and geographical maps would need modification in the event that all could arrive at a common system of representing longitude from an agreed initial prime meridian. In his view this was not an insuperable difficulty. Most pressing was the question of where that prime meridian should be: all IGCs since 1871 had, he claimed, simply restated the problem. Although he emphasized the importance of accuracy in science, geodesy, and astronomical measurement, Hirsch's most important criterion—flagged to Siemens in January of that year and aired with others in late 1882—was that the world's prime meridian be determined at an astronomical observatory. This appeal to precision and to the prestige associated with established observatories had the effect of eliminating several prime meridians from consideration. Ferro was ruled out: there was no observatory, using an island was no guarantee of accuracy (Pico de Tenerife was likewise excluded), and historical precedence counted for nought. The Bering Strait option—indeed any proposed prime meridian aligned in the oceans as the antithesis or near-antithesis of an established European observatory—was problematic. The proposal of his countryman, Henri Bouthillier de Beaumont, was rejected out of hand: Where would an observatory be built to serve that scheme? On similar grounds Hirsch set aside the Great Pyramid at Giza and the proposal regarding Jerusalem.

In dismissing these options, Hirsch was emphasizing the requirements of accuracy and the demands of science over political neutrality. Adherence to a single standard on scientific grounds was important—not any requirement over neutrality that if overlooked might, to again cite Sandford Fleming, give "offence to local prejudice or national vanity." The problem over longitude was similarly easily solved in principle: longitudes should be given in one direction only, starting at 0° at the initial meridian and increasing eastward around the globe to 360° (at the initial meridian). Hirsch's dismissal of oceanic prime meridians and those without an astronomical observatory left four principal prime meridians: Greenwich, Paris, Berlin, and Washington. On scientific grounds there was nothing to choose between these four possibilities. Because of this Hirsch proceeded to argue on practical grounds, much as Fleming had: Which prime meridian was the most used outside its native country and for what purpose? Which choice would require the least change in the production of topographical maps, marine charts, nautical almanacs, and geographical literature? Unlike Fleming, who had chosen Greenwich's antimeridian on the grounds of its neutrality, as did Struve in 1875, Hirsh's choice of Greenwich reflected the prime meridian most used in marine charts and in commerce, as well as in the best-selling nautical ephemeris (Britain's *Nautical Almanac and Astronomical Ephemeris* significantly outsold either the *Connaissance de temps* or *The American Ephemeris and Nautical Almanac*). Although he did not make the connection, Hirsch's 1883 recommendations in Rome over the prime meridian paralleled those of Struve in 1870.

On the question of universal time, Hirsch reasoned that if all ephemerides and almanacs were based on one initial meridian, then the time of that meridian would form the basis of a single standard time for the world at large. Retaining national times was not an option. Fleming's combination of local time with cosmopolitan time was acceptable in principle to Hirsch (but Fleming's notion of standard regional time within these twenty-four-hour sectors was not). There needed to be a single global time that would coexist with local or national time. To this end Hirsch argued that what he called "universal time" should be counted from zero to twenty-four hours, with local time to be continued with the day divided into two groups of twelve hours respectively designated as "A.M." and "P.M." The issue of where to begin the universal hour and the universal day was complicated by the difference between the civil day, which began at midnight, and the astronomical day, which commenced from the following noon. Hirsch knew that it would not be easy to convince astronomers to change the start of the astronomical day from noon to midnight nor to have civil society change the civil day from midnight to noon. He suggested that universal time should be regulated by the meridian 180° from Greenwich—Fleming's 1879 "Proposed Common Prime Meridian" (Figure 4.5)—such that the universal day would start at midnight along that meridian. The universal and the astronomical day would thus coincide. This choice of universal time mirrored navigators' practice of changing their date when crossing the 180° meridian. After some debate but following collective agreement (unlike the IGC), the IGA brought forward these propositions as "Resolutions . . . Concerning the Unification of Longitude and of Time" (Table 4.4).

Like the discussions on which they were based, these resolutions reflect a mix of scientific recommendations and diplomatic maneuvers. The Swedes, led by the astronomer Hugo Gyldén, offered a different scheme for universal time but found no support. The Belgian delegates urged acceptance of them all. The French did not agree with the recommendation of Greenwich. Reporting upon the Rome meeting, William Christie pointed out that this was a personal decision by France's delegates, not the view of the French government, "which is disposed to accept the propositions of the Geodetic Congress." Government and individual views on the matter

Table 4.4 The resolutions of the International Geodetic Association meeting, Rome
1883, concerning the unification of longitude and of time

Resolution	Substance of the recommendations proposed
I	The unification of longitude and of time is desirable . . . recommended to the Governments of all States interested . . . that hereafter one and the same systems of longitudes shall be employed in all the institutes and geodetic bureaus
II	Notwithstanding the great advantages . . . [of] decimal division . . . co-ordinates . . . and in the corresponding time divisions . . . it is proper to pass it by
III	The Conference proposes to the Governments to select for the initial meridian that of Greenwich . . . [which] fulfils all the conditions de-manded by science . . . and . . . presents the greatest probability of being generally accepted
IV	It is advisable to count all longitudes, starting from the meridian of Greenwich, in the direction from west to east only
V	The Conference recognises for certain scientific wants and for the internal service in the chief administrations of routes of communication . . . the utility of adopting a universal time, along with local or national time
VI	The Conference recommends, as the point of departure of universal time and of cosmopolitan date, the mean noon of Greenwich, which coincides . . . with the commencement of the civil day, under the meridian situated 12 hours or 180 degrees from Greenwich. It is agreed to count the universal time from 0h to 24h
VII	States which . . . find it necessary to change their meridians, should introduce the new system of longitudes and of hours as soon as possible
VIII	The Conference hopes that if the entire world . . . [agrees to] the meridian of Greenwich . . . Great Britain will find in this fact . . . a new step in fa-vour of the unification of weights and measures, by acceding to the Metre Convention [Convention du Mètre] of the 20th May 1875
IX	These resolutions will be brought to the knowledge of Governments and recommended to their favourable consideration

Source: Philosophical Society of Washington, "Resolutions of the International Geodetic
Association," *Bulletin of the Philosophical Society of Washington* 6 (1884), 107–109; Ian R. Bartky,
One Time Fits All: The Campaigns for Global Uniformity (Stanford, CA: Stanford University Press,
2007), table 6.IA, 88.

were in fact finely nuanced: as Christie further noted, Hervé Faye, one of
the French delegates, "voted in favour of the adoption of Greenwich time
though he opposed the choice of the Greenwich meridian for longitudes."
    In their turn Christie and his colleagues were sympathetic to the idea
of metrological uniformity and aware that the metric system in Britain
was widely established in "many branches of Science and Manufactures."

Christie counseled the British government that while Resolution VIII in Rome "implies no obligation whatever to extend the use of Metrical Weights or Measures beyond what is thought expedient in the interests of Great Britain," other countries had given up much in accepting the recommendations regarding Greenwich (Resolutions III and VI). "It must be remembered," observed Christie, "how great a concession it is proposed that the States of Europe should make in adopting the Meridian of Greenwich and Greenwich time, thereby giving up what many of those States have long and naturally regarded as matters concerning their national independence and importance."[51]

The recommendations in Rome (Table 4.4) foreshadowed and clearly influenced the debates in Washington in 1884. It is important to understand, however, that the proposals made in Rome were shaped by the resolutions advanced not only in Rome in 1883 but also in Venice in 1881 (Table 4.1). In the 1883 IGA resolutions, the solution to the prime meridian problem was seen to lie in questions of science and accuracy. As Struve and Fleming considered it, political neutrality alone could not determine the world's shared initial meridian. Bartky has observed how after Rome 1883, "the chance that any initial meridian other than Greenwich would be adopted was virtually nil."[52] Yet Rome left the "meter question" hanging. Some (the French in particular) thought Britain might accede to it if Greenwich should come to rule the world in Washington 1884.

○   ○   ○

Between 1870 and 1883, the question of a single prime meridian for the world became a distinctively international issue; a problem recognized within and shared among nations. It was debated in international forums. Its solution was understood as something that would benefit all nations. From the start of this period in 1870, in Saint Petersburg from the view of a Russian astronomer, through Rome in 1883, from the view of a Swiss astronomer-metrologist, many leading scientists advanced arguments that the world's initial meridian should be Greenwich. But their reasons for doing so were not the same. Other prime meridians were presented as candidates for the world's baseline: Jerusalem; Ferro; Kamchatka; the mid-Atlantic; the Bering Strait; and Greenwich's antimeridian.

The prime meridian in this period did not cease to be of national importance. Many nations continued to use one or more prime meridians for different purposes even as the problem assumed transnational dimensions.

Some nations strengthened the defense of their meridian in the face of the perceived threat that a universal first meridian presented to their established usage—unless their prime meridian was the one being advanced for global acceptance. Some proposals stated that the world's initial meridian should be politically neutral and no one nation's. In Rome in 1883, the argument over neutrality as a criterion—forcefully put by Sandford Fleming—was overturned by the imperatives of science.

The issues of a single prime meridian for the world and of universal time became international for several reasons. These were problems whose solutions required transnational political agreement. Scientific bodies with international reach saw these issues as part of what defined those bodies and their science as global. The internationalization of the prime meridian was shaped by a few individuals and articulated in particular scientific spaces—but not with the same first meridian in mind. The route from Struve in 1870, from de Beaumont in 1876, and from Fleming in 1879 to Hirsch in 1883— from Saint Petersburg, Paris, Venice, and Rome to Greenwich—was not a direct one.

From a North American perspective, the problem of universal time arguably outweighed that of the single prime meridian: the United States, after all, had two. For Sandford Fleming, cosmopolitan time reckoning depended upon the choice of the prime meridian. Considered from Paris, a number of proposals were accommodated, sometimes reluctantly, or as in Rome in 1883, in the hope that giving up one's long-established national meridian might prompt others to accept a different metrology. Arguments about Jerusalem or Ferro foundered on their lack of a scientific basis. From Britain's point of view—the point of view, that is, of the Royal Geographical Society, the Admiralty, the Royal Observatory, the Royal Society, and the Science and Art Department of the government—Greenwich was the only choice.

5

Greenwich Ascendant

Washington 1884 and the Politics of Science

THE INTERNATIONAL MERIDIAN CONFERENCE, organized "for the purpose of fixing upon a meridian proper to be employed as a common zero of longitude and standard of time-reckoning throughout the globe," began at noon on October 1, 1884, in the Diplomatic Hall of the U.S. Department of State in Washington, DC.[1] Thirty-five delegates were present—a mix of diplomats and scientists—representing twenty-one countries. A further seven delegates from six countries had not yet arrived (some expected delegates, among them Adolphe Hirsch from Switzerland, never appeared). Eventually, forty delegates were present, representing twenty-five countries. One month later, on November 1, 1884, the conference closed, its business complete and its proceedings—which had "not been free from difficulty," as the conference president put it—characterized "by great courtesy and kindness, and by a conciliatory spirit."[2]

This chapter examines the workings of the 1884 Washington meeting and the resolutions arrived at there with respect to a single prime meridian and universal time. Rather than focus only on what was said and upon the resolutions agreed to, I consider why and how it was said by addressing both the tone of the discourse and the essence of the arguments offered by different delegates. In addressing the intentions of individual speakers, the chapter continues and refines the narrowing of focus that was a feature of Chapter 4. There, we looked at debate across continents, within nations, and in international meetings over a fourteen-year period; here, we consider debate over the nature, wording, and the implications of seven resolutions in seven sessions within one month. Where others have primarily focused on the substance of the resolutions arrived at in Washington, what follows examines not just what was said but how and why it was said, in a particular "speech space," the stateroom in the U.S. Capitol.[3]

The fact that the 212-page published report of the sessions was titled *Protocols of the Proceedings* is, I contend, not incidental. The setting, the Diplomatic Hall of the Department of State—and the fact that the invitation had been issued by the president of the United States—helped produce a quasi-judicial space in which authoritative views could be advanced and discussed. Here, for the first time, delegates had the authority to vote on behalf of their home countries on the basis of what was heard. This politicized setting produced what cultural historian of science Thomas Gieryn terms a "truth-spot"—a location, such as the witness box in a courtroom, where authoritative claims were aired, argued over, and agreed upon on the basis of the evidence presented.[4]

There are problems, of course, in working with printed records of spoken evidence, such as trust in their completeness and the dangers of inference over pauses and gaps. Not every instance and utterance was captured. Each session was punctuated by periods of recess; voting on minor procedural matters was by voice, not formal enumeration. Days passed between sessions in Washington—days in which delegates discussed the matters at hand, yet we have no way of knowing what was said and by whom. Nevertheless, the printed *Protocols of the Proceedings* is a rich source that may be read closely as a written record of spoken discourse, its analysis supplemented by other sources.

In order to understand what was resolved, why, how, and by whom, it is necessary to turn first to the delegates themselves. In Washington in 1884, as in Rome in 1883, delegates came to debate the prime meridian and universal time—in a conciliatory spirit perhaps—but with their minds, and those of their governments, already made up.[5]

Agreeing to Proceed: Delegates and Instructions

Reporting in October 1883 upon the International Geodetic Association (IGA) conference in Rome, the *Times* (London) considered it memorable for "having contributed towards the simplification of an important department of science." The paper recognized also that the meeting had "no power to enforce its recommendations, since it is a gathering of men of science, not of plenipotentiaries. It remains for an International Convention to give official sanction to its proposals and to carry out the practical reforms it has suggested." As the *Times* reporter understood and as was more widely recognized, the resolutions proposed in Rome effectively provided both the rationale and the agenda for Washington: the latter would have the power to decide upon what had been proposed in Rome (see Table 4.4). As the *Times* reporter put it, "There may be no reason to doubt that an International Convention will ratify what the Conference has done, but at the same time we may congratulate ourselves upon the evidence of progress afforded by this anticipation." He might have been too sanguine. In fact, there would be considerable reason to doubt, and on more than one occasion, that Washington would ratify Rome regarding Greenwich.

In Paris, London, and elsewhere, the knowledge that the Washington meeting would probably confirm the Rome resolutions raised the stakes for everyone concerned. Scientific organizations, national governments, and the media all expressed concerns about not only who the delegates should be but also what the delegates were permitted to say and how they should vote.

In Russia, a commission consisting of representatives of the war and navy departments, the Imperial Academy of Sciences, and the Imperial Geographical Society was brought together to examine the resolutions from Rome, to identify the persons to be delegated to Washington on Russia's behalf, and "to submit and draw up instructions for their guidance." Otto Struve was not one of the three Russians selected, although a relative, Carl von Struve, Russia's ambassador to the United States, was chosen by virtue of his geographical work in Central Asia. The instructions drawn up by this commission stated that Russia's delegates to Washington should follow the resolutions of the Rome 1883 meeting.[6]

In Paris, Hervé Faye reported to a December 1883 meeting of the Académie des Sciences on his actions in Rome, where he had supported the proposal on Greenwich and universal time but rejected Greenwich's possible adoption as the universal prime meridian. In London, the Council of the Royal Astronomical Society curtly recorded the French decision to vote against those articles "which involve the adoption of the meridian of Greenwich."[7]

To the editors of *Science*, the journal of the American Association for the Advancement of Science, the United States needed to be represented by "men of the highest authority." Two men had already been identified: Frederick A. P. Barnard and Commander William T. Sampson, assistant to the superintendent of the Naval Observatory. Sampson was unknown outside the Naval Observatory but uncontroversial. Doubts were expressed over

Barnard, but because of his deafness, not his scientific credibility. As we have seen, Barnard had been engaged with the question of the prime meridian since at least 1850 following his membership in the American Association for the Advancement of Science's committee on the matter. As recently as December 1883, he had forcefully rejected the claims of Piazzi Smyth and other pyramidologists over the Great Pyramid of Giza (see Chapter 2). Barnard never did bring his expertise to bear in Washington, however: he resigned a little over a week before the conference was due to start, citing the pressure of work from Columbia. His place was taken by Rear Adm. Christopher R. P. Rodgers, former superintendent of the U.S. Naval Academy: Rodgers would become president of the International Meridian Conference. The other U.S. delegates were William F. Allen and Cleveland Abbe, both involved in standardizing time on America's railways (the latter in correspondence with Sandford Fleming), and the amateur astronomer Lewis M. Rutherfurd.

Britain appointed four delegates, three of them distinguished scientists. John C. Adams was an astronomer and director of the Cambridge Observatory. Capt. Sir Frederick J. O. Evans, hydrographer to the Royal Navy, had, as vice president of the Royal Geographical Society between 1879 and 1881, discussed the prime meridian with Lefroy and with Airy. The third delegate was Lt. Gen. Richard Strachey, who had planned Britain's engagement with the 1883 Rome IGA meeting. Strachey was involved because of his role as a member of the Council of India and his involvement with the Great Trigonometrical Survey: he could speak from considerable "on-the-ground" experience to Britain's imperial interests. As Strachey was appointed "to represent India," Sandford Fleming was appointed to represent the Dominion of Canada. The possibility of a representative to attend the Washington meeting on behalf of the Australian colonies never materialized.

To judge from correspondence between the government in Canada and Britain's Foreign Office and Colonial Office, Fleming's name was the first to be advanced. The Canadian authorities named Fleming as Canadian delegate to the proposed "International Conference for Common Prime Meridian" as early as May 1883—that is, in the immediate wake of the invitation from the Italian government and its geographical society. This colonial presumption led to confusion in London. Officials in the Colonial Office exchanged letters with the Canadians and their associates in Whitehall about "whether it is at Rome or Washington that the Dominion of Canada desire to be represented, and if it be at the latter whether the Canadian Government have received any intimation as to the date at which the Congress is to take place."The tensions that these letters reveal between London, the metropolitan center of Britain's imperial interests, and Ottawa and Toronto, where colonial officials had been pushing Fleming's interests and dispatching his publications, would become further apparent in dealings on the conference floor between Fleming and his fellow British delegates.[8]

The coordination of the British delegation to Washington—as with Britain's delegation in Rome—was the responsibility of Col. John Donnelly, secretary to the Treasury, on behalf of the government's Science and Art Department. This body took its instructions from the Committee of Council on Education, from the Treasury, and from Britain's foreign secretary, Granville Leveson-Gower. As a result of the involvement of these several branches of government and the intrusiveness of the Canadian authorities, Donnelly found himself unclear on who was doing what in identifying Britain's delegates. As he wrote to William Christie, the Astronomer Royal, in February 1884: "Through the action of the Foreign Office referring to the Treasury after they had told us [the Science and Art Department] they expected us to do that, there is a pleasing confusion—I am trying to unravel it and find out who is to take [the] next step and appoint delegates &c &c." In response, Christie suggested that the Science and Art Department contact the India office in order "to send a delegate to Washington to represent India," thus setting in motion Strachey's involvement. By late March 1884, the situation was no less confused since the Royal Society had become involved despite being unsure whether it was expected to appoint delegates or not. By then, however, Christie had identified Sir Frederick Evans and John Adams as "good men" and recommended them to Donnelly. The British took preparing for Washington seriously, at least in terms of their delegates, but they did not do so in a straightforward manner.[9]

The prime meridian problem was clearly understood. Donnelly's instructions to Strachey in July 1884 over how he and colleagues should proceed echo those issued to William Christie eleven months earlier for Rome:

My Lords understand that the one and only question announced for discussion is that of a common first meridian. It is presumed that any decisions of the Conference will be "ad referendum." But it is advisable that you should make it clear that the Government, though willing to consider any opinions that may be expressed, does not undertake to be bound in any way by the decision of the majority of the Conference if this should be to alter the first meridian at present in use in this country.

It has been suggested that the question of the adoption of the metrical system of weights and measures may be brought before the Conference. Should this occur, I am to request that you will point out that as the question does not arise on the invitation, you have no instructions with regard to it, and that you can hold out no hopes that the Government will take any measure towards the general adoption of the metrical system in this country. Any part, therefore, that you may take in the discussion should simply be regarding it as a scientific question.[10]

As for Rome and even before, Britain's official position in advance of the Washington meeting was unequivocal: engage with other participants but stand fast on Greenwich; adoption of the metric system was not to be a quid pro quo in the event that some nations might have to accept a prime meridian other than their own.

For the French these issues would be crucial given the resolutions advanced in Rome. France's position was initially discussed within the Académie des Sciences by its astronomy, geography, and navigation sections, which prepared a joint report of their recommendations. Drawing upon this report, France's minister of state for public instruction and fine arts established an expert commission—La Commission de l'Unification des Longitudes et Des Heures—to guide France's delegates in Washington. Hervé Faye, who had been in Rome in 1883, was appointed president of this twenty-one-person group. Other members with experience debating the prime meridian included geographer Antoine d'Abbadie, who, with others in the Paris Geographical Society thirty years previously, had called for a global political solution to the prime meridian problem, and Alexandre-Émile Béguyer de Chancourtois, France's inspector general of mines, who had spoken on the matter at consecutive International Geographical Congresses (IGCs) in 1875 and 1881. Other members were scientists, personnel of the French Navy, and a director of a railway company—constituencies with interests in how the questions of a single prime meridian and universal time would be approached. The commission's report on the matter, published in August 1884, was prepared by Édouard Caspari, hydrographer to the navy.[11]

The French commission understood that four issues were to be discussed in Washington: the utility of the adoption of an initial meridian with respect to geographical longitude; the utility "d'une heure universelle"—that is, universal time—from scientific and practical points of view; the convenience, also from these points of view, of the choice of Greenwich as the world's initial meridian; and the possible adoption of the decimal system with respect to global geodetics and time's measurement. The phrasing in Caspari's *Rapport fait au nom de la Commission de l'Unification des Longitudes et des Heures* regarding Greenwich—that is, "du choix du meridian de Greenwich" and not any proposition or resolution concerning that choice (for the language used in reporting upon the outcomes of the Rome meeting, see Table 4.4, Resolution III) is suggestive. The French knew they had to counter a situation that in their view was in danger of being decided upon without further debate. In this respect the attitude of the French to the proposals in Rome—and not just to resolutions over Greenwich—would prove to be important to how the Washington meeting proceeded.

The commission proposed four resolutions, effectively notes of instruction on content and protocol, to the nation's delegates. On the principal issue, the prime meridian, the most important attribute for the French was that it be genuinely international in nature and neutral in character. This meant that it should not divide a major continent (either Europe or the Americas). Two possibilities were entertained as candidate prime meridians: one that ran through the Bering Strait or one that ran through Ferro (even though this latter idea had been dismissed by Hirsch in Rome in 1883). The French believed that, even if it were generally agreed upon, individual nations should have the right to judge the benefit of the single prime meridian in terms of national imperatives. This was undoubtedly, as Bartky has it, "a hedge to allow selective noncompliance": an "opt-out clause" in the event that France's national prime meridian, Paris, should be replaced by another. The commission was dismissive of the resolution over universal time, noting that it would not oppose its take-up if the same conditions of neutrality were adhered to and if both universal and local time were noted in telegraph messages. By contrast, the French saw "the occasion of this top-level meeting" as an opportunity to press for the application of the decimal system. Should this not be resolved in Washington, a further conference could be held to debate the matter. Armed with these instructions, France's delegates to Washington were appointed: Jules Janssen, the astronomer and director of the Meudon Observatory and a member of the commission behind Caspari's report, and the diplomat Consul General Albert Lefaivre.[12]

In effect, the issues up for debate in Washington had already been weakly determined by those proposed at the Venice IGC in 1881 and more strongly decided by those ratified under Hirsch's guidance in Rome in 1883. It is precisely because these issues had been signaled beforehand—even to a degree rehearsed—that each of the twenty-five nations formally represented in Washington had prior instructions from their respective governments on how to proceed once there. Delegates in Washington had access to the resolutions drawn up in Rome, which were presented almost as a formal agenda, clearly making some delegates feel constrained in their actions. As Otto Struve observed:

A great number of the Delegates were not provided with special instructions in regard to particular questions, but had only received as a rule of conduct that they should hold to the Resolutions of the Congress at Rome. . . . These delegates evidently did not feel themselves at liberty to depart from what had been laid down at Rome, even when their own personal views in the course of the discussions at Washington rather inclined them to the prevailing direction of the Resolutions there brought forward, on the ground of common utility and their conformability to the requirements of the case.[13]

As had been the case in Rome, delegates to Washington had no authority to commit their nations to the decisions proposed. The fact that nations had different numbers of representative delegates was unimportant: unlike in Rome, where voting had been by individuals, voting in Washington was by nation. In order to understand Washington, we need to recognize the salience of Struve's remarks: what happened in the American capital depended greatly upon the degree to which delegates felt themselves bound, or not, by events in the Italian capital one year previously. And for nations with three or more delegates—Britain and the United States being particular cases in point—differences among those members over how to proceed and what to say complicated things even more.[14]

Proceeding to Agree: The Work of the 1884 Washington Conference

Before business could begin, procedural matters had to be attended to. Most modern commentators have tended to overlook these, referring, if at all, only to the meeting's "preliminaries." Yet what could be said in the conference's speech space was made possible only by agreement at its start over the protocol to be followed, the secretaries to be appointed, the additional expertise to be included or excluded, and the procedures to be put in place for discussion.

Delegates were formally welcomed on October 1, 1884, by Frederick T. Frelinghuysen, the U.S. secretary of state. Before the practical arrangements were discussed, Rear Admiral Rodgers made two remarks of note as the meeting's president. The first outlined the cosmopolitan aim in view: "Happy we shall be, if, throwing aside national preference and inclinations, we seek only the common good of mankind, and gain for science and for commerce a prime meridian acceptable to all countries, and secured with the least possible inconvenience." The second declared his own nation's disinterest— namely, that despite the United States covering 100° of longitude, being traversed by railways and telegraph lines, and being "dotted with observatories," its delegation "had no desire to urge that a prime meridian shall be found within its confines."

Rodgers's initial remark reflected the intellectual and political language of universal common good in which discussions of the prime meridian were usually couched. His latter declaration, over America's position with regard to the prime meridian, is interpretable in different ways. It is, in one reading, an expression of neutrality: the host nation stating that it had no expectation of being favored. The choice of an American prime meridian was, if unlikely, still a possibility: Hirsch and others had recognized Washington's credentials in reviewing the prime meridian on scientific grounds in Rome. Read another way, Rodgers's remarks are a disguised but profoundly non-neutral declaration from a country that had been formally using two prime meridians for the last thirty-four years and that within the past year had added to its long-established use of Greenwich for navigational purposes by affirming it as the baseline for the nation's railway times. The everyday experiences of living and working with Greenwich, together with the tenor of the resolutions at the 1881 Venice IGC and the 1883 Rome IGA, meant that the Americans had nothing to gain by placing the world's prime meridian within their bounds—provided, of course, that Washington would ratify the resolutions advanced in Rome.

In addition to appointing Rodgers as conference president, the first two sessions were taken up with other organizational matters: the appointment of secretaries; views over the proceedings being recorded in two languages, English and French; discussion of whether the meeting should be open to the public; "the propriety of inviting distinguished scientists, some of whom are now in Washington, and who may desire to be present at the meetings of this Conference, to take part in the discussion of the questions pending"; and the best way to consider and incorporate into the proceedings, if necessary, any written submissions on the matters at hand received from outside parties.

Some matters were quickly resolved. Secretaries were unanimously appointed: Luis Cruls, director of Brazil's Imperial Observatory; Jules Janssen from France; and Lt. Gen. Richard Strachey. Proceedings were to be recorded in English and in French and made available in those languages for consultation the next day. The meeting would not be open to the public. The issue of further scientific input was less easily resolved. Commander Sampson proposed that the conference invite several figures then in Washington to attend: Prof. Simon Newcomb, superintendent of the Nautical Almanac Office; Prof. Julius Hilgard, superintendent of the U.S. Coast and Geodetic Surveys; Prof. Asaph Hall, mathematician and astronomer at the U.S. Naval Academy; Prof. Karl W. Valentiner, director of the observatory at Karlsruhe; and Sir William Thomson, the British engineer and physicist. Discussions went to and fro over the nature of their involvement: To attend only but not to offer an opinion? To present written views? Should they be allowed to represent their governments? On this last matter, a vote was taken and lost. Strachey's amendment to Sampson's proposal allowing the conference president "to request an expression of the opinions of the gentlemen invited to attend the Congress on any subject on which their opinion may be likely to be valuable" was unanimously adopted. Abbe's suggestion that the conference acknowledge the receipt of communications from outside parties but "abstain from any opinion as to their respective merits" was rejected. Rodgers had already received some in advance of the formal proceedings, one arguing that Jerusalem should be the world's prime meridian. It was agreed to review such proposals at an appropriate point during the proceedings. These preliminaries reveal that the meeting was permeable: voices other than the delegates' were heard; written submissions by persons not present were considered. Newspaper reporters publicized news of the sessions. As the meeting progressed, these things would matter.[15]

Greenwich, Neutrality, and Matters of Principle

The first significant item of business was proposed on the afternoon of October 2, 1884, by American delegate Lewis Rutherfurd: "That the Conference proposes to the Governments represented the adoption as a standard meridian that of Greenwich passing through the centre of the transit instrument at the Observatory of Greenwich." Delegates would have immediately understood that this meant endorsing Resolution III of the Rome meeting (Table 4.4). From France, Albert Lefaivre at once considered this "out of order." He did so from an understanding that the conference should express only a "collective wish" to be submitted "to the approval of our respective Governments"—theirs was a mission of "lofty international bearing," as he put it, but not one of decision making. Their authority lay in proposing recommendations that would later be ratified by their different governments. Lefaivre held to the view that delegates at Washington were not in thrall to the Rome proposals from 1883: "We are in nowise bound by the decisions of the Conference held in Rome." This, he emphasized, was a matter of some importance: "I will say even more than this: The results of the Conference held at Rome are by no means regarded as possessing official authority by the Governments that have accredited us; for if those results had been taken as a starting point, there would be no occasion for our Conference, and our Governments would simply have to decide with regard to the acceptance or rejection of the resolutions adopted by the Geodetic Congress at Rome."

Jules Janssen followed up his French colleague's observations to make two points. Rutherfurd's proposal was in his view presumptive: "It is wholly undesirable that a proposition of so grave a character ... should be put to the vote while our meeting has scarcely been organized, and before any discussion relative to the true merits of the questions to be considered has taken place." Consideration needed to be given to the principles behind the issue, not to practical proposals: "Before discussing the question of the selection of a meridian which is to serve as common zero of longitude for all the nations of the world, (if the Congress shall think proper to discuss that point,) it is evident that we must first decide the question of principle which is to govern all our proceedings; that is to say, whether it is desirable to fix upon a common zero of longitude for all nations."[16]

The French position regarding principle threatened to derail the meeting before it had effectively begun. In response Strachey, Rutherfurd, and Fleming made reference to the initial letter from Frelinghuysen in October 1882 from Chester Arthur, president of the United States, and his invitation of December 1, 1883, regarding the Washington meeting. These missives made it clear, they stressed, that the purpose of the meeting was to agree upon a prime meridian and universal time. The French were, at the least, being evasive in ignoring the tenor of the invitation and the guidance afforded to Washington over the practical implications of the resolutions proposed in Rome. Janssen reiterated his objection nonetheless: "Mr. Lefaivre, my honorable colleague, and I are of the opinion that the mission of this Congress is chiefly to examine questions of principle. I consider that we shall do a very important thing if we proclaim the principle of the adoption of a meridian which shall be the same for all nations." Janssen even raised the possibility of a further congress: "The principle having once been adopted, our Governments would subsequently convoke a conference of a more technical character than this, at which questions of application would be more thoroughly examined." Rutherfurd again stressed the wording of the initial invitation—"not to establish the principle that it is desirable to have a prime meridian, but to fix that prime meridian; that that was the object of the meeting."

Courtesy demanded that his and others' disquiet with the French position on principle should be phrased with care. To Rutherfurd it seemed "that there must be some misapprehension on the part of the learned gentleman from France in thinking that this Conference has not the power to fix upon a prime meridian." He supposed that every delegate had studied this matter before coming to Washington (although he admitted that he was not in possession of the instructions the French delegates had received from their government). The responses of the delegates from Italy, Spain, Sweden, Russia, Germany, Mexico, Brazil, and Strachey on behalf of Britain confirmed his point, in his view: delegates were there to debate and to vote upon the resolutions advanced and then to recommend them to their respective governments. Divided between adherence to matters of principle and the actual wording of the initial invitation with its clearly practical implications, the session of October 2 adjourned. No decisions were made.[17] The *Times*, receiving its news via telegraphic cables, covered the business to date by reporting on the importance of a single prime meridian and announcing that no progress as of yet had been made to that end. The *New York Times* reporter seemed to have inside knowledge as to what would happen: "Officers to the United States Bureau of Navigation express the opinion that the Conference will either adopt the meridian of Greenwich or come to no agreement whatever, as neither Great Britain or the United States will consent to any other."[18]

The meeting recommenced on October 6. Rutherfurd's initial proposal, with the words "as the additional meridian for longitude [referring to Greenwich]" added as an amendment, remained on the table. By then he had exchanged words with the French delegates, and possibly others, out of formal session. Rutherfurd temporarily withdrew his motion in order to allow the French to submit a further motion: "That the initial meridian should have a character of absolute neutrality. It should be chosen exclusively so as to secure to science and to international commerce all possible advantages, and in particular should cut no great continent—neither Europe nor America." In advancing what was their national commission's initial instruction to them as delegates, Janssen and LeFaivre set in motion prolonged discussions about neutrality and how it was to be understood. In doing so, they threatened to undermine the scientific bases on which earlier discussions in Rome and elsewhere had been based and to dismiss the candidacies of Paris, Greenwich, Washington, and Berlin.

To Britain's Frederick Evans, this was a matter of major importance. It was crucial not to lose sight of the scientific bases by which possible prime meridians had been identified in Rome. Science should come first—that was his key principle. Only then, all things being equal among the candidate sites, should delegates consider practical criteria: "This Conference should be particularly guarded, looking at the question from a scientific point of view, not to depart from the conditions laid down by the Conference at Rome." From the U.S. delegation, Sampson concurred. His remarks betray a continuing impatience with the French delegates' position: "If, then, we are of one mind as to the desirability of a single prime meridian, and if we are fully empowered to make the selection, which may be taken as another way of saying that we are directed by our respective Governments to make the selection [here referring to the wording of the initial invitation and to its clearly practical implications], we may proceed directly to the performance of this duty." It was essential, Sampson continued, that the world's prime meridian should run through an astronomical observatory; "no scientific or practical advantage is to be secured by adopting the meridian of the great pyramid." For him, myriad practical matters and the "additional consideration of economy"—the costs involved for different countries in reprinting maps and navigation charts in relation to the decided-upon first meridian should it not be their own—meant that it had to be Greenwich. Natural pride had made individual nations establish "their own prime meridian within their own borders." The United States had been led into "this error" about thirty-five years before (alluding to the decision to adopt both Washington and Greenwich in 1850), and now was the time to set aside national prejudice and practice.[19]

The French countered that their argument was not about Paris. Had it been, Paris "would not lose by the comparison [with Greenwich]." In defense of his claim, Janssen cited "the latest observations of the differences of longitude made by electricity [telegraphy]," which "have given very remarkable results of great accuracy." In fact, the determination of the longitude of the Paris and Greenwich observatories was to be a source of continuing difficulty (as I show in Chapter 6), as it had been in earlier periods, but Janssen was never going to admit to that in public. The key issue for the French was neither Paris nor claims as to its accurate longitudinal fixing. The important thing was the matter of global neutrality for the prime meridian selected. If there was no commitment to that, cautioned Janssen, there was little point in continuing: "What we ask is, that after the general declaration of the second session as to the utility of a common prime meridian, the Congress should discuss the question of the principle which should guide the choice of that meridian. Being charged to maintain before, gentlemen, the principle of the neutrality of the prime meridian, it is evident that if that principle was rejected by the Congress it would be useless for us to take part in the further discussion of the choice of the meridian to be adopted as the point of departure in reckoning longitude."

Janssen's prolonged defense of "neutrality" drew together several issues. There was, he reasoned, a difference between the scientific and the moral bases to the prime meridian. The first was a matter of long-term imprecision and indecision over where the prime meridian had and should run. The second concerned the issue of nations' preferences. The French recognized the difference between a geographical and an observed prime meridian. They were, of course, anxious over the necessity of changing the country's maps and charts and the *Connaissance des temps* should a prime meridian other than Paris be selected. Importantly, too, they considered neutrality a matter of geography. Addressing the delegates in general and Sampson and Rutherfurd in particular, Janssen remarked how "we [the French] do not advocate any particular meridian." This was a controversial point. What the French wanted—and wanted others to accept as a matter of defining principle—was the adoption of the world's prime meridian not on the grounds of science as its initiating determinant but on the grounds of natural geographical disposition: "Upon the globe, nature has so sharply separated the continent on which the great American nation has arisen, that there are only two solutions possible from a geographical point of view, both of them very natural. The first solution would consist in returning, with some small modification, to the solution of the ancients, by placing our meridian near the Azores; the second by throwing it back to that immense expanse of water which separates America from Asia, where on its northern shore the New World abuts on the old [the Bering Strait]."

This claim about neutrality as a consequence of geography—that nature offered an optimally positioned global prime meridian that was no one nation's—denied that long-established view, understood since at least the early modern period and stated by the author of *Sayling by the True Chart* in 1703, that there was "nothing in nature on the Earth ... which obliges us to place it in one place rather than another." The French were neither denying the logic of science nor dismissing the importance of accuracy, upon which science depended. In Rome, delegates had assumed that the key to knowing the precise location of the prime meridian would be to locate it at one of four observatories under consideration. In Washington, Janssen and LeFaivre argued that science would allow a neutral prime meridian to be accurately positioned not by placing it at one of these existing observatories but by working to calculate it in relation to these observatories: "The position thus obtained would be connected with certain of the great observatories selected for the purpose from their being accurately connected one with another, and the relative positions thus ascertained would supply the definition of the first meridian." Exactly how this was to be done and which of the two solutions should be adopted—near the Azores or in the Bering Strait—was, argued Janssen, unimportant: "I do not propose to dwell upon the details of the establishment of such a meridian. We have only to advocate before you the principle of its acceptance. If this principle be admitted by the Congress, we are instructed to say that you will find in it a ground for agreement with France."[20]

The French arguments prompted a variety of responses. For John Adams, Janssen's remarks rested "almost entirely on sentimental considerations" and not upon "the aggregate convenience of the world at large." If there was to be a neutral meridian as the French proposed, this demanded an observatory on the point selected—an argument Hirsch had rebutted in 1883 for Ferro, Giza, the Azores, and Jerusalem. Or, noted Adams, "we should merely have a zero of longitude by a legal fiction, and that would not be a real zero at all." Suppose, continued Adams, referring to one of the two prime meridian lines proposed by the French in their "neutral" model, "they took a point for zero twenty degrees west of Paris, of course it would be really adopting Paris as the prime meridian; that it would not be so nominally, but in reality it would be." Cleveland Abbe pressed the French to declare what a neutral meridian was: "On what principle shall the Conference fix upon a neutral meridian, and what is a neutral meridian? Shall it be historical, geographical, scientific, or arithmetical? In what way shall it be fixed upon?" After all, observed Abbe, France had given the world a "neutral system for weights and measures," but in reality the metric system was not a neutral system but a French system. In the same way, "The expression 'Neutral system of longitude' is a myth, a fancy, a piece of poetry, unless you can tell precisely how to do it." Janssen reiterated the geographical basis to the principle of neutrality: "Our meridian will be neutral if, in place of taking one of those which are fixed by the existing great observatories, to which, consequently, the name of a nation is attached, and which by long usage is identified with that nation, we choose a meridian based only upon geographical considerations, and upon the uses for which we propose to adopt it."[21]

At this point Rodgers invited Simon Newcomb to speak—the first nondelegate to voice an opinion. Newcomb's position in the Nautical Almanac Office and his growing reputation lent his words authority even if, as we have seen, he was especially protective of America's interests and not at all supportive of Fleming's proposals over universal time. Newcomb admitted that it was not in fact necessary to have a fixed observatory as the basis for the prime meridian (so snubbing Sampson, who was present, and Hirsch, who was not) and that he had some sympathy for the French position over principle. But pragmatics determined otherwise. In his view it was "impracticable under any circumstances to have an absolutely neutral prime meridian; that the definition of the prime meridian must practically depend upon subsidiary considerations, no matter where it might be located." Neutrality was simply not possible because of science's requirements for accuracy of observation and record. Matters of practicality (the fact of existing observatories) trumped matters of principle (the capacity of astronomers and others to fix a new prime meridian in relation to those observatories). Drawing upon his experience of topographical survey in India and the Himalayas, Strachey concurred: "It is absolutely essential for fixing an initial meridian for the determination of longitude that it should be placed at an astronomical observatory which can be connected with other places by astronomical observatories and by telegraph wires, and that the idea of fixing a neutral meridian is nothing more than the establishment of an ideal meridian really based upon some point at which there is located an observatory." Adams took issue with Newcomb's seeming endorsement of the French position on principle but agreed with him over the practical grounds on which the first meridian should be chosen.[22]

Having spent the day considering questions of principles and neutrality, the session ended with a recess in order to review, in print, the details of what had been discussed. But here there was a hitch: no French stenographer had been available. Janssen, who had been speaking in French and who was anxious that his meaning might have been lost, called for an adjournment of the vote on the principle of neutrality. When the conference met in its fourth session one week later, on October 13, 1884, the French-language version of the matters discussed on October 6 had still not been completed. There is no evidence to suggest that the French delegation expressed concern at this state of affairs, but the fact remains that the arguments in French concerning a neutral prime meridian were available for checking much later than their English-language equivalent. Arguments over principle were delayed by the practicalities of translation and of printing.

As reporters carried news of the meeting, so delegates wrote to friends and family of their roles in it. The *Times* reported only briefly on Janssen's lengthy address and his call for a neutral meridian through the Bering Strait or the Azores and expressed doubts as to whether the conference would have any result whatsoever. Strachey wrote to friends in America between sessions. One replied to express the hope "that our testy French cousins may become more reasonable in their demands, agree upon Greenwich as the only proper common meridian and thus enable you to see something of our Country before your return to England." Janssen wrote to his wife in Paris. His speech on the neutral prime meridian was, he said, "a truly terrible session. . . . They went for me, one after the other, trying to demolish my meridian by all possible arguments." There was even, he hinted, an element of premeditation from certain parties: "They had planned to attack me one after the other, each on different grounds. I was alone against forty, among whom were the principal English and American scientists, but the struggle aroused me so much that ideas and expressions came effortlessly." If this hints at an element of extemporization from Janssen as he spoke, the recollection of his speech also afforded him an opportunity to reflect upon its reception and his own sense of achievement and to privately express a sense of resignation as to the final outcome that he could not entertain in public: "Besides, almost everyone congratulated me, and were kind enough (especially the gallery) to say that the session had been magnificent and that they had been enchanted to see a scientist who was an orator. The moral success belonged to France. This was all we could hope for, because everyone had precise, formal instructions to vote for Greenwich. But our attitude needed to be a noble one, and it was." His efforts in the crowded and hot space of the Diplomatic Hall, he wrote, had left his shirt wringing wet: "It took two days to dry."[23]

The fourth session of the conference began at one o'clock on October 13. Aside from the expression of unanimity on October 2 that there should be a single prime meridian, nothing had yet been agreed upon. No votes had been cast on the French resolution of the week before. Publically, reaction to the meeting was cool. In London, the *Times* noted that American newspapers were reporting the French delegates' behavior to date as "puerile." The first to speak on October 13 was Sandford Fleming. Drawing from his written accounts on universal time and the prime meridian, Fleming stressed the cosmopolitan common good, urging his fellow delegates to "set aside any national or individual prejudices we possess, and view the subject as members of one community—in fact, as citizens of the world." Reminding the assembled delegates that the French resolution "undoubtedly involves the selection of an entirely new meridian," Fleming made three points. The principle of a neutral prime meridian was, he averred, not wrong, "but to attempt to establish one would, I feel satisfied, be productive of no good result. A neutral meridian is excellent in theory, but I fear it is entirely beyond the domain of practicality." To lend support to his view that the world's prime meridian should relate to the initial meridians in use but not be one of them, he read out the table on relative tonnage in relation to these meridians that he had published five years earlier (Table 4.3). This pointed to Greenwich as the primary candidate—"but Greenwich is a national meridian, and its use as an international zero awakens national susceptibilities." All this was leading to his own scheme, laid out in *Longitude and Time-Reckoning*, that the world's initial meridian should be Greenwich's antimeridian (Figure 4.4): Fleming again cited Otto Struve's 1880 paper in support of his case. There was, in short, no need for a neutral meridian on the principles outlined by the French: "The Pacific meridian referred to would soon be recognized as being as much neutral as any meridian could possibly be."[24]

Fleming's was the last effective contribution in the prolonged discussion on neutrality. As one of the meeting's secretaries, Luiz Cruls summarized the issues at hand before Rodgers restated the French resolution over "absolute neutrality" for the prime meridian and subjected it to a vote. Only Brazil, France, and San Domingo voted yes; twenty-one other countries voted against. This was a clear rejection of the matter of principle, at least as interpreted by the French. It was not yet a decision on the prime meridian's actual position. It allowed Rutherfurd to return to his earlier resolution and to the agenda set for Washington by the matters agreed upon in Rome. But before Rutherfurd's resolution regarding "the centre of the transit instrument at the Observatory of Greenwich as the initial meridian for longitude" could be taken further, Sandford Fleming proposed an amendment that the meridian should be "at a great circle passing through the poles and the centre of the transit instrument at the Observatory at Greenwich." That is, Fleming advanced as a formal amendment his own 1879 scheme (and that of Otto Struve in 1880) regarding the antimeridian of Greenwich. In doing so, Fleming exposed a key difference of opinion among the British delegation and contradicted the British government's position on the matter. John Adams rose at once to express his British colleagues' dissent from Fleming's views and to state that they would vote against it: "The proposition to count longitude from a point 180 degrees from the meridian at Greenwich appears to them not to be accompanied by any advantage whatever. . . . If there is objection to the meridian of Greenwich on account of its nationality, the meridian of 180 degrees from Greenwich is subject to the same objection. The one half is just as national as the other half." Fleming's amendment was put to a verbal vote and defeated.

At this point what the British government had feared came to the fore—namely, the issue of the metric system in relation to the possible adoption of Greenwich. Juan Valera, from Spain, informed the delegates that he and his colleagues had been instructed by their government "to accept the Greenwich meridian as the international meridian for longitudes . . . in the hope that England and the United States will accept on their part the metric system as she [Spain] has done herself." Rodgers intervened to insist upon protocol: "The question of weights and measures is beyond the scope of this Conference." Richard Strachey reported that although the issue was not on the agenda at Washington, he could report to the meeting that Great Britain had resolved to join the Convention du Mètre: "The arrangements for that purpose, when I left my country, were either completed, or were in course of completion, so that, as a matter of fact, Great Britain henceforth will be, as regards its system of weights and measures, exactly in the same position as the United States" (Britain had formally joined on September 11, 1884). In response, Albert Lefaivre—who like most delegates could see the formal vote upon Greenwich fast approaching—reminded delegates of Hirsch's view in Rome: should the adoption of Greenwich proceed—"a sacrifice for France"—Britain should respond "by favouring the adoption of the metric system." The meridian at Greenwich, insisted Lefaivre, "is not a scientific one . . . its adoption implies no progress for astronomy, geodesy, or navigation; that is to say, for all the branches and pursuits of human activity interested in the unification at which we aim." Adoption of the metric system might ease France's anxiety over her principled defeat. "However, it is with great pleasure that I heard our colleague from England declare that his Government was ready to join the international metric convention, but I notice, with sorrow, that our situation is not as favourable as that of Rome, since the total abandonment of our meridian is proposed without any compensation."[25]

Two more delegates spoke before Rodgers returned the meeting to Rutherfurd's much-delayed resolution. Sir William Thomson urged the French to accept the resolution on practical grounds. Frederick Evans presented detailed statistics on the worldwide sales of Britain's Admiralty charts, based on Greenwich, and on the sales of *The Nautical Almanac and*

Astronomical Ephemeris (well in excess of fifteen thousand per annum in each of the preceding seven years). Together with Fleming's tonnage and meridian statistics, Greenwich's dominance in the world of mapping and navigation was undeniable.[26]

Rodgers reiterated the resolution proposed by Rutherfurd: "That the Conference proposes to the Governments here represented the adoption of the meridian passing through the transit instrument at the Observatory of Greenwich as the initial meridian for longitude." Delegates knew Rutherfurd's resolution was crucial. Discussions of the question had occupied three full conference sessions without agreement. The printed record of them takes up nearly half of the *Protocols of the Proceedings* of the Washington meeting. Upon formal vote, twenty-one countries were in favor, with one against (San Domingo). There were two abstentions (France and Brazil): Salvador's delegate was absent with an illness as the votes were cast. By a considerable majority, but not unanimously, Greenwich was decided upon as the world's prime meridian for longitude (Table 5.1, Resolution II). This, reported the *Times* with only half an eye to the truth, was as it should be: "Nothing but extremely sensitive national feeling could have stood in the way of the adoption of the English line, which map makers have in general long consented to use." Even so, the reporter admitted that "the question at issue has not proved so simple as it might at first blush seem to be."[27]

This detail behind the resolution that Greenwich be adopted as the world's prime meridian is vital to show why and how it was taken up, not just that it was. The Washington meeting was not in any strict sense a contest between competing prime meridians. Proposals in Washington over Greenwich's candidacy were first advanced by America's delegates, not by Britain's, and adhered to an agenda established in Rome in 1883 that was itself shaped in Venice in 1881. Rodgers's declaration at the outset over the possibility of the world's prime meridian being within the United States was designed to rule it out of contention (the Washington Observatory was one of four possibilities identified in Rome on scientific grounds). It had the additional effect of limiting the possibilities for discussion.

French resistance to the Rome resolutions, to Rutherfurd's proposal, and to Greenwich's candidacy was not based upon the promotion of Paris. Where the British disagreed among themselves over the implications in practice of different prime meridians, the French argued collectively on matters of principle over neutrality and science. Initially, neutrality meant a denial of the bases on which the meeting had been called (that is, Washington 1884 should not be bound by Rome 1883). It also meant rejecting any

Table 5.1 The resolutions of the International Meridian Conference held at Washington, DC, for the purpose of fixing a prime meridian and a universal day

Resolution	Substance of resolution	Voting patterns on resolution
I	That it is the opinion of this Congress that it is desirable to adopt a prime meridian for all nations, in place of the multiplicity of initial meridians which now exist	Unanimously adopted
II	That the Conference proposes to the Governments here represented the adoption of the meridian passing through the centre of the transit instrument at the Observatory of Greenwich as the initial meridian for longitude	Ayes = 22[a] Noes = 1 Abstentions = 2
III	That from this meridian longitude shall be counted in two directions up to 180 degrees, east longitude being plus and west longitude minus	Ayes = 14 Noes = 5 Abstentions = 6
IV	That the Conference proposes the adoption of a universal day for all purposes for which it may be found convenient, and which shall not interfere with the use of local or other standard times where desirable	Ayes = 23 Abstentions = 2
V	That this universal day is to be a mean solar day; is to begin for all the world at the moment of mean midnight of the initial meridian, coinciding with the beginning of the civil day and date of that meridian; and is to be counted from zero up to twenty-four hours	Ayes = 14[b] Noes = 3 Abstentions = 7
VI	That the Conference expresses the hope that as soon as may be practicable the astronomical and nautical days will be arranged everywhere to begin at mean midnight	Carried without division
VII	That the Conference expresses the hope that the technical studies designed to regulate and extend the application of the decimal system to the division of angular space and time shall be resumed, so as to permit the extension of this application to all cases in which it presents real advantages	Ayes = 21 Abstentions = 3

Source: International Conference Held at Washington for the Purpose of Fixing a Prime Meridian and a Universal Day. October, 1884. Protocols of the Proceedings (Washington, DC: George Brothers, 1884), 199–203.

[a]The Salvador delegate, Mr. Antonio Batres, was ill on the day of the vote, October 13, 1884. His vote in favor of Greenwich was accepted by the conference at its session the following day and added to the "proper entry in the Protocol"—the printed record kept by the stenographers—bringing the total in favor of Greenwich to twenty-two.

[b]In initial voting upon this resolution, the delegate from Turkey, Rustum Effendi, had voted yes. At the final session of the conference on November 1, 1884, Rustum Effendi indicated that he wanted to change his vote to no. This was accepted without objection by conference delegates, but the printed record was not correct, as it is here.

established observed prime meridian on the grounds of its national associations. The French additionally understood neutrality to be geographical: a prime meridian placed in relation to nature's geography, not sited at a single astronomical observatory but fixed by accurate measurement between several. Empirical evidence of commercial practice—Fleming's on tonnage statistics and different initial meridians; Evans's on sales of maps, charts, and *The Nautical Almanac and Astronomical Ephemeris*—all worked to deny French arguments over principle. So too did the fact that many countries' delegates had been advised to vote in favor of Greenwich even before they arrived in Washington. To the editors of *Science,* the meeting was more politics than science: "The time has been mostly taken up with political diplomacy and sentiment."[28]

Reckoning Longitude

The wording of the resolution in Washington over longitude—proposed, again, by Rutherfurd—effectively stated that longitude was to be "counted in two directions up to 180 degrees [from the chosen prime meridian], east longitude being plus and west longitude minus." This was a departure from the resolution agreed to in Rome, which had suggested that "longitude be counted from the meridian of Greenwich in one direction only, from east to west" (Table 4.4, Resolution IV). To some delegates this was only a matter of detail. To others it was the detail that mattered. From Sweden, Count Carl Lewenhaupt proposed in amendment to Rutherfurd's resolution that Washington should adopt the Rome proposal. This was not to everyone's liking. To Sampson the initial resolution introduced by Rutherfurd would be in "perfect conformity with the habits of the world." The habits that really mattered, stressed Frederick Evans, were those of seamen. To Evans the "proposed dislocation of the methods of seamen by reckoning longitude in one direction only"—that is, to accept the Rome resolution and not Rutherfurd's—would "be extremely inconvenient" even though, as he freely admitted, "it is the easier method." Two of his fellow British delegates were of the same mind. According to Strachey, delegates should reject the recommendations advanced in Rome for fear of the social dislocation that would follow: "The result of the system which was proposed in Rome would be to cause the break of dates to take place at Greenwich at noon, so that the morning hours of the civil day would have a different universal date from the afternoon hours, and this would be the case all over Europe. But if the universal day be made to correspond to the civil day of Greenwich, and the longitude is counted east in one direction and west in another direction to the 180th meridian, these difficulties would be overcome, and a perfectly simple rule would suffice for converting local into universal time." To Adams the different expressions being proposed for the world's longitude—measuring east to west, west to east, from Greenwich as 0 to 360—mattered little: "It amounts, mathematically speaking, to the same thing."[29]

For Sandford Fleming, however, it did matter: what really was at stake was the question of universal time. Washington afforded him further opportunity to reflect upon the modernity of the age. Where once there had been no need for a common system of time reckoning and longitude, argued Fleming, the modern world required a new universal scheme to be adopted. In his view, "The application of science to the means of locomotion and to the instantaneous transmission of thought and speech have gradually contracted space and annihilated distance." Any such scheme should define and create a universal day common to the whole world, establish universal time on the same basis, and produce "a sound and rational system of reckoning time which may be eventually be adopted for civil purposes everywhere, and thus secure uniformity and accuracy throughout the world." This was a matter of keeping time and of reckoning longitude: they could be expressed by a common notation. In defense of his reasoning, Fleming put forward proposals titled "Recommendations for the Regulation of Time and the Reckoning of Longitude." The printed protocol is effectively a summary of these. Fleming had his longer arguments brought together in pamphlet form. To judge from its title page, which notes "Respectfully submitted," these were made available to delegates during the meeting.[30] Central to his recommendations were the ideas of a "Cosmic Day" and of "Cosmic Time," both reckoned not from Greenwich but from "the meridian twelve hours from the Prime Meridian." Longitude, suggested Fleming, should "be reckoned continuously towards the west, beginning with zero at the Anti-prime meridian, twelve hours from Greenwich."[31]

Fleming's insistence upon returning to plans he had laid out in 1879 in *Longitude and Time-Reckoning* was rooted in well-intentioned cosmopolitan principles, but it went against established practice and the views of his fellow British delegates. "It departs from the usages and habits now existing," noted Adams. "That to my mind, is a very great and insuperable objection, and I do not see any countervailing advantage." Evans expressed "dismay" at Fleming's remarks. Strachey noted simply that the current system was a "rational and symmetrical method." No vote, formal or viva voce, was taken on Fleming's recommendations, and the discussion returned to Rutherfurd's initial resolution. The differences within the British delegation over how longitude should be measured is less significant than the reasoning behind them: defense of the status quo, adherence to established maritime practice, and rejection of the resolution from Rome.

Yet what suited sailors over longitude's reckoning did not necessarily suit others. Sampson spoke to just this question as the vote approached: "So far as I have been able to learn, many of the delegates have come here instructed to favor the resolution adopted by the Rome Conference" (Table 4.4, Resolution IV). But, he continued: "It is my own opinion that the recommendation to count longitude continuously from the prime meridian from west to east, as recommended by the conference at Rome, is not so good as the proposition now before us [Rutherfurd's resolution to the effect that longitude be counted in two directions from Greenwich]." This, however, conflicted with his own preference for astronomical practice: "Personally, however, I would prefer to see it counted continuously from east to west, as being more in conformity with present usage among astronomers." This he knew to be unlikely, as was the possibility of adopting the recommendations from Rome: "But, as it appears that so many delegates are instructed by their Governments to favor counting in the opposite direction, and as, if this Congress adopts any other plan than that proposed by the Conference at Rome, they will have to lay before their Governments as the action of this Congress something that will be opposed to the recommendation of the Roman Conference, and as these two recommendations would naturally tend to neutralize each other, I would favor the proposition which is now before us as being the most expedient."[32]

Rutherfurd's resolution over longitude was supported by fourteen votes in favor, with five against. Six countries abstained (Table 5.1, Resolution III). The voting on the resolution concerning longitude reflected what Lewenhaupt, Sampson, and several others recognized as "great differences of opinion" on the matter among delegates. These varied opinions spoke to differences of practice among and between seamen and astronomers, not to distinctions between nations. The resolution in Washington went against the recommendations of the Rome meeting and privileged the customary usage of one community, mariners, over that of astronomers, cartographers, and others. That this was a resolution rooted in expediency should not disguise the fact that it was not expedient for all.

Regulating Time

The third resolution on delegates' printed circular concerned universal time: "That the Conference proposes the adoption of a universal day for all purposes for which it may be found convenient, and which shall not interfere with the use of local time where desirable. This universal day is to be a mean solar day; is to begin for all the world at the moment of midnight of the initial meridian coinciding with the beginning of the civil day and date of that meridian, and is to be counted from zero up to twenty-four hours."[33]

This was considered too complex for a single resolution and so was broken into component clauses, beginning with the adoption of a universal day. Drawing upon his experiences in standardizing time on American railways, Cleveland Abbe spoke at length upon the advantages of a universal day and of universal time: practical convenience was what mattered. In this context, as Albert de Foresta from Italy noted, what was being discussed was little different from the fifth resolution in Rome (Table 4.4). His proposed amendment to adopt the Rome wording was defeated, but his and others' interjections prompted an alteration to the initial clause with the addition of the words "or other standard" (with respect to local time). This was passed by a vote of twenty-three in favor, with two countries abstaining (Table 5.1, Resolution IV).

This left hanging the remaining part of the resolution: the issue of the universal day and whether it would coincide with the civil day, as well as how time was to be counted. For Juan Valera from Spain, this was of "very great importance," yet it was not, he confessed, an issue on which he felt authorized or competent to judge: "I acknowledge that my mission is already fulfilled. The Government of Spain had directed me to admit the necessity or the usefulness of a common prime meridian, and also to accept the meridian of Greenwich as the universal meridian. I have attended to these directions." No other nation's delegates were as open in their self-assessment of competence and responsibility over this issue. It is illustrative, nevertheless, of the variable extent to which delegates felt able to engage with the details of the resolutions, with the science involved in these decisions, and with the nature of the implications. At this point several other matters conspired to delay affairs. Discussion on the question had begun toward the end of the day, and there was little time left for full debate. French involvement with the proceedings was hindered because, as Janssen reported, "preparation of the protocols is very much behind-hand, and it is desirable that the members of the Conference be kept fully acquainted with all the discussions." The meeting adjourned until October 20.[34]

The first order of business at the sixth session was dealing with the written communications, seventeen in all, received from nonparticipants by Rodgers and scrutinized by a committee under Adams's chairmanship. Some were immediately discounted: proposed patents for timepieces; claims about measuring systems; a suggestion that Bethlehem be adopted as the initial meridian. Others received more attention. Even before the meeting, Béguyer de Chancourtois, writing from Paris, had pressed his case for a mid-Atlantic initial meridian (first advanced by him during the Paris IGC in 1875) by sending each delegate a copy of his proposals (see Figure 4.2). These proposals, "nearly identical with those which were so ably laid before the Conference by Professor Janssen, but which failed to meet with their approval," were rejected not because they advocated such a meridian—they were after all very similar to one of the options advanced by the French delegation—but because they did not fall "within the limits indicated by the instructions which we have received from our respective governments." This was not wholly true since the French arguments over principle had entertained the possibility of the Azores as one "neutral" prime meridian. Letters on the metrology of the Great Pyramid sent through Professor Hilgard, who was present but not a delegate, were engaged with and set aside.

No collective action was taken on these written ideas. What matters are not the views advanced but the views of the conference members regarding the inadmissibility of written evidence received from outside their authoritative speech space (despite initial claims to the contrary) and the timing of its review. Had these proposals been examined earlier, before important decisions were made, it is possible that some—Béguyer de Chantcourtois's correspondence, for example—might have influenced the discussion.

In thus reaffirming their epistemic and political authority by collectively discounting the evidence of others (and in Valera's case, despite admissions of individual fallibility), delegates turned to the remaining part of the resolution proposed six days earlier. This stated that the universal day should be a mean solar day, should begin "for all the world at the moment of mean midnight of the initial meridian coinciding with the beginning of the civil day and date of that meridian," and should be counted from zero up to twenty-four hours. Immediately an amendment was put forward, from Count Lewenhaupt of Sweden, to the effect that the initial point "for the universal hour and the cosmic day" should be the mean midday of Greenwich, not mean midnight. Before this could be discussed, Ruiz del Arbol from

Spain advanced a further amendment arguing that there already was a meridian in use for reckoning universal time. Here again it is important to understand the protocols of the Washington meeting. As Rodgers noted, "The Chair must pursue the principle on which it has acted hitherto, taking the amendments in the order in which they are offered, and presenting them inversely for the action of the Conference." More than any other resolution in Washington, with the exception of the majority decision over Greenwich as the world's initial meridian, the resolution over universal time was arrived at not by the collective scrutiny of its intrinsic merit but through the rejection of others.

Ruiz del Arbol's interjection centered upon two matters. The first was his rather confused argument that the world in fact already had a single universal meridian for time—namely, the antimeridian of Rome. When the Christian world brought the Julian and the Gregorian calendars together and made two systems of reckoning days into one, so Rome, he argued, had become an agreed base point. Now was the opportunity to use its antimeridian, a point 180° from Rome, as the modern world's initial meridian for time. The second was a plea to choose no initial meridian at all with respect to universal time, leaving it to be decided upon by railway and telegraph companies, postal authorities, and individual governments. Adams's terse response hints at considerable irritation: "I cannot consent to that proposition. It appears to me to be wanting in every element of simplicity, which should be the chief aim in this Conference." The wording of the printed *Protocols of the Proceedings*—"The Chair would politely suggest," which are Rodgers's words to the Spanish delegate—indicates a degree of tension over the issue. As Rodgers reminded the meeting, what was under discussion was Lewenhaupt's amendment—that is, the adoption of the proposition recommended by the conference at Rome that universal time begin at midday at Greenwich (Resolution VI, Table 4.4). Ruiz del Arbol stood his ground and raised, as Janssen had earlier for France, the issue of a yet further meeting: "My proposition is to abstain from the adoption of any one meridian, and that we leave the matter to some other Congress, organized with the special object of regulating this question."[35]

The Spaniard's argument was not so much about the rejection of any meridian (for that had already been discussed and agreed upon) as it was a poorly articulated opinion on an issue that had yet to be voted upon—namely, where the universal day should commence. As del Arbol's fellow Spanish delegate Juan Pastorin pointed out, the proposition was about universal time and the idea of an antimeridian, 180° from an established prime meridian, whether it was the antimeridian of Rome, Paris, or Greenwich. This issue, Pastorin observed, was not a new one "in spite of its having been modified in the course of our sitting," and he cited those upon whom his case rested:

The works of our eminent colleague and indefatigable propagandist, Mr. SANDFORD FLEMING, the resolution of the Conference at Rome, the valuable opinions of Messrs. Faye, Otto Struve, Beaumont de Bouthillier, Hugo Gyldén, the scientific work of Monsieur Chancourtois, and the report which M. Gaspari has just presented to the Academy of Sciences of Paris are the text upon which I base the simplest and most practical method of solving the problem, namely, to adopt as the prime meridian for cosmic time and longitude a meridian near the point at which our dates change, and to reckon longitude from zero hours to twenty-four hours towards the west, contrary to the movement of the earth.[36]

Of importance here is the appeal to knowledge that was circulating before Washington and to those documents that influenced, even constrained, the voting patterns of delegates before their arrival. Some, such as Caspari's report, which had directed the French on how to proceed, were not available to the delegates of other countries. Other material, Fleming's and Chantcourtois's work most evidently, had been circulating since the IGC; in the case of the latter, because the author had sent each delegate in Washington a copy of his work. Delegates, in short, had prior knowledge of the questions—the resolutions of the Rome meeting—but some judged the Washington resolutions and the Rome recommendations differently because they had gained insight into the question of the prime meridian as a result of attending the IGC or the 1883 Rome IGA or because of their own scientific study of the issue. The propositions of Ruiz del Arbol and of Juan Pastorin were separately put to a voice vote. Both were rejected. We should not see the failure of their views as in any sense a "diversion" from successful individual resolutions. This was how agreement was reached in Washington's Diplomatic Hall. To examine these proceedings is to reveal how these questions were shaped by the contingencies and protocols of a meeting whose conduct was spoken yet whose corroborative evidence in print was not equally available or understood by everyone.

Rodgers directed the delegates back to Lewenhaupt's amendment that the universal day and cosmic time should begin at midday at Greenwich (the resolution advanced in Rome), not mean midnight as proposed by Rutherfurd. Adams spoke against the amendment, emphasizing that it was "most natural" that the universal day "should begin and end at the instant of mean midnight on the initial meridian." One difficulty, as he conceded, was that astronomers measured time differently: "Astronomers, instead of adopting the use of the civil day, like the rest of the world, are accustomed to employ a so-called astronomical day, which begins at noon." In emphasizing the differences in customary usage within those communities debating the prime meridian and universal time, Adams bluntly proclaimed the solution: "It is plain that it is the astronomers who will have to yield." To support his view, he cited a letter from Professor Valentiner, who had been present for the opening sessions in Washington but who had since left the meeting and the city. Valentiner was equally to the point: "I am of the opinion that it is the astronomer only that must give way. For all purposes of civil life one cannot begin the day in the middle of the day-light—that is to say, in the middle of that interval during which work is prosecuted." Strachey called upon Professor Hilgard, who similarly favored midnight at Greenwich as the beginning of the universal day. Evans and Sampson further endorsed this view as the basis to the nautical day.

In the face of these expressions of practical convenience and an evident readiness by astronomers to accommodate the new basis to calculate the universal day, Lewenhaupt's amendment was put to formal vote and defeated: fourteen countries voted against, six voted for, and four countries abstained. Brief exchanges on universal time and local timekeeping practices followed before delegates turned to vote upon Rutherfurd's resolution. This was initially passed with fifteen votes for, two against, and seven abstentions: one delegate later reversed his decision (Table 5.1, Resolution V). The issue of the proposed alignment of the astronomical and nautical day "as soon as may be practicable" was immediately turned to and carried unanimously (Table 5.1, Resolution VI).[37]

Regulating Space: Metrological Uniformity

Following the votes upon universal time and the intention to standardize the astronomical and the nautical day, the agenda for the Washington meeting—as planned in the initial invitation—was completed. For Janssen, however, one issue remained—"the desire that studies relative to the application of the decimal system to the division of angular space and time should be resumed in order that this application may be extended to all cases." Having earlier opposed the view that the resolutions proposed in Rome should as a matter of course guide debate in Washington, Janssen and Lefaivre turned to the 1883 recommendation regarding the decimal measurement of astronomical calculations, in addition to its wider implementation in linear and volumetric measurement. To Rodgers this was "beyond the scope of the Conference," and the subject should be dropped. But, he inquired of Janssen, did he wish to appeal this decision, "to take the sense of the Conference upon it?" This, replied Janssen, was indeed the case: by a majority of votes in his favor, Janssen's appeal was upheld.

This—the only time in the Washington meeting in which Rodgers's management was countermanded—afforded Janssen an opportunity to stress the importance of the end in view: extending the use of the metric system without anyone identifying the solution. Nobody voted against the resolution and, by a vote of twenty-one in favor with three abstentions, the final and most open-ended resolution was agreed upon (Table 5.1, Resolution VII), despite never being a formal part of the proposals to be discussed at Washington.

Two more resolutions were aired, both by Strachey. The first concerned the possibility of reckoning local civil time at successive meridians at ten-minute intervals. As Count Lewenhaupt remarked, this was more or less the scheme that had been proposed earlier by his countryman Hugo Gyldén; Gyldén's paper was later printed for delegates' benefit and included in the formal *Protocols of the Proceedings*. The second resolution suggested that the arrangements for the universal day be left to the consideration of the International Telegraph Congress. Neither was put to a vote. At the session on October 22, Strachey withdrew his resolutions: as he and Rutherfurd pointed out, nothing further would be gained by them. Sandford Fleming tried to address the conference "before any action is taken" but, as Rodgers pointed out, with the withdrawal of Strachey's resolutions, there was no business to conduct. Rodgers met with Cruls and Janssen at his home on the evening of October 31 to prepare for the meeting's end. The "final act" of the International Meridian Conference on November 1, 1884, was to ratify the seven resolutions proposed (Table 5.1).[38]

o   o   o

The "solution" arrived at in Washington regarding Greenwich as the world's prime meridian was neither an inevitability nor an agreed-upon, world-ruling decision.

What happened in Washington over a period of one month beginning on October 1, 1884, was greatly shaped by the events in Rome in September 1883. In the wake of the IGA meeting, different national governments instructed or advised their chosen delegates how they should proceed to vote upon arriving in Washington. An analysis of the printed *Protocols of the Proceedings* from the Washington meeting discloses a more complicated route to the choice of Greenwich than adherence to this earlier set of resolutions. For the French in particular, what is revealed is a quite different set of expectations and a capacity in Washington to both recognize and to reject Rome's agenda-setting resolutions. Decisions reached in Washington were the result neither of arguments over national imperatives regarding observed prime meridians nor scientific considerations that had previously recognized the prime meridians based on the observatories at Paris, Berlin, Washington, and Greenwich as equal contenders. Decisions rested upon practical convenience and the eventual dismissal of French arguments over principle concerning the neutrality of the world's initial meridian.

French delegates did not advance the candidacy of the Paris prime meridian. They argued for matters of principle. The French view of the principles that should direct the choice of the prime meridian—what Jules Janssen saw as the "moral basis" to the issue—was established by the Académie des Science's commission and the Caspari report produced two months before Janssen and Lefaivre spoke in Washington's Diplomatic Hall. In his lengthy articulation of those principles, Janssen revealed them to be a combination of things. The most important was not the appeal to a natural disposition for the world's prime meridian that might allow, by sleight of hand, the baseline to be either the Bering Strait or the Azores. Nor was it the promotion of Paris and the rejection of other national candidates. It was, rather, the emphasis they accorded an altogether new global prime meridian on the basis of its scientific positioning in relation to existing observatories: the sciences of astronomy, geodesy, and geography could, if they were accurate enough, establish an altogether new initial meridian. As the British and American delegates in particular knew and made clear, this flew in the face of the resolutions from Rome and denied practical common sense. Once Greenwich was recommended as the world's prime meridian for longitude, resolutions over the universal day and the coincidence of the astronomical and the nautical day were more easily arrived at. What is also clear is that two of the resolutions—over the reckoning of longitude and the unification of the astronomical and the nautical day—rested on assumptions made about the

user communities of astronomers and practical navigators: their preparedness to reconcile the differences in reckoning their respective days and,
for sailors, the practice of measuring longitude both east and west from
Greenwich.

The resolutions agreed upon at Washington were just that—
recommendations. The fact that they were not—because they could not
be—binding upon the governments of the delegates, upon other nations, or
upon the citizens of the world would lead to particular and geographically
uneven consequences in the wake of this meeting.

PART III

GEOGRAPHICAL AFTERLIVES

6

Washington's "Afterlife"

The Prime Meridian and Universal Time, 1884–1925

WHEREAS THE PREVIOUS CHAPTER considered the "what" and the "how" of the 1884 International Meridian Conference in Washington, DC, this chapter considers the responses—the "So what?"—of that meeting. The argument is structured around two related themes. The first concerns Resolution VI, the proposed unification of the astronomical and nautical (and civil) day. The resolution had seemed unproblematic. It was debated promptly and passed unanimously on October 20, 1884. Diverse scientific communities, however, would disagree about it for decades. Most notably, leading naval astronomers in the United States refused to act on it. For the United States, dissent over this resolution was one reason—but only one reason—behind the country's failure tout court to adopt the recommendations of the Washington conference nearly five years after the event. Sandford Fleming was also at odds with the recommendations of the meeting, given the failure to specify exactly how universal time might be measured locally and applied globally in relation to a twenty-four-hour system of time reckoning. The Washington meeting had led to two recommendations in this respect: a universal day should be adopted, but it should not interfere with the use of local time (Resolution IV), and the universal day would begin for all the world at mean midnight of the initial meridian (Greenwich) and be counted from zero up to twenty-four hours (Resolution V) (Table 5.1). Fleming's continued agitation over the proposed alignment of the astronomical and civil day between 1886 and 1897 reflected both a particular concern with Resolution VI and its connection to the prime meridian and his long-term interests in time reform and modernity.

The second theme concerns the continuing call for the use of a prime meridian other than Greenwich as the benchmark for universal time. After Washington, countries that had already used Greenwich for topographical, nautical, or astronomical purposes continued to do so. Others, like Japan, adopted Greenwich for these purposes for the first time—but did not do so for the reckoning of time itself. Still others, like France, made no change at all. And all the while, many scientists and others around the world kept arguing that for the purposes of measuring time, a prime meridian other than Greenwich should be used, one better suited to telegraphic communication and railway timekeeping and not to the demands of global metrological regulation. This view, principally articulated in international geographical congresses from 1889 and in other scientific bodies, returns us to those spaces of scientific debate discussed in Chapter 4. The case made for Jerusalem between 1888 and 1891 is unequivocal proof that the recommendation proposed in Washington's Diplomatic Hall of Greenwich as the world's prime meridian was not taken up straightforwardly, at once, and by everybody.

In closing, the chapter returns us to two earlier questions. The late nineteenth century was characterized by the standardization of time, not just in the United States but in European countries as well. This move— what we can think of as the international regulation of civic time—was broadly independent of the Washington conference, although that meeting lent impetus to the internationalization of timekeeping and reflected the work of Allen and Dowd in establishing time zones for America's railways based upon Greenwich (see Figure 4.3). As Europe's nations increasingly employed synchronized systems of railway and civic time from the 1880s, other mechanisms of modernity were working to unify the world in terms of timekeeping: one, the telegraph, would be used to justify the post-Washington proposal over Jerusalem as the world's prime meridian in relation to universal time.

The early twentieth century witnessed the advent of radio time signals (radio telegraphy) as a new means of time reckoning. The fact, however, that radio time signals sent from different recording stations could vary from one another by several seconds was a problem when the position of those stations was used as a base point for metrological accuracy in astronomical calculation and in topographical surveying. This was a matter of importance in the continuing efforts to "fix" the position of the Greenwich and Paris observatories. As we shall see, the use of telegraphy in the late nineteenth and early twentieth centuries to measure the distance between the observatories of Paris and of Greenwich encountered long-standing difficulties, which had implications for the "certainty" of the world's prime meridian.

Geographies of an "Afterlife"

As Derek Howse has written, the implementation of the recommendations proposed in Washington varied by nation and by resolution: "Whereas the adoption of Greenwich as Longitude Zero on maps and charts proceeded slowly but surely after the Conference's completion, the recommendations connected with Universal Time were adopted much more slowly, and that to do with the decimalization of angles and time was implemented hardly at all. The principal impact of the Washington conference on the man-in-the-street was the adoption, country by country, of a time-zone system based on Greenwich."[1]

These remarks mask considerable complexity. Washington's "afterlife," particularly as it concerned the resolutions over Greenwich and universal time, can be read in ways other than as diverse and delayed and as responses at a national level. We may distinguish, for example, between popular reactions and scientific responses. Only three weeks after the conference ended and half a world away, the editors of the *Australian Town and Country Journal* pronounced definitively upon the meeting and its results: "All such congresses are of little or no significance. . . . No possible congress can devise a system of uniformity for different countries. . . . The congress must be, as every predecessor has been, a failure." In Britain, popular reaction to the proposed adoption of Greenwich as the world's baseline, on maps and "on the ground," was muted: Greenwich was already in widespread use, and no one in any position of scientific or political authority contemplated the possibility of another prime meridian being selected. In terms of time-keeping, the response was more engaged, publically and scientifically. On January 1, 1885, William Christie, the Astronomer Royal, changed the hands on the public clock outside the Royal Observatory at Greenwich in accordance with the resolution passed in Washington that the astronomical day, traditionally reckoned from midday, should conform to the civil day, which began at midnight. This also conformed to Resolution V over the nature of the universal day. He was not prompted to do so by the government: the Washington resolutions were recommendations only. In Britain, Christie's move toward the proposed change in reckoning time was the subject of much newspaper reporting. In the United States, the

Philadelphia Inquirer of January 2, 1885, reported upon Christie's action and the system proposed of numbering the hours consecutively from one to twenty-four. "Certainly," that paper observed, "the new system has gotten the best start it could have. If adhered to at Greenwich, which might be called the world's chronometric pole, it can scarcely fail of adoption sooner or later by seafaring men throughout the world, and from the sea it will spread gradually over the land."[2]

We may also consider the reactions of leading participants to the Washington meeting and their view of its significance. In Paris, Jules Janssen reported to the Académie des Sciences that the result, simply, "is considerable." But, he continued—reiterating an earlier refrain as he did so—"its significance lies mainly rather with the principles that the Conference enunciated, rather than with the solutions it adopted." In Cambridge, John Adams expressed his feelings on the meeting in a letter to a friend in November 1884: "I am perfectly satisfied with the results of our Congress at Washington in which I took a more prominent part than I expected to do." In Ottawa, Sandford Fleming gave a lengthy account of the meeting and of his role within it while reporting to Canadian officials on December 31, 1884: "The question of the Prime Meridian was not settled without argument and divergence of view. It was one indeed, on which some national sensitiveness was to be looked for. . . . It was not from any national reason that the Meridian of Greenwich suggested itself as the one to be chosen. It was because of its convenience and its general use by the great majority of sea-going ships." Fleming also tersely recounted the stated differences of opinion in Washington between himself, Adams, and the other British delegates over longitude and universal time: "My own colleagues from Great Britain were not in accord."

From Washington there was little to report. In June 1885 George Wheeler informed Fleming "that so far no official adoption by any single Government has followed the action of the Washington Conference but possibly it is yet too early to expect it." Two years later, by which time Japan had become the first nation to formally adopt Greenwich as the world's prime meridian as a result of the Washington meeting (Figure 6.1), the United States had still not acted upon any of the resolutions proposed. On December 15, 1887, Rear Admiral Rodgers, former conference chair, confessed to Fleming at being "much annoyed by the failure of our Government to give effect to the action of the Conference." By March 1889 Rodgers could "not put in words the annoyance I felt and feel" over the United States' continued failure to effect the conference's resolutions.[3]

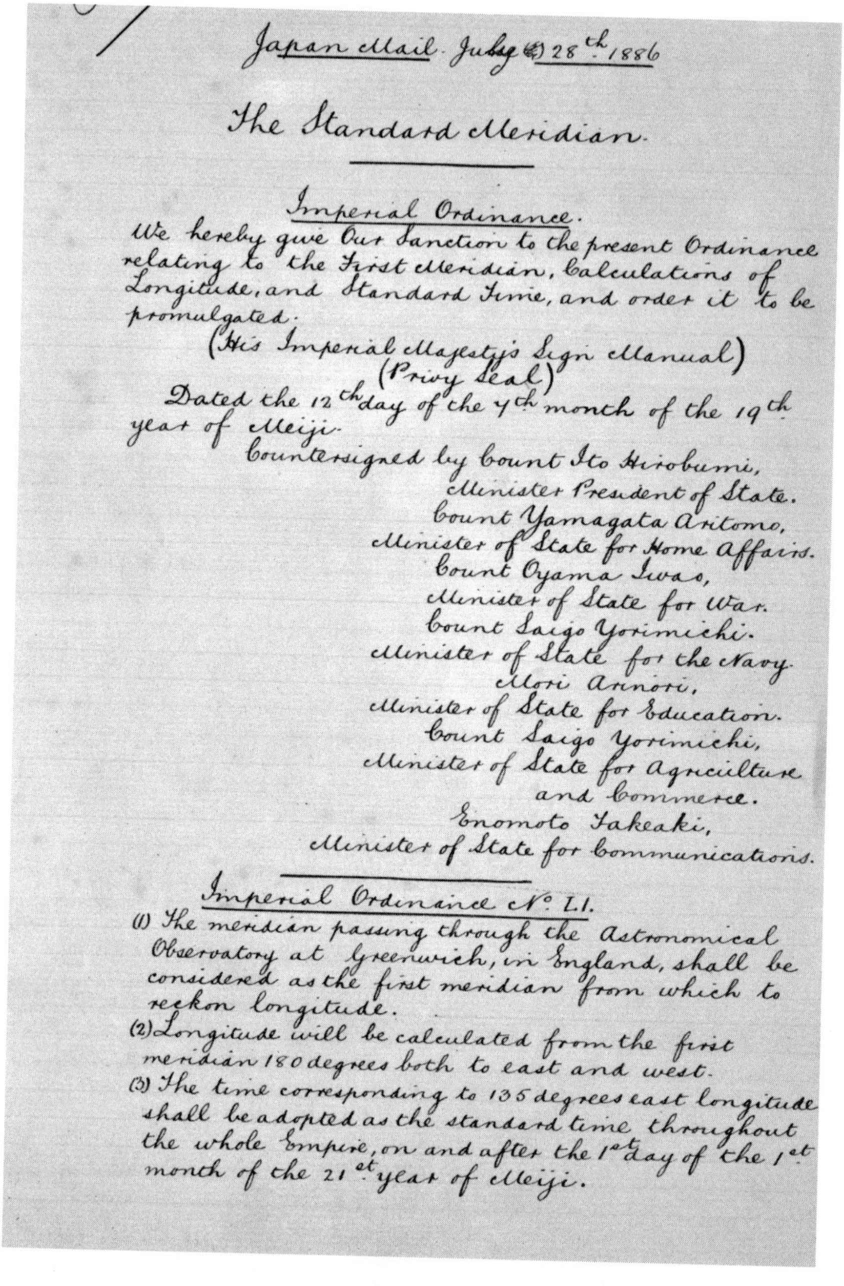

Japan Mail. July 28th 1886

The Standard Meridian.

Imperial Ordinance.

We hereby give Our Sanction to the present Ordinance relating to the First Meridian, Calculations of Longitude, and Standard Time, and order it to be promulgated.

(His Imperial Majesty's Sign Manual)
(Privy Seal)

Dated the 12th day of the 4th month of the 19th year of Meiji.

Countersigned by Count Ito Hirobumi,
Minister President of State.
Count Yamagata Aritomo,
Minister of State for Home Affairs.
Count Oyama Iwao,
Minister of State for War.
Count Saigo Yorimichi.
Minister of State for the Navy.
Mori Arinori,
Minister of State for Education.
Count Saigo Yorimichi,
Minister of State for Agriculture
and Commerce.
Enomoto Takeaki,
Minister of State for Communications.

Imperial Ordinance No. 51.

(1) The meridian passing through the Astronomical Observatory at Greenwich, in England, shall be considered as the first meridian from which to reckon longitude.

(2) Longitude will be calculated from the first meridian 180 degrees both to east and west.

(3) The time corresponding to 135 degrees east longitude shall be adopted as the standard time throughout the whole Empire, on and after the 1st day of the 1st month of the 21st year of Meiji.

FIGURE 6.1 Japan's adoption of the Greenwich meridian "as the first meridian from which to reckon longitude" (and standard time) was confirmed by Japanese Imperial Ordinance No. 51 on July 28, 1886. Reproduced by kind permission from the Syndics of Cambridge University Library, RGO 7/146.

Knowing there was a difference between popular and scientific audiences and how individual delegates reacted is interesting but does not explain why and how the resolutions recommended in Washington were taken up, or not, by different nations. To understand that, we must recall that the Washington conference did not culminate in a single proposal but in different resolutions with distinct yet related histories and the potential for distinct and different consequences. Reaction to the Washington meeting was expressed neither solely at the level of the nation nor as a verdict on the conference overall—"success" or "failure," as the Australian periodical cast it—but was directed at the effect of individual resolutions upon different scientific communities.

Matters of Practice: Dealing with Resolution VI, 1884–1925

In examining the evidence relating to Resolution VI and before turning to review international exchanges on the matter, it is helpful to distinguish between the responses of scientists, mariners, and government officials in the United States, Europe, and Britain as they debated the issue. A note of clarification is also necessary. The specific wording of Resolution VI refers to the astronomical and the nautical days; other resolutions to the civil day (see Table 5.1). The civil day begins at midnight and is twenty-four hours in duration. It is usually divided into two twelve-hour sections. The twenty-four-hour astronomical day began at noon after the start of the civil day. In Britain from 1805 and in the United States from 1848, entries in ships' logs used the civil day: in effect the nautical and the civil day were from these respective dates equivalent. In this regard the wording of Resolution VI was unfortunately imprecise, even unnecessary, since what was being linked to the start of the proposed universal day was the civil day. The onus to act, then, lay with astronomers—as Frederick Evans had said in Washington in 1884—whose system of timekeeping was at odds with others'. But the dividing lines among and within the scientific communities were never absolute. It was after all the astronomers who had established the predictive bases to the nautical ephemerides used by mariners. Differences within scientific communities were evident at national and international levels: some astronomers saw every reason to unify the two time systems as delegates at Washington had proposed. Others argued just as forcefully that they had no need to change since in their view different time systems could coexist without inconvenience.

American Indecision, December 1884–April 1885

The *Protocols of the Proceedings* of the Washington conference was brought to the attention of America's politicians in a message from their president, Chester Arthur, in December 1884. In February 1885, via Secretary of State Frelinghuysen, President Arthur let it be known that he "had done all that is necessary to bring the matter again within the jurisdiction of Congress (where the project originated), and that it is open to that body to signify its wish as to whether the conclusions reached by that Conference shall be brought by this Government formally to the notice of the other Governments, with an invitation to adopt them for universal use by means of a general international convention to that end." Arthur, who had issued the initial invitation to attend the International Meridian Conference, was of the view that responsibility for acting upon the conference's resolutions rested with several offices of the U.S. government in terms both of discussion within America and in liaison with other governments.

But he also believed that another international convention was required to discuss the means to adopt the Washington resolutions universally. Related proposals were advanced with this end in view. The Senate passed a resolution "authorizing the president to communicate with all countries regarding the resolutions adopted at the International Meridian Conference and to invite their accession." Concurrently, proposals were drafted to repeal the act of 1850 that had formalized the use of two prime meridians in America: the meridian of Greenwich was set to be adopted "for all nautical and astronomical purposes."[4] All was in place for the adoption of the 1884 recommendations and the replacement of the United States' twin prime meridians.

And nothing happened. Politics intervened. In March 1885 Chester Arthur was succeeded as president by Grover Cleveland. The second session of the Forty-Eighth Congress, much engaged with other matters, took no action on the Washington resolutions. Cleveland's administration had little interest in advancing the proposals. As Rodgers recounted to Fleming in December 1887, "When the new administration came in, I found no great interest in what its predecessors had begun, and such a pressure of new business, that its attention to old affairs, was not easily to be had." Although the Senate approved the proposals, Congress as a whole dithered, reported Rodgers, "through the negligence of the Chairman of the House Committee on Foreign Relations" (a Mr. Blaine). Rodgers suggested to Cleveland that as incoming president he should call the attention of the new Congress to what Rodgers termed "our obligations to the different countries which had, at our invitation sent delegates to the Conference at Washington," but, as he told Fleming, "As yet, I have had no reply."[5]

Here we have one conjoint reason for the failure of relevant authorities in the United States to take forward the Washington resolutions: personal ineptitude, political circumstances, and administrative inertia. Yet this accounts only for the failure to adopt the resolutions in the first few months after the Washington meeting. A further significant and longer-lasting reason was what Bartky has called the "firestorm of controversy" among astronomers, and between astronomers and others, over Resolution VI.

On the day that Chester Arthur transmitted the official *Protocols of the Proceedings* to the House of Representatives, Samuel Franklin, superintendent of the U. S. Naval Observatory—and one of the U.S. delegates at the 1884 meeting—issued General Order No. 3. This directed that "on or after the Ist of January, 1885, the astronomical day shall be considered as beginning at midnight, corresponding to the civil date." This unilateral proclamation, a direct consequence of the hoped-for alignment in time reckoning to which Franklin had been party, was opposed by several leading American astronomers. Edward Holden, director of the Washburn Observatory at the University of Wisconsin–Madison, considered Franklin's move presumptive since as Holden understood it, the "final cut of the Meridian Conference"—its recommendations—did "not bind any nation, not even the United States, to adopt its provisions." It was doubtful, continued Holden, whether some nations would adopt any of the resolutions within a decade or more: wait, he reasoned, "till a majority of the nations represented at the Conference have at least legalized this new day before introducing it in their official publications."[6]

Another leading opponent was Simon Newcomb, superintendent of the U.S. Nautical Almanac Office. Newcomb, we should recall, had been an informal participant in the Washington meeting. Newcomb's opposition was rooted in three concerns: whether astronomers had any need to adopt this change at all since they were not formally obliged to do so; whether any such change would affect the long-term production of the U.S. Navy's *American Ephemeris and Nautical Almanac;* and whether such a change should be made to any nautical almanac until there had been international agreement on the date for the change in time reckoning. In the absence of legislative direction given the change in government, Franklin's General Order No. 3 could not be formally effected. Franklin dispatched a circular to America's leading astronomers soliciting their views. Of the eleven who responded, most favored the shift and the timing that Franklin had indicated, but several pointed out that *The American Ephemeris and Nautical Almanac* for the years 1885, 1886, and 1887 was already in print since, like all nautical almanacs, it was a predictive text.

Early in 1885 Newcomb renewed his opposition to the proposed alignment of time systems even as he lauded the proposals for the creation of a universal day and its link to Greenwich civil time. The issue in hand, Resolution VI, had not been a source of "trouble or confusion in the past"—by which he meant that astronomical and civil time schemes did not conflict with one another. Should the move be thought necessary, however, it should be by international consensus among astronomers and at an agreed date far enough ahead to make the required alteration to nautical almanacs manageable.[7]

In late April 1885, William Whitney, secretary of state to the U. S. Navy, commissioned the National Academy of Sciences and its president, Prof. Othniel Marsh, to report upon the implications of shifting the start of the astronomical day in *The American Ephemeris and Nautical Almanac*. In reporting favorably upon the issue—that is, recommending the adoption of Resolution VI—Marsh's committee advocated international consensus: "The change should be made as soon as sufficient concert of action can be secured among the leading astronomers and astronomical establishments of the civilized world." In stating this, Marsh and his academy committee merely reiterated the views of Samuel Franklin and ignored Newcomb's authoritative position. In any case they lacked the means to effect such consensus. Moreover, whereas Newcomb had not been an official participant at the 1884 Washington meeting, we should recall that Asaph Hall, the leading naval astronomer of the U. S. Naval Observatory, had been. Hall published a pointed critique of the National Academy of Sciences' report on the grounds that its recommendation reflected the views of the committee only, not of the academy as a whole, whose astronomical members were largely opposed to Resolution VI and who would have said so had they been asked. Marsh and his fellow academy committee members, and perhaps Samuel Franklin too, were conscious of an obligation to the wider political and scientific community: as Marsh noted, "It has been said that, inasmuch as the Meridian Conference was assembled on the call of the United States, there is a certain fitness in a prompt acquiescence on our part in the recommendation of that body." Newcomb and Hall likewise felt obligated but to a different constituency—those astronomers in the United States and elsewhere who saw no reason to adopt Resolution VI. America's scientists were firmly divided over the resolution.[8]

Differences in European Opinion, 1884–1885

Astronomers in Europe, as in the United States, had divergent views. In Berlin's observatory, Wilhelm Foerster, who had been a delegate at the 1883 Rome International Geodetic Association (IGA) meeting, argued that the astronomical day and the civil day should indeed coincide but with the universal day starting at noon, as had been proposed in Rome (see Table 4.4, Resolution VI). In his view the universal day was of little concern to the general public in its day-to-day time reckoning: the customary use of "local time" would not be affected; neither should astronomers bother themselves with civic practices. In Vienna the astronomer Theodor von Oppolzer, who with Adolph Hirsch had overseen the Rome IGA and had voted there to unite the universal and the astronomical day, strongly supported the Washington resolution: that the universal day should, at mean midnight, conform to the civil day at the meridian of Greenwich. From the Imperial Astronomical Observatory in Pulkova near Saint Petersburg, Otto Struve in 1885 offered a lengthy analysis of the prime meridian question in historical context—referring to the "positive evil" caused by the use of different national initial meridians—before reviewing the Washington meeting and its seven resolutions.

As Struve recognized, the first resolution—that there should be a single global prime meridian—"was manifestly purely formal." France's views on the neutrality of that initial meridian had foundered in the face of the vote for Greenwich on practical grounds (Resolution II). Noting that Resolutions III, IV, and V were related to one another (and to Resolution II), Struve turned his attention to Resolution VI and to his support for it (and in passing took aim at Foerster's elitist stance on astronomers and astronomical time): "At a period when everything tends to the simplification of reciprocal relationships, it must appear to us desirable that the numeration of date differing from the rest of the world, must also be abandoned by astronomers, and indeed for the greater reason, that in modern times the mission of many observatories is not simply to subserve scientific purposes, but also to unite with them matters of practical utility." Here was a clear statement by one of the world's leading astronomers: the astronomical community should accede to Resolution VI.

Failure to do so might threaten all that had transpired in Washington, Struve wrote: "It is now asked by everybody whether there is any prospect that the Washington Resolutions will come into operation, and by what means that result may be attained?" His answer was in the affirmative. And his proposed means were simple: "Therefore, naturally it comes to be preeminently a duty for those who in the different countries are in the position to exercise influence in this direction, to make this influence felt in the sphere of their labours." Of all the countries concerned, he noted, Great Britain "has the greatest reason to be satisfied with the Washington Resolutions, for, in her case, there is the greatest accomplishment of her wishes, with a minimum of discomfort and sacrifice. The cartography of the whole Kingdom and its Colonies is already based on the Meridian of Greenwich, and the notation of time in commercial relations in civil life in England and Scotland is determined by mean Greenwich Time, which hereafter also will be recognized as Universal Time." This minimum of disruption was also a responsibility. Struve spoke of Britain's "moral obligation to exert herself to carry out earnestly the wishes expressed at the Washington Conference, namely, the establishment of accord between ordinary Astronomical and Nautical time notation."[9] This was to prove easier to say than to do.

British Views and Reactions, April 1885–February 1886

In Britain, as in the United States, sharply held differences created an obstacle to action on Resolution VI. Turning Struve's moral imperative into metrological consequences for the world would not be easy. In Cambridge, as letters reveal, John Adams supported the proposed shift in astronomical time. In December 1884 a correspondent apologized for having missed Adams during a brief visit to Cambridge, "especially as I had hoped to ask you for some information as to what England is going to do about the resolutions of the conference." Adams's correspondent reveals a seeming acceptance that the resolutions at Washington would be adopted given time and an awareness of the protagonists debating Resolution VI: "Of course they will be accepted I suppose: but I learn from the Astronomer Royal that Profs Forster and Newcomb object to the change in the astronomical day, & that he [Christie] is therefore going to defer taking any public action though he will adopt the change in the Obs^y itself. Could you give me any further information? Will you change at Cambridge? Will the Admiralty accept the change in the nautical day?" Here was an intimation that Christie's forthcoming adjustment of the observatory clock at Greenwich on January 1, 1885, would be without either wider public application or government endorsement. It is also evidence that while these leading astronomers were in favor of the change, Britain's naval authorities could not be presumed to be.[10]

In April 1885 the British established a committee to advise the government and relevant departments upon the resolutions proposed in

Washington. Three of its five active members (Adams, Evans, and Strachey) had represented Britain in Washington. The committee, which never had a formal name, was overseen by Major Gen. John Donnelly, who had been coordinating Britain's institutional and government responses to the prime meridian question since 1881. Now permanent secretary to the Department of Education and Science, Donnelly was assisted by G. F. Duncombe, undersecretary of state at the Foreign Office. Its fifth member was Dr. John Russell Hind, president of the Royal Astronomical Society and the person responsible for preparation and publication of the Admiralty's *Nautical Almanac and Astronomical Ephemeris*.

Donnelly and Duncombe directed that inquiries be made of telegraph companies, learned societies, and government departments regarding the resolutions, principally those relating to universal time and to the proposed alteration of the astronomical day. The Eastern Telegraph Company, in London's Old Broad Street, duly reported that it had long used the twenty-four-hour system. In June 1885 the Royal Society debated the issue in a meeting of its Prime Meridian Committee chaired by Sir Frederick Evans, Evans having been approached to do so by Donnelly. The society recommended the adoption of Resolution VI on the understanding that other nations should do likewise at a convenient time: "If the change of time-reckoning be made in the Nautical Almanacs of all Nations for 1890, the Committee recommend that year for the change to be made." This view—even its timing—was shared by the Royal Astronomical Society. Britain's postmaster general was of a mind that introducing the twenty-four-hour system of time notation "should much depend on popular feeling in the matter." Farther afield in British India, Lord Kimberley, secretary of state for India, was reported to have "no remarks to make on the subject . . . beyond expressing his conviction that the Government of India will be perfectly prepared to accept whatever conclusions may be arrived at by Great Britain."[11]

In January 1886 this committee reported upon the results of its consultation to the effect that "the first five resolutions of the Washington Conference have met with unanimous approval." These required "no action on the part of this country." Resolution VII, on the possible extension of the decimal system in measuring angles and time, was not mentioned. In the committee's view, the problem lay squarely with Resolution VI, the proposed unification of the astronomical and nautical day "as soon as may be practicable," as the *Protocols from the Proceedings* stated. "As regards the sixth resolution . . . it appears that the opinion of the United Kingdom is generally in favour of this change in the mode of reckoning astronomical time; and that the

Lords Commissioners of the Admiralty have expressed their willingness to take the necessary steps to give effect to this resolution of the Conference by introducing civil reckoning into the English Nautical Almanac, if other maritime nations are prepared to adopt the proposed method of reckoning astronomical time."[12]

The wording of this last clause is important. The adoption of this resolution depended not upon individual nations' decisions but, as with the Americans, upon international consensus. In order to effect this, the committee directed Britain's Foreign Office via the British ambassador in Washington to inquire of the United States, as convenor of the Washington meeting, how ready the Americans were to secure "the adhesion of other maritime States to the sixth resolution of that Conference."[13]

Trans-Atlantic Scientific Exchange and Disagreement, 1886–1889

From mid-February 1886, despite ongoing scientific and political disagreements in both countries, officials in Britain and the United States began to negotiate over how to respond to the Washington resolutions, in particular Resolution VI. The United States' response to Britain's request for these negotiations was overseen by George Belknap, Samuel Franklin's replacement as superintendent of the U.S. Naval Observatory, and Simon Newcomb from the Nautical Almanac Office. Belknap's management of the issue was more inclusive, showing little of his predecessor's autocratic style. Newcomb had already made his views clear. So had Asaph Hall, backed in his opposition to altering the start of the astronomical day by his astronomical colleagues at the U.S. Naval Observatory and Professors Harkness, Eastman, Frisby, and Brown of the naval office in Annapolis.

Belknap and Newcomb advanced several reasons to justify this rejection of Resolution VI. One was the view, widely but not unanimously held, that there was in principle no necessity for the astronomical day to align with the civil day. As was further pointed out, Resolution VI was not directly linked to the preceding resolution—a universal day based on the civil day at the meridian of Greenwich—but to "local astronomical [time] and nautical days, as used by astronomers in their several observatories, and by ships on the ocean." If the resolution was adopted, "Washington astronomers will count their days from Washington midnight, the Berlin astronomers from Berlin midnight, &c. . . . this change will not in any manner conduce to uniformity of time reckoning upon different meridians." Far from promoting a common standard, namely universal time, the adoption of Resolution VI would promote the singularity of local time reckoning. Another objection concerned the fact that the principles of nautical astronomy would have to be redrafted, and the astronomical and nautical textbooks would have to be rewritten—and not just ephemerides such as *The American Ephemeris and Nautical Almanac* and Britain's *Nautical Almanac and Astronomical Ephemeris*. Further, Belknap and Newcomb made a semantic distinction between the resolutions. They termed Resolutions I to V "proposals" and Resolutions VI and VII "simple expressions of hope" that should be judged on their own merits. Concerning VII at least, they echoed the views of Britain's Royal Society Prime Meridian Committee, which considered Resolution VII to be "the expression of a hope that the subject might be further studied, to which of course there could be no objection."

At the same time, the views of those American astronomers who had supported the adoption of Resolution VI in answer to Franklin's inquiry of December 1884 were dismissed: "The views given therein were expressed without discussion and without consultation." Marsh's National Academy of Sciences' report was similarly rejected: "The report has never been laid before the Academy itself for examination and discussion, adoption or rejection." Doubt was cast both on the agenda of the Washington meeting and the expertise of its delegates in failing to discuss the merits of telegraphy in determining longitude, from which straightforward calculations could be made to effect a universal time in relation to the observatories most involved: Greenwich, Paris, Washington, and Berlin. The "inconveniences" to astronomers, as Belknap and Newcomb termed them, were in essence twofold. First, they felt strongly that the proposed change could not easily be made given the existing literature on the subject, which both their contemporaries and astronomers in the future would consult. As a result these groups would have to work with two distinct methods of reckoning time. Second, they perceived that no easy means would be found of making the change globally effective and "securing a general consensus of action in making the change."

The Belknap-Newcomb report did not call for rejecting all of the resolutions made in Washington some fifteen months earlier: "We do not conceive ourselves recommending any action adverse to the recommendations of the Meridian Conference." Rather, it carefully turned the wording of those recommendations into a justification for selective dissent: "So far as the universal day is concerned . . . the Conference did not recommend its adoption in the Nautical Almanac, but very wisely restricted it to cases in which it should be found convenient. We do not deem it convenient to introduce it into astronomy at present." The language of the 1884 Washington conference was turned back on itself. In closing, however, Belknap and Newcomb seemed to leave space for the resolution's later possible adoption: "But if, despite the views and considerations herein expressed, this Government, as the convenor of the Prime Meridian Conference, deems it best to adopt and conform to the suggestions embodied in the sixth resolution of the Conference, then the undersigned beg leave to suggest that the date of 1900 may be fixed upon for such purpose."[14]

At the same time this report was effectively proscribing the formal adoption of Resolution VI within the United States or at least delaying it until century's end, Sandford Fleming was continuing to press his own thoughts on the resolution on Canada's politicians, Europe's astronomers, and Britain's shipmasters.

Sandford Fleming, Modernity, and Universal Time, 1884–1896

Sandford Fleming had several reasons for taking up the issue of the proposed alteration of the astronomical day in the wake of the 1884 International Meridian Conference. Reporting in December 1884 upon the Washington meeting to the secretary of state for Canada, Joseph-Adolphe Chapleau, Fleming stressed that with the resolution concerning Greenwich carried, "the establishment of a system of Universal Time became possible." As Fleming was not slow to point out, some years before, he had outlined the principles of universal time accepted in Washington (in his *Time-Reckoning and the Selection of a Prime Meridian to Be Common to All Nations* and *Longitude and Time-Reckoning*). Fleming's purpose in reporting was also to provide for Canadian officialdom a sense of that country's contribution to the debates: "I trust it will not be an impropriety on my part thus dwelling upon the important part Canada has played in the establishment of Universal Time, and in the determination of an initial Prime Meridian for the world." By implication, of course, he was lauding his own endeavors in this respect. So when Fleming again turned to proselytize upon universal time in his new *Time-Reckoning for the Twentieth Century*, he was reviving his old personal mission by drawing upon the strong support of Canada's politicians and scientists and his own interpretation of the Washington meeting, where he believed some of his ideas had been ignored not because of their intrinsic flaws but because of his fellow British delegates' opposing views.[15]

Time-Reckoning for the Twentieth Century is of a piece with Fleming's earlier arguments, with the significant addition of his reflections upon the Washington meeting and with heightened appeal to the value of universal time reckoning in a rapidly modernizing world. Alluding to Christie's adjustment of the public clock at Greenwich in 1885 and to the earlier work of the American Society of Civil Engineers' committee on standard time, Fleming emphasized the importance of "a new measure of time entirely non-local" that would be based upon Greenwich. But for universal time— "Cosmic Time," as he also termed it—to be genuinely global, the astronomical and the civil days needed to be aligned. That is, Resolution VI needed to be taken up everywhere.

Fleming drew upon others' authority to support his case. He cited William Christie's lecture of October 1885 that had reported that Thomas Henry Huxley, chairman of the Board of Visitors of the Royal Greenwich Observatory, had recommended adoption of Resolution VI and the alteration of Britain's *Nautical Almanac and Astronomical Ephemeris* in 1891 despite dissent from the Admiralty and opposition from the United States. The testimony of Otto Struve was again called upon—"that all astronomers throughout the world should simultaneously abandon Astronomical Time and bring their notation into harmony with the civil reckoning." In Fleming's eyes, no one now "can question that the change of the century is an appropriate period for effecting the complete unification of time, and doing away with all the errors of our present mode of reckoning.... The proceedings of the Washington Conference have given the movement an immense impulse."[16]

Given the evidence stated, this last phrase was more an expression of hope than a statement of fact. Although the suspicion must remain that he was promoting his own long-running interests in universal time rather than advancing Resolution VI per se, Fleming continued to work on having the proposed alteration in the astronomical day adopted. In November 1889 he prepared a "Memorandum on the Movement for Reckoning Time on a Scientific Basis." This was based on *Time-Reckoning for the Twentieth Century* but supplemented by reference to the letters sent to Donnelly's committee from those companies and academic bodies who in 1885–1886 had declared themselves in favor of the resolution and of the global adoption of the twenty-four-hour scheme of time reckoning. In January 1890 this memorandum was sent from the Canadian Institute in Toronto to the governor general in Ottawa and from there to Britain's Colonial Office and to Donnelly in the Science and Art Department. In July 1890 Donnelly instructed that the memorandum be sent to Britain's colonial administrations for their views. Donnelly's directive addressed only the possible "adoption of the hour zone system in reckoning time generally, and of the 24 hour notation for railway time tables": it did not specifically identify the alteration of the astronomical day or seek views upon this. The reasons for this selective omission, if not simple oversight, are not clear.

One fact is certain: from July 1890 Donnelly and his Science and Art Committee had their attention drawn elsewhere. At just the moment they were dealing with the Colonial Office, the Foreign Office informed them that a communication had been received from Italy's ambassador in London "respecting a projected Congress to be held at Rome, to discuss the question of a Prime Meridian, and asking if Her Majesty's Government would be disposed to be represented thereat."[17] The matter of this proposed congress is examined below. Debates over universal time and Resolution VI must have seemed less significant to Britain's interests and the authority of the Greenwich meridian than the threat raised by this new message of moves to consider an alternative prime meridian in the wake of Washington.

Back in Canada, Fleming established a joint committee late in 1892 of the Canadian Institute and the Astronomical and Physical Society of Canada, with himself as chair. He did so following communication with Christie over Resolution VI and meetings with Captain Wharton, a former hydrographer to the navy, and with Donnelly during a visit to London in July 1892. Interestingly, although Wharton was not personally in favor of the astronomical day being changed now, "He is quite willing & he says the Admiralty will be willing to favour the change provided Astronomers give their assent." This was not straightforward. How, asked Fleming, "are we to reach the astronomers?"[18]

Recalling his earlier schemes to promote universal time, Fleming's joint committee distributed a circular in April 1893 to the world's astronomical community asking for their views on altering the astronomical day. Answers were sought to one question only: "Is it desirable, all interests considered, that on and after the first day of January 1901, the astronomical day should everywhere begin at mean midnight?" Although only 171 replies were received from almost one thousand circulars sent, the results further illuminate the diversity of international opinion on the issues enshrined in Resolution VI (Table 6.1). The low level of return, especially from certain countries, cautions against our placing too much weight on this as unequivocal evidence of national sentiments with regard to Resolution VI. But one point may be made. These were the views of the civic astronomical community—interested individuals, private astronomical observatories, and some national establishments—more than of official or naval astronomers. This was not exactly the "preponderating weight of opinion amongst astronomers themselves that a change should be made in the Astronomical Day," as

Table 6.1 The response of the world astronomical community to Sandford Fleming's 1893 question on the proposed alteration of the astronomical day, with effect from January 1, 1901

Nation	Total responses	Yes	No	Majority
Austria	12	7	5	In favor
Australia	2	2	0	In favor
Belgium	6	6	0	In favor
Canada	5	5	0	In favor
Colombia	1	1	0	In favor
England	20	16	4	In favor
France	4	4	0	In favor
Germany	38	7	31	Against
Greece	1	1	0	In favor
Holland	1	0	1	Against
Italy	11	8	3	In favor
Ireland	4	4	0	In favor
Jamaica	1	1	0	In favor
Madagascar	1	1	0	In favor
Mexico	5	5	0	In favor
Norway	1	0	1	Against
Portugal	1	0	1	Against
Romania	1	1	0	In favor
Russia	11	6	5	In favor
Scotland	1	1	0	In favor
Spain	2	2	0	In favor
Switzerland	4	2	2	Equal
United States	38	28	10	In favor
Total	171	108	63	

Source: "Proposed Reform in Time Reckoning," *Forty-Second Report of the Department of Art and Science of the Committee of Council on Education, with Appendices* (London: Her Majesty's Stationery Office, 1895), app. A, 20.

Fleming put it in May 1894 when inviting Canada's governor general to forward the results to the British authorities. But it was sufficient for Fleming to argue that the Belknap-Newcomb report should not stand for the United States as a whole: "With all the facts before us, it is impossible to consider that this adverse report signed by three officials of the United States' Naval Observatory, fairly represents the mind of the United States Government, of Congress, or of the people of the United States."[19]

Fleming misjudged the authoritative position of the U.S. naval astronomers. Nor did he appear to understand what the response in Britain of the Admiralty and of Donnelly's Science and Art Committee would be, given earlier evidence. The Admiralty rejected the Canadian committee's report on the grounds of its lack of evidential authority: in their eyes, the list of respondents "omits nearly the whole of the names of leading Astronomers in the world." They reiterated their view, first articulated in 1886, that until such time as those other countries issuing nautical ephemerides (France, Germany, the United States, Austria, Spain, Portugal, Brazil, and Mexico) were agreed on the matter and on the timing, no action would be taken. The response of Donnelly's Science and Art Committee was no more encouraging, which was surprising given that three of its five members had agreed to the resolutions put forward in Washington. They did suggest that, via the Foreign Office, the British should raise the matter with the governments of the mentioned countries "with a view to ascertaining whether there is such a general agreement on the subject and would admit of concerted action in regard to the introduction of the change proposed in the VIth resolution of the Washington Conference, at the beginning of the next century." Here, internationalism was a disincentive to action on Resolution VI. The spirit of internationalism was a shared end in view of these debates over a single prime meridian but was not itself sufficient to ensure action on the issue should consensus be forthcoming.[20]

International consensus was gestured toward but never realized. Appeals for agreement revealed only national differences over whether to adopt Resolution VI and if so when to take it up and make the necessary alterations in respective ephemerides. What was expressed was at best a preparedness to act only if and when others did. Yet nothing motivated anyone enough to make that happen. Neither the German nor the Portuguese authorities replied. In Austria the Academy of Sciences in Vienna did not think it "especially expedient, as it would hardly afford any practical results," although Austria's Royal Navy and merchant marine had no objection to the proposal. Both Spain and Brazil saw merit in the adoption from 1901 if, as the Spanish put it, "the majority of the Ephemerides Offices which will regularly issue Nautical Almanacs are in favour of it." The Mexican government blithely assumed that other countries' astronomers had already approved Resolution VI. The United States bluntly declared itself "decidedly opposed to any change in the existing mode of reckoning astronomical time and therefore recommend that no departure be made from the present system."[21]

In Washington, Cleveland Abbe—who voted yes in response to Fleming's 1893 circular—understood the implications of the United States' steadfast refusal to adopt Resolution VI. As he wrote to Fleming on May 21, 1895, "If this be taken as the final decision of the United States, it assuredly means that the reform will never be made in any part of the world. But I cannot bring myself to believe that a country so advanced as the U.S. is in everything, will in this matter stand in the way of a reform proposed at the Conference by U. States delegates." In a letter three days later, Abbe advised Fleming that attempts to coerce naval and nautical almanac authorities, in the United States or in Britain, were not likely to be successful:

> I doubt whether it is best to spend much time trying to force the American Nautical Almanac to adopt the time reform that is to say trying to force it by order of higher authority. When you address the Naval Observatory or the Hydrographic Office, or the Secretary of the Navy the reply is, of course, suggested by the Nautical Almanac Office which comes under this organization.
>
> ... The real question now is to what extent do the navigators of the world desire the abolition of the astronomical day and the exclusive usage of the civil day and date on sea and land. The astronomers proper and the nautical almanac makers are not personally troubled by this duplicate system of reckoning and if they wish to keep it up in their observations and ephemerides and astronomical tables we, outsiders, can hardly prevent it.[22]

Fleming took Abbe's guidance as a prompt to renewed action. He dispatched a further circular in July and August 1895 directed not at the "navigators of the world" but toward the shipmasters using British ports. The circular asked four questions (Figure 6.2). Of the 243 replies received, the responses to all four questions were overwhelmingly positive, the yeas being 237, 234, 233, and 223 "and the nays 5, 8, 7, and 19, with 1, 1, 3, and 1, questionable or doubtful answers to the queries 1, 2, 3, and 4, respectively."[23]

This 1895 circular should be understood in several ways. In his first question, Fleming sought to solicit the views of one significant maritime community, Britain's shipmasters, upon a resolution regarding Greenwich that had effectively already been adopted (Resolution II). This was in order to generate action on a resolution that almost eleven years after the Washington conference had still not been adopted (Resolution VI). Fleming already knew, or thought he did, what the answer would be. As long ago as 1879 and his *Longitude and Time-Reckoning*, he had shown that the Greenwich prime meridian was the initial meridian most widely used by the world's—and Britain's—mercantile community (Tables 4.2 and 4.3). Practical navigators had not changed established practices in the interim. Fleming understood

Please give an answer "Yes" or "No," signing your name, with the name of your vessel and her employment, Foreign or Home-trade, to the four following questions:—

UNIFICATION OF TIME.

1. Are you in favour of the Greenwich Meridian being universally recognised as the prime or first Meridian by all Maritime Nations?

2. Are you in favour of the Unification of Time as reckoned from such prime Meridian, and extended to all nations irrespectively?

3. Are you in favour of the Unification of Time as applied to the Civil, Nautical, and Astronomical days, and is it desirable in the interest of all concerned that such days should each commence at Greenwich Mean Midnight?

4. Are you in favour of reckoning the day by the 24 hours system, counting the hours for each Civil, Nautical, and Astronomical day from O, or Mean Midnight at Greenwich throughout the 24 hours to Midnight again? This will do away with the old a.m. and p.m., but will make Mean Noon at Greenwich the 12th hour for all three systems of time reckoning alike?

"Yes" or "No."	Name of Master.	Name of Vessel.	Employment—Foreign or Home.

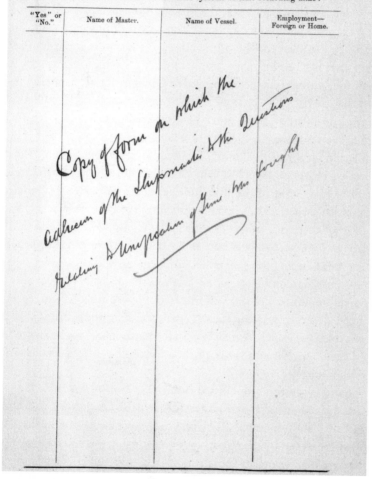

FIGURE 6.2 Sandford Fleming used a standard questionnaire to determine the attitudes of British shipmasters toward both Greenwich as the universal prime meridian and the unification of the civil and astronomical days, as this copy of the form shows. Reproduced by kind permission from the Syndics of Cambridge University Library, RGO 7/146.

this. He was, of course, garnering evidence in support of his work on universal, "cosmic," or "cosmopolitan" time. He was hoping to use evidence regarding mariners' practice to counter what he saw as astronomers' prejudice.

In October 1895 he sent William Christie—"for your private information"—an advance copy of the Canadian joint committee report of October 1894 that was shortly to be sent, via officials in Canada, to Donnelly's committee in London. "The report," wrote Fleming, "speaks for itself. You will notice that we have been constrained to allude to a well known gentleman in Washington [Newcomb] who has always assumed a hostile attitude and unfortunately is at the present moment in an official position to stop the wheels of progress if he be listened to, or unless the bureaucratic platform be demolished. The prime object of the report is not to place our friend in a ridiculous light but simply to carry a scientific reform for which we have long striven." As the report moved through official channels, Fleming hoped that Christie, as someone close to the Science and Art Committee and as an influential astronomer in his own right, would have time to examine the evidence "in support of what I mention to call an irrefutable argument in favour of the United States as an assenting nation." Replying to Fleming in March 1896, William Christie expressed his frustration over the matter: "It is unfortunate that Prof Newcomb has taken up such a position as his obstruction is a serious difficulty. We have obstruction too on this side but there is, I believe, a sufficiently strong feeling in favour of the change amongst astronomers here to break down the 'official' obstruction." In response, Fleming considered—with some justification given what we have seen of Newcomb's attitude—that Newcomb had "been at war with the general movement of time reform from the beginning." The real worry, continued Fleming, was the "fear ... that, speaking from his official position, his voice [Newcomb's] may be mistaken in England for the voice of the United States & that as a result of the mistake the desired change will be stopped for another century."[24]

What does all this amount to? In his *One Time Fits All: The Campaign for Global Uniformity,* Bartky observes that if Resolution VI and altering the start of the astronomical day was effectively a dead issue in the United States within a year of the Washington meeting, it was the country's naval astronomers who wrote its "official obituary." This judgment is correct for the United States but less applicable to Britain. Given the authoritative weight of Belknap and Newcomb's report and the support lent to it by senior naval astronomers such as Asaph Hall, there was little possibility of the United States officially adopting Resolution VI, despite Franklin's early presumptive efforts and the supportive views of some in the U.S. astronomical community. Rodgers's displeasure in 1887 and in 1889 at his nation's failure to adopt the recommendations from the Washington meeting doubtless reflected a sense of personal slight and frustration at the United States dragging its heels. But the root cause lay with individual leading astronomers in the American naval community and their persuasive rhetoric that, when not in outright opposition, always sought to push the possible adoption of Resolution VI into some unspecified future and only on the condition of further international accord. The failure to adopt Resolution VI did not rest with astronomers as a whole, with tardy politicians, or with changes of administration.[25]

In Britain and later in France, the question of altering the astronomical day to commence at midnight (at the initial meridian) remained a topic of scientific and political concern. The evidence of various national committees in the United States, in Canada, and in Britain, which from December 1884 debated the issue, and of Fleming's work on universal time between 1886 and 1897 shows just how complicated the "afterlife" of Resolution VI was. What is revealed are lingering "last rites," not a universal obituary. In Britain, negotiation of the issue centered upon Donnelly's committee in the Science and Art Department. Between 1885 and 1896, it was the focus for coordination with the British Admiralty, British companies and shipmasters, Fleming's Canadian joint committee, Britain's colonial administrations, and foreign governments. In Britain the common response to Resolution VI—from the Admiralty, the Royal Society, and other bodies—was not absolute rejection of the proposed alteration of the astronomical day. Rather, it was a preparedness to admit the change would be logical but only if other countries changed too. In the absence of an agreed method to coordinate such a move, the failure to adopt Resolution VI was inevitable—but not absolute. In advising Donnelly's committee, Christie was of a mind to accept the proposal and Adams was too; Hind was not. Sanford Fleming showed, to his satisfaction at least, that many private astronomers were prepared to adopt Resolution VI. Replies from Britain's shipmasters to his questions of 1895 confirmed that attitudes to Resolution VI varied at the level of nations within scientific and other user communities and even within government committees.

Fleming's schemes may not have been successful in effecting agreement over Resolution VI. But his fear that the issue would be set aside for a hundred years was without foundation. Newcomb's death in 1909 removed one obstacle to the adoption of Resolution VI. And the French had earlier been making helpful moves toward an accord. In November 1896 France agreed in principle to "substitute officially in France the meridian of Greenwich for that of Paris" and to adopt Resolution VI, citing in reply to further queries from Donnelly's committee what was a common response—France would agree to the reform if other countries would do likewise. In fact, France did not adopt Greenwich mean time until 1911. In Britain the secretary to the Admiralty wrote in December 1917 to the Astronomer Royal and to the Royal Astronomical Society in the wake of a conference on timekeeping at sea held in June of that year. He recommended adopting the change from noon to midnight in the reckoning of the astronomical day (Resolution VI). Within a year, the society reported in favor of such (as, in principle, it had in 1885). In June 1919 the Admiralty issued instructions that this change should be incorporated in Britain's *Nautical Almanac and Astronomical Ephemeris* and take effect in 1925. Only from that date may we think of Resolution VI as formally adopted in Britain. This drawn-out alignment of the astronomical and civil day took place alongside continuing debates over the possibility of a prime meridian other than Greenwich and as part of international moves to standardize time.[26]

Not Greenwich: A Different Global Prime Meridian after Washington?

In his 1885 overview of the Washington conference, Otto Struve considered its principal job done: "We may, therefore, regard the chief object of the Washington Conference, namely, the establishment of the First Meridian, from which all the remaining questions are more or less natural consequences, as satisfactorily solved."[27] Nothing could be further from the truth.

Existing prime meridians other than Greenwich did not cease to be used after 1884 simply because delegates in Washington proposed Greenwich as the world's initial meridian. Because the resolutions advanced were recommendations, not requirements, different national prime meridians remained in use: Greenwich for navigational purposes by the majority of the world's maritime nations; Greenwich and other national prime meridians such as Paris for geographical publishing—in atlases, for example; and numerous national prime meridians for in-nation topographical surveying and astronomical inquiry. The Washington meeting had proposed Greenwich as the world's initial meridian in relation to universal time. But universal time was slow to be adopted—as Sandford Fleming's activities before and after 1884 make clear and as we will see as well in the next section. As usage continued of prime meridians other than Greenwich, it was thus the case, at least in principle, that one initial meridian could be used in association with universal time and another, Greenwich, might be employed for longitude, astronomical calculation, mapmaking, or oceanic navigation. Proposals for just this arose not once but twice after Washington: from Geneva, the idea of a global meridian based on the Bering Strait was put forward; from Bologna and Rome, others advanced a case for Jerusalem.

The Bering Strait Revisited

In April 1888 in a paper to the Geographical Society of Geneva, Henri Bouthillier de Beaumont proposed that the prime meridian be aligned on Cape Prince of Wales on the eastern shore of the Bering Strait. In one sense this was old news—the reiteration of an earlier concern and one shared by others, albeit with variations. De Beaumont had advanced it in 1876, his argument attracting the approbation of Charles Daly of the Geographical Society of New York, among others (Chapter 4). In another sense this was new work. Unlike in his 1876 paper, de Beaumont explicitly associated his 1888 proposal for a Cape Prince of Wales / Bering Strait prime meridian with the requirements of universal time and of map projections, which in his view would allow the clear presentation of hour-based meridians around the globe. Every 15° would correspond to one hour. De Beaumont reiterated his views in August 1888 in an address to a congress of Swiss geographers. His arguments were met with "considerable sympathy and loud applause."[28]

It is impossible to know what this reception of de Beaumont's revivified argument for a prime meridian other than Greenwich was based on: his status and standing as a senior figure within the geographical body in question, the utility or nature of his proposed map projection, or his simple perseverance in presenting these views over a dozen or more years. Certainly, his advocacy of a prime meridian other than Greenwich, four years after Washington, flew in the face of that conference's attempt at achieving global metrological uniformity. Yet in advancing an alternative to Greenwich, de Beaumont was not alone.

The Bologna Proposal and the Jerusalem Option

Also in August 1888, Cesare Tondini de Quarenghi, an Italian scholar, addressed the mathematical section (Section A), of the British Association for the Advancement of Science (BAAS) at its meeting in Bath. By vocation a Catholic missionary (in Croatia) and the author of works on the liturgical calendar and religion and the state in Russia, de Quarenghi was on this occasion representing the Academy of Sciences of the Institute of Bologna. "His" paper—it originated with the mathematicians of the academy and in events held in June 1887 to mark its eighth centenary—was titled "A Suggestion from the Bologna Academy of Science towards an Agreement on the Initial Meridian for the Universal Hour."

De Quarenghi proposed a case for a new prime meridian, not Greenwich, to be used in association with universal time. Airy's remarks in 1879 on the prime meridian and universal time were cited as evidence that Airy did not at that time axiomatically accord global primacy to Greenwich: "Nearly all navigation is based on the Nautical Almanack, which is based on Greenwich observations and referred to Greenwich meridian.... But I, as Superintendent of the Greenwich Observatory, entirely repudiate the idea of founding any claim on this" (see Chapter 4). De Quarenghi drew attention to the continuing failure of America's administration to endorse the resolutions from Washington, citing the published records of the U.S. Congress on January 9, 1888, that recommended the U.S. government "take action to approve the resolutions passed in 1884 and to invite the Powers to accede to them." The resolutions, he noted, are "not approved yet, nor have the Powers [other leading nations] acceded yet to them." The work of Struve and of Fleming was invoked, de Quarenghi using the latter's interest in universal time to stress "the urgency" of the question. Finally, he argued that Caspari's report of August 1884 to the Commission de l'Unification des Longitudes et des Heures made the case for the simultaneous use of more than one prime meridian. All of this was justification for the Bologna academy's proposal regarding the world's initial meridian: "That, navigators and astronomers being at liberty to go on using their own initial meridians, another truly international meridian be chosen for all other purposes for which the unification of time is required. That, moreover, since the Jerusalem meridian has already the suffrage of scientific authorities, its appropriateness to serve as the universal initial meridian be seriously taken into consideration."[29]

This proposal shows just how little Washington had really resolved. Resolution II in favor of Greenwich had not been acted upon, with disagreement assuming a different character in different places. Scientists in Bologna were sufficiently aware of this state of affairs to be able to cite from American congressional minutes, Airy's correspondence, Sandford Fleming's and Otto Struve's work, and French government reports. Technically, the Italians were correct to state that navigators and astronomers were free to continue using their own prime meridians despite the recommendations made in Washington and Greenwich's predominance as a global prime meridian before 1884 (Tables 4.2 and 4.3). The proposal, that Jerusalem be the world's prime meridian and that it was neutral, albeit having particular significance to Christians, had been made before, initially at the International Geographical Congress (IGC) in Paris in 1875. In Rome in 1883, however, Adolphe Hirsch had rejected the possibility, given Jerusalem's lack of an astronomical observatory. In Washington in 1884, correspondence proposing Jerusalem had come to the delegates' attention as they sat in session, although the case for the city was never formally entertained there.

What was novel about the 1888 Bologna proposal was the link made to universal time. Since universal time—in the form of a twenty-four-hour system of time reckoning based on a recognized zero as midnight—had not yet been adopted everywhere, it was possible to have a base prime meridian for time other than Greenwich (or any other prime meridian then in use). Despite an earlier lack of support, the continued advocacy of Jerusalem, albeit unsuccessful, was sufficient evidence for the Italians to claim that the city had "the suffrage of scientific authorities." Again, technically at least, they were correct to claim that universal time might be measured from any initial meridian. There was, in short, a legitimacy to the Bologna proposal, despite Greenwich's primacy on pragmatic grounds.

The group that de Quarenghi addressed, the BAAS, was in the nineteenth century Britain's leading body for the discussion of science in a civic context. As such it had discussed questions related to the prime meridian and metrology on several occasions, not least in the advice of a BAAS committee over the advantages to science of the metric system (see Chapter 3). As a corresponding scientific society, academicians in Bologna received the BAAS's printed annual report and so were aware of the association's role and influence. The BAAS committee, which in the early 1860s had come out in favor of the metric system in the interests of science and which included Airy among its members, reiterated its recommendations in the early 1870s as debate intensified over the metric and imperial systems. In 1876 Sandford Fleming, in keeping with his other work on modernity and universal time, had proposed a paper to the BAAS's annual meeting in Glasgow that stressed the "desirability of adopting Common Time over the Globe for Railways and Steam-Ships," but there is no evidence that he actually delivered his talk. Ten years later—his work on universal time being by then more widely known—Fleming received an invitation from leading geographer Ernst Georg Ravenstein on behalf of the geographical section (Section E) of the BAAS: "I am instructed by the Committee for Section E . . . to ask you, to read a paper at Birmingham [the location of the BAAS meeting in 1886], dealing with the system of universal time, originated by you, and adapted on the Canadian and American railways. The subject may not be strictly geographical, yet it is one in which geographers naturally take an interest." Fleming could not accept Ravenstein's invitation, being en route to Canada from Liverpool by the time he received it, though he found it "very gratifying" given his "intense desire to bring my views before the British public through the medium of the influential association which you represent." Fleming took the invitation to indicate that "the authorities of the British Association now consider the question of sufficient importance to occupy time of its respective meetings." Ravenstein was similarly abreast of current topics: his paper to the 1887 Manchester meeting, "A Plea for the Metre," left no doubt where he stood with regard to questions of metrological uniformity.[30]

Given this activity within the BAAS, de Quarenghi and his Bologna colleagues had every reason to approach that body and to expect that their proposal over Jerusalem would be taken seriously. The BAAS's response was to form a committee—or to try to. Of the four men invited to consider the Bologna proposal, neither the Astronomer Royal, William Christie, nor the mathematician, Dr. Longstaff, had been present at the 1888 Bath meeting. Both later declined to serve. One year later, at the Newcastle meeting in September 1889, the BAAS committee of two, a Mr. Glaisher and Sir Robert Stowall Ball, the Royal Astronomer of Ireland, delivered their report on the Bologna proposal. Their view was unequivocal: "The question of a universal prime meridian is one that cannot usefully be considered by a Committee of the British Association at the present time."[31]

The records of the BAAS are unfortunately silent on the reasons behind this judgment. For the British, of course, the question of a universal prime meridian had already been decided de facto—it was and would be Greenwich. At issue here was a different prime meridian for a different purpose, not a single prime meridian for all global metrology. The obduracy of the BAAS, however, did not signal the end to the Italian proposal. As he waited for the BAAS committee to report from Newcastle, de Quarenghi again brought forward the Jerusalem option for others' scrutiny, this time to delegates in group I, Mathematical Geography, at the fourth IGC in Paris in August 1889.

The organizing committee for the Paris IGC included several figures with long-standing interests in the prime meridian: geographer Antoine d'Abbadie, who in 1851 had urged international consensus on the matter; Henri Bouthillier de Beaumont; Charles Daly; and Jules Janssen, in his position as president of the Academie des Sciences in Paris and vice president of the Paris Geographical Society. The president of group I was Hervé Faye, who had participated in the 1883 IGA in Rome and had presided over the 1884 French Commission de l'Unification des Longitudes et Des Heures. Cesare Tondini de Quarenghi's paper was the subject of lengthy debate on the morning of Saturday, August 10, 1889. Resolution VII of the Washington meeting was also discussed, but the Jerusalem option was the key item.

The case for Jerusalem, it was stressed, lay in its proposed utility for universal time compared to Greenwich and in telegraphic communication. Jerusalem had an advantage over Bouthillier de Beaumont's Cape Prince of Wales option by virtue of its proximity to the centers of European population. Washington, it was argued, had not been a disinterested meeting and had always favored the British and the Americans. In support of this view, reference was made to the statements in Washington of Ruiz del Arbol, one of the Spanish delegates, whose comments there on the prime meridian and universal time and the role of the railway and telegraph companies—points central to the Bologna proposal—had been dismissed by John Adams and Rear Admiral Rodgers. "That may have been true in Washington in 1884," concluded de Quarenghi in his presentation to group I delegates, "but it is no longer true for us. It is never too late to acknowledge that one acted without any right—scientific or historic—by voting for the universal day to begin at Greenwich midnight instead of Jerusalem midnight."[32]

When the vote was taken, twelve delegates were in favor of the proposal and twelve were against. Without achieving the unanimity required, Faye and group I could not advance the proposal over Jerusalem to the general session of the Paris congress. The Jerusalem question was not taken forward.[33]

The Italian proposal over Jerusalem as a world prime meridian did not, however, end with its dismissal in 1889 by the BAAS in Newcastle and the IGC in Paris. De Quarenghi spoke to the issue again at the International Telegraph Conference in June 1890 in Paris and also used the occasion to address the Paris Geographical Society. By July 1890 the Italian authorities were sufficiently persuaded of the merits of the case to request that their ambassador in London, Giuseppe Tornielli, write to the British government over the prospects for a further international meeting. It was the correspondence about this invitation that had arrived on Donnelly's desk as he was liaising with the Colonial Office over Fleming's "Memorandum on the Movement for Reckoning Time on a Scientific Basis."

Tornielli's letter to Britain's foreign secretary, the Marquis of Salisbury, was pointedly direct: "The Academy of Science of Bologna have made accurate studies to discover a solution, which might be accepted by all States, of the question of adopting an initial meridian in order to obtain unity in the calculation of time, and therefore a uniform hour all over the world." On this basis, he continued, "I am directed by my Government, who have decided to support these proposals, and to summon a Congress to be held in Rome by the representatives of the States interested." The British turned to the Science and Art Committee for a decision. Duncombe's reply to the Foreign Office in early October 1890, for onward transmission to the Italians, was similarly direct: "I am directed to state, for the information of Lord Salisbury, that the Lords of the Committee of Council of Education have consulted the Astronomer Royal, and that they concur in the opinion expressed by him that no practical result is likely to follow from the Congress, the question of the choice of a Prime Meridian having already been fully discussed at Rome in 1883 and at Washington in 1884." In these circumstances, continued Duncombe, "my Lords do not feel that they would be justified in applying to the Treasury to pay the expenses of delegates to attend the Congress, which, they are informed, is to consider whether the meridian of Jerusalem should be adopted for time reckoning for telegraphs and railways, but not by astronomers, sailors, or geodesists."[34]

In late October 1890, Tornielli tried again. His tone was diplomatic and measured, at once consistent with others' entreaties over the prime meridian, modernity, and science. He took what he saw as the nonbinding outcomes of Washington as a reason to advance the case for Jerusalem:

The intelligence has been received at Rome with regret that Her Majesty's Government, considering that the question of the Prime Meridian was discussed in 1883 and 1884 without satisfactory result, are not disposed to send delegates to a new Conference charged with the re-examination of the matter. The subject is one both of great scientific interest and of practical utility in some of its direct applications. If it is true that the progress of science is continuous, and that the relations of different nations with each other are being developed more and more, it must be admitted that those questions which are not only of scientific interest but also of practical utility in their bearing on the relations between different countries, ought never to cease to be considered fit subjects for study, even if their previous examination may not have resulted in the conclusion of agreements which may have been considered desirable with a view to the solution of those questions. . . . The initiative taken by my Government, while it is intended to carry out these recommendations, is not meant to trench in any way on existing rights, since it does not in any way interfere with the use of national meridians in navigation, astronomy, topography, or local cartography.

By late November 1890, Tornielli's letter had been reviewed by Donnelly, Christie, and others. Christie was trenchant in response, pointing out historical parallels and the inadequacies intrinsic to the Jerusalem proposal:

I do not see any reason for reconsidering the refusal to take part in the proposed Conference. It would be inconsistent with the progress of science that a place should be selected for a Prime Meridian at which there is no observatory, and at which an efficient observatory is not likely to be established. The meridian of Jerusalem would become purely fictitious like that of Ferro, which, after being adopted by several European countries as Prime Meridian, had to be defined as 20°W. of Paris, thus becoming virtually the meridian of Paris in an inconvenient form. I look upon the proposal to adopt the meridian of Jerusalem as utterly unscientific, and the Italian Ambassador, in urging the progress of science and the scientific interest of the question of a Prime Meridian, has simply furnished strong arguments against the Italian proposal.

The choice of a Prime Meridian which could not be used for astronomy, geodesy, or navigation would hardly tend to the unification of time, as it leaves out of account the most important scientific applications.[35]

Further Italian entreaties to the British early in 1891 fell on deaf ears, with Donnelly pointing out that the Italians' proposal "is not received with any favour by a representative international body especially qualified to judge of it." Even as Tornielli had been writing to the British, the Italians had contacted the IGA for its view on the proposal and the benefits of another international congress to discuss the prime meridian. The IGA saw no virtue in either and considered "the essentials" of the prime meridian question to have been dealt with in Rome in 1883 and in Washington in 1884. Undaunted, de Quarenghi turned again to the IGC, this time in Berne, in August 1891.

The business of group I at the fifth IGC was thus dominated by Henri Bouthillier de Beaumont and Cesare Tondini de Quarenghi. De Beaumont offered a variant to his Cape Prince of Wales suggestion. Arguing that national meridians had no place in the modern world—but without mentioning either Greenwich or the Washington conference—de Beaumont returned to his 1876 idea of the global prime meridian as "the mediator." Rather than press for Cape Prince of Wales or a 180° "antimeridian" to Greenwich as earlier proposed, de Beaumont now advanced Rome as the prime meridian for universal time. De Quarenghi pressed the case for Jerusalem. He saw no contradiction in retaining a different prime meridian for astronomy, geodesy, and navigation if Jerusalem were to be adopted for timekeeping purposes. Although he did not mention them by name, he countered the arguments of Hirsch in 1883 and of Christie in 1890 by observing that Jerusalem did have a telegraph bureau, making it possible to communicate with the principal observatories and so be used as the baseline for the worldwide measurement of time. Neither proposal was advanced for the general approval of the Berne congress. De Quarenghi is listed as speaking of his proposal when the BAAS met in Leeds in the autumn of 1891, but he did not deliver his talk there.[36]

The Bologna proposal over Jerusalem makes clear that the issue of a single global prime meridian had not been settled by the 1884 Washington meeting. Greenwich was widely used and had received the endorsement of delegates at that international meeting—it was in effect primus inter pares from a short-list of four (even if, as the Italians hinted, recommendations arrived at there with perhaps undue haste might later be overturned). The candidacy of Jerusalem as prime meridian was subject specific—to be used in relation to civic timekeeping and thus for the railways and for telegraphy but not, as Christie pointed out, for astronomy, geodesy, or navigation. Thematic prime meridians to replace national prime meridians were a possibility widely recognized as an implication of the Jerusalem proposal. It is hard to know how much contemporary writing on the prime meridian—Airy's declaration, Struve's lectures, Fleming's papers on universal time, and so on—was easily available to scientists and others around the world. But the Jerusalem option certainly received widespread scrutiny from mathematicians, astronomers, and geographers in the BAAS in Bath and Newcastle; from geographers in Paris and in Berne; from telegraphists in Paris; and from astronomers, hydrographers, and career civil servants in London before it faded from view in the face of British intransigence and American indecision.

Continuing Complexity: The Prime Meridian in Geographical Congresses, 1891–1904

Among the audience at Berne in 1891 to hear de Beaumont and de Quarenghi was Thomas Holdich, the British military surveyor and geographer (and, later, president of the Royal Geographical Society). Holdich knew firsthand the importance of agreed-upon standards in mapmaking: he had that year been appointed superintendent of Britain's frontier survey in Afghanistan, India, and the Himalayas. In his report on the Berne meeting to the Royal Geographical Society, Holdich made clear the indecision currently prevailing over the prime meridian:

The meridian question, although it is apparently as far from solution as it was previously to the Washington Congress, has certainly advanced far enough that all English maps should possess a common origin for longitude. At present this is not so, for maps of India and of parts of the bordering countries are published with a longitude value based on an incorrect assumption of the position of the Madras observatory, differing about two and half miles from the true Greenwich value; so that, as our mapping extends westward through Persia and eastwards through Burma we become involved in awkward discrepancies. I would venture to suggest that the opinion of the Surveyor-General of India should be consulted as to the advisability of adopting the Greenwich meridian in future for all Indian mapping. I am quite aware of the nature of the reasons which have prevented its adoption hitherto, but since attending this Congress I have come to the conclusion that a continuance of the present system is a grave disadvantage if we wish to persuade other nations to adopt Greenwich as their longitude origin, and that this disadvantage outweighs previous considerations.[37]

In addition to the multiplicity of prime meridians that Holdich observed on British imperial topographic maps, there was of course great variety on maps produced by other countries. In 1885 fifteen different national prime meridians were used in large-scale topographic mapping throughout the world. By 1898 all bar one were still in use for national topographic purposes (only that of Warsaw, used as a prime meridian in some Russian topographic mapping, had been dropped in the interim).[38]

This enduring problem and the prevalence of different scales and standards in national topographic mapping was addressed in Berne in 1891 by the German geographer Albrecht Penck, who announced there a scheme of world mapping at a scale of one to one million. Scholars of the so-called Millionth Map Project have shown its importance and that it developed only slowly and unevenly. They have overlooked its significance with regard to the IGC and the prime meridian. Penck's scheme was linked to the resolution passed in Berne that the "English prime meridian" (Greenwich) should be universally adopted (the meeting also expressed the hope that Britain should adopt the metric system).[39]

Penck returned to his scheme at the sixth IGC in London in 1895. In absentia, Henri Bouthillier de Beaumont returned to his own scheme on time, with a paper read by another delegate. Other contributions included an address from M. de Rey-Pailhade of Paris on Resolution VII from Washington, the decimal measurement of time and of angles, and a talk from Enrico Frassi of the Italian Geographical Society on time reform and a system of hour zones. Most of the papers, except Frassi's, were short. This may have been deliberate. John Donnelly was one of the organizers of the 1895 London congress, as was the explorer Edward Delmar Morgan, who had represented the Royal Geographical Society in discussions on the prime meridian at the Paris IGC in 1889. It is possible that they, Donnelly especially, limited the time available for non-Greenwich options, much as Germain had tried to steer the agenda of the 1875 Paris IGC. Donnelly would certainly not have warmed to Frassi's appeal to assembled delegates in 1895 over the need for yet another international meeting: "I beg you, Mr President [of the Congress], in my own and country's name, to present a request to the General Meeting, to nominate an International Commission ... to effect an agreement between the views of Rome and Washington as to a universal meridian."[40] Resolutions over Greenwich, now over a decade old, were still not being acted upon.

There was no discussion of an alternative prime meridian at the seventh IGC, held in Berlin in 1899, but the meeting unanimously recommended that in relation to Penck's scheme, Greenwich be used as the base 0° and the meter for topographical measurement. In fact, this recommendation had first come forward in 1895. In his paper in Berlin on the advantages to geography of the metric system, British geographer and climatologist Hugh Robert Mill shed light on the discussions at the London IGC. There the French delegates had proposed that heights on Penck's proposed one-to-one million world map should be given in meters and that the prime meridian for longitude should be Greenwich. This "very happy international compromise ... might before long be extended to all maps and geographical writings," noted Mill. After all, "the objection to the metric system in English-speaking countries is no more strongly based than the objection to accepting the meridian of a foreign observatory as the zero of longitude is in France. The vast benefit of international uniformity in standards should outweigh all other sentiments."[41]

By 1904 the IGC had taken heed of Mill's views. At the Washington IGC that year—twenty years after Greenwich's status in this respect had been voted upon and agreed—the IGC affirmed Greenwich as the world's prime meridian in relation to time: "In view of the fact that a large majority of the nations of the world have already adopted systems of standard time based upon the meridian of Greenwich as prime meridian, that this congress is in favour of the universal adoption of the meridian of Greenwich as the basis of all systems of standard time."[42]

Standard Time, Telegraphy, and the Paris and Greenwich Meridians, c. 1884–1912

The take-up by different countries of standard time, or zone time, based on midnight at Greenwich as the initial meridian—not to be confused with universal time, the use in time reckoning of the zero to twenty-four-hour system—has been documented at length elsewhere. Belgium and Switzerland made the change to Greenwich time in 1892, for example. In Germany, the system of middle, or central, European time, in which clocks and public timekeeping were set in relation to Greenwich, came into force in 1893, as it did in Austria. In Australia, resolutions "that there should be one time in Australia & that the meridian should be the 135th, i.e., 9 hours East of Greenwich" came into effect on February 1, 1895 (Figure 6.3).

The move to standard time was not without inconvenience, in Australia and elsewhere: on the eve of the change, the *Sydney Morning Herald* observed how Adelaide time had been predominant hitherto, except for post and telegraph offices, which observed Sydney mean time, and that hoteliers, publicans, and banks would all have to adjust their hours of opening, to the public's "undoubted confusion." In South Africa, where the astronomer and geodesist David Gill coordinated the move to Greenwich time in 1892 (and who liaised with the Cape authorities over their response to Fleming's proposals over universal time), Gill's advocacy had to overcome considerable "local prejudice."

By 1905 most countries in the world were using Greenwich 0° as their prime meridian and Greenwich mean midnight as the basis to their

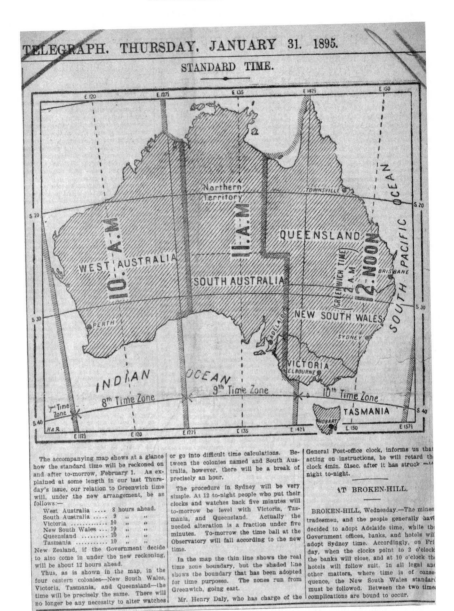

The accompanying map shows at a glance how the standard time will be reckoned on and after to-morrow, February 1. As explained at some length in our last Thursday's issue, our relation to Greenwich time will, under the new arrangement, be as follows:—

West Australia .... 8 hours ahead.
South Australia .... 9 ,, ,,
Victoria .... 10 ,, ,,
New South Wales ... 10 ,, ,,
Queensland .... 10 ,, ,,
Tasmania .... 10 ,, ,,

New Zealand, if the Government decide to also come in under the new reckoning, will be about 12 hours ahead.

Thus, as is shown in the map, in the four eastern colonies—New South Wales, Victoria, Tasmania, and Queensland—the time will be precisely the same. There will no longer be any necessity to alter watches

or go into difficult time calculations. Between the colonies named and South Australia, however, there will be a break of precisely an hour.

The procedure in Sydney will be very simple. At 12 to-night people who put their clocks and watches back five minutes will to-morrow be level with Victoria, Tasmania, and Queensland. Actually the needed alteration is a fraction under five minutes. To-morrow the time ball at the Observatory will fall according to the new time.

In the map the thin line shows the real time zone boundary, but the shaded line shows the boundary that has been adopted for time purposes. The zones run from Greenwich, going east.

Mr. Henry Daly, who has charge of the

General Post-office clock, informs us that acting on instructions, he will retard the clock 4min. 51sec. after it has struck mid night to-night.

AT BROKEN-HILL.

BROKEN-HILL, Wednesday.—The miners tradesmen, and the people generally have decided to adopt Adelaide time, while the Government offices, banks, and hotels will adopt Sydney time. Accordingly, on Fri day, when the clocks point to 2 o'clock the banks will close, and at 10 o'clock the hotels will follow suit. In all legal and other matters, where time is of conse quence, the New South Wales standard must be followed. Between the two times complications are bound to occur.

FIGURE 6.3 This map of Australia's zone times based on Greenwich was published in the *Sydney Telegraph* on January 31, 1895, to alert readers to the changes in time that would go into effect the following day: "There will no longer be any necessity to alter watches or go into difficult time calculations." Reproduced by kind permission from the Syndics of Cambridge University Library, RGO 7/146.

timekeeping. Countries that fell into line after that date include France—which conformed to the recommendation from Washington and adjusted Paris mean time to Greenwich mean time (GMT), less nine minutes and twenty-one seconds, in 1911—Portugal (1912), Brazil and Colombia (1914), Greece (1916), Turkey (1916), Ireland (1916), Russia (1924), Argentina and Uruguay (1920), and the Netherlands (1940). Liberia adopted Greenwich mean time only in 1972 (Figure 6.4).[43]

The move toward greater uniformity in public timekeeping occurred alongside those disputes in the scientific world over the unification of the astronomical and civil day discussed previously. By and large, however, it was untouched by the specialist concerns of astronomers and navigators over projected adjustments to nautical ephemerides and the demands for international agreement on Resolution VI. The use of standard time zones based on the Greenwich meridian and on Greenwich midnight was a reflection of uniformity in timekeeping's widespread benefits to science and, chiefly, to the public. As the *Sydney Morning Herald* put it in February 1885, "The extensive applications of scientific observations to the weather and to navigation, by the electric telegraph, by railways, and by extremely interesting and valuable astronomical researches, all tend towards the practical course adopted. And there can be no question as to the advantages of all countries having a universal basis for scientific observations and calculations."[44]

Recognizing that contemporaries placed great importance upon telegraphy in establishing this sought-for "universal basis for scientific observations and calculations"—a single prime meridian being both the cause and the effect of such metrological universality—it is instructive to remember that telegraphy's use in the calculation of longitude and of time was problematic. Telegraphy was a central feature of the proposal for Jerusalem as a prime meridian. Proponents argued that in the absence of an astronomical observatory an initial prime meridian could be "fixed" by wire. We should recall too that this point about positioning by proxy had formed part of Janssen's arguments in Washington over the neutrality of the prime meridian. As Henry James found in the late 1850s, using the telegraph to determine longitude did not lead to certainty, especially when different linear units were employed among countries. Metrological variety was not the only problem. Overseeing longitudinal measurements in the Egyptian desert as part of the 1874 transit of Venus expedition, Airy recounted the calculative adjustments necessary given human error: "The Greenwich-Mokattam longitude requires correction for the personal equations between the observers who

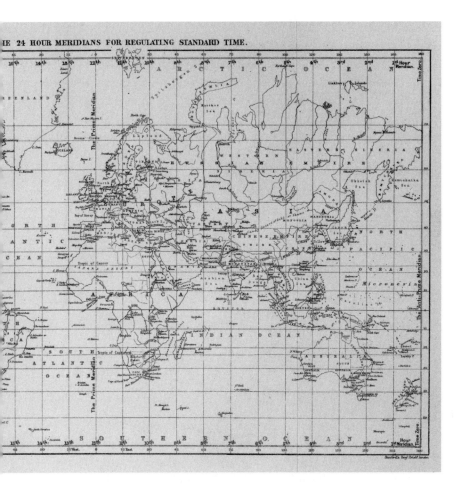

FIGURE 6.4 This world map of one-hour time zones, with Greenwich as the prime meridian, was produced by the American Society of Civil Engineers as part of its promotion of universal and standard time. Note the labeling of "The Anti-Prime Meridian"—the international date line in modern parlance—at 180° (twelve hours) from Greenwich. Reproduced by kind permission from the Syndics of Cambridge University Library, RGO 7/146.

gave and received signals, and for the difference of the manner in which the observers at Greenwich and Mokattam determined clock-error."[45]

This problem was not a consequence of the transit expedition, or of the location, or of the instruments used. It was—and is—intrinsic to the measurement of phenomena, longitude and the prime meridian included, as was illustrated by the Paris and Greenwich meridians in the late eighteenth century and was reiterated for these meridians in the late nineteenth century.

In June 1888, as debates over the unification of astronomical and civil time were taking place on both sides of the Atlantic, as Rodgers was lamenting American inaction over Washington, and as de Beaumont and de Quarenghi were preparing their alternatives to Greenwich, William Christie reported that provision had been made for "the expense of a re-determination of the difference of longitude between Greenwich and Paris." Reporting a year later, Christie documented the procedure followed:

Observations were made in four groups of three nights each (or the equivalent in half nights). An English and a French observer were stationed at each end, each with a separate instrument and chronograph, and the pairs of observers were interchanged twice, to eliminate any change in the personal equations during the progress of the work. The pairs of English and French instruments were similar, and the signals as well as the star transits were recorded on similar chronographs. On a full night each observer recorded about forty star transits, reversing his instrument three times, and exchanged signals twice (near the beginning and the end of the evening) with his compatriot at the other end of the line, and once with the other observer. . . . The actual stations were the Front Court of the Royal Observatory and the Observatory of the Service Géographique de l'Armée at Paris, the position of which with reference to the Paris Observatory has been accurately determined.

Despite such operational management, the calculation and publication of the results—and from that the latest declaration of the relative positions of two leading national prime meridians—was delayed by what Christie termed "a curious discrepancy between the results obtained for clock error at one of the stations . . . by the two instruments mounted side by side, and used by the French observer and the English observer respectively." In June 1891 Christie reported that "the English definitive result of the difference of longitude between the Greenwich Transit Circle and Cassini's

Meridian is $9^m 20^s \cdot 86$, while the French result (not yet published) is about $0^s \cdot 15$ greater." Although this "discordance" was only about half what it had been, it was in Christie's estimation still so large that no alternative was left but to repeat the work in its entirety. This "anxious consideration" mattered with respect to Paris and Greenwich and to Greenwich in relation to other locations then being linked globally by telegraph (Montreal) or, as for Waterville in Ireland, as part of the European arc of longitude pioneered by Wilhelm and Otto Struve.

Modifications designed to reduce the errors and the differences calculated between Paris and Greenwich were introduced into the procedures in 1892: two clocks instead of one were placed at each end of the telegraph line and all clocks were placed in rooms kept at nearly constant temperature. Observers were interchanged. The mean of the results, from Greenwich, was a difference of longitude between there and Paris of $9^m 20^s \cdot 82$, against the figure calculated in 1888 of $9^m 20^s \cdot 85$. But from Paris, French observers confirmed exactly their previous result—$9^m 21^s \cdot 06$. Greenwich, it seemed, was farther away from Paris than the French capital was from Greenwich. Differences in longitudinal measurement continued to be found in the years following. By 1905, however, the difference reckoned was judged so slight that, to the British at least, the Greenwich-Paris problem was no longer a concern.[46]

This evidence concerning Paris and Greenwich affirms earlier problems in fixing these two observatories and thus of determining these two national prime meridians. Its significance after 1888 lies in its relation to the foregoing discussions over Resolution VI and an alternative prime meridian and what it reveals about the complex afterlife of the Washington conference. Arguably, the concerns expressed by France's astronomers over accuracy, longitude, and the positioning of Paris as a site of metrological authority were part of that country's rather halting acceptance of Greenwich's primacy. It is hard to explain the French engagement with questions of time unification in the early twentieth century if not as a belated realization of the importance of international science as a form of global politics. The promotion of radio telegraphy in timekeeping was led by the French. The French employed radio signals telegraphed from the Paris Observatoire via the Eiffel Tower from 1910 (Figure 6.5). France's revised legal time—*l'heure anglaise de Greenwich,* as the French press dubbed it—came into effect on March 11, 1911. In that year, the first International Congress on Astronomical Ephemerides met in Paris and formally adopted the Greenwich meridian, including its use, "completely and without reserve," as the base meridian in the *Connaissance des temps.* The first International Conference on Time, in Paris

FIGURE 6.5 This 1913 French newspaper conveys the role of Paris—and of the Eiffel Tower—as the early twentieth century's time capital for the world based on radio telegraph signals transmitted to Greenwich. *Source: Excelsior,* October 25, 1913.

in 1912, was additionally significant in making the French capital an important "truth-spot" for the standardization of time.[47]

France's adoption of Greenwich mean time and thus its acceptance of the primacy of the Greenwich meridian for timekeeping purposes (as opposed to France's in-principle recognition in 1896 of Greenwich as the world's prime meridian in relation to longitude, astronomy, and nautical literature) was covered in the British press. The *Times* (London) report ended on a note that was conciliatory in tone yet aware of the historical and geographical context in which the now-displaced Paris prime meridian was to be understood: "France, however, having changed her prime meridian more than once in the past—it has at various times been reckoned from St. Michael in the Azores, from Tenerife, and from Ferro, the westernmost of the Canary Islands—is, perhaps, all the more inclined once more to accommodate herself to the advance of science and the concord of the nations. Nevertheless, we do not underrate the sacrifice of national sentiment that she has made in abandoning her historic meridian, and we rejoice to think that in these days it is easier for her to do so in favour of Greenwich than it could have been in favour of any other meridian."[48]

o o o

This chapter has shown the importance of a thematic, resolution-by-resolution approach in understanding the afterlife of Washington 1884. Of the seven resolutions advanced in Washington, effectively four, and in substantial terms only two, were the subject of attention after 1884. Resolution I was a formality. Resolution III concerning longitude's east and west reckoning was largely ignored. Resolution VII continued as the preserve of a few in scientific meetings. Resolutions IV and V, which concerned the universal day and universal, standard, and local time, were bound up with discussions over the unification of time and over Greenwich's place as the baseline for the world's timekeeping. The most contentious was Resolution VI, the unification of the civil and astronomical days. The adoption of Greenwich as the world's prime meridian, Resolution II, was the other main focus of attention. In Britain, Resolution VI was not formally agreed upon until 1925. The international geographical community ratified Greenwich as the world's prime meridian in relation to standard time in 1904. In France, Greenwich's place as the world's meridian was recognized by the state as an "urgent matter" in 1886, but Greenwich was not formally adopted until 1911, the year in which the French sanctioned its use in France's nautical almanac. In the United

States, the twin prime meridians of Washington and of Greenwich were not replaced in law until 1912.

In Britain, in France, and in the United States in particular, the drawn-out adoption of Resolution VI was not a national matter. It divided scientific and political communities within these nations and across national boundaries. It is impossible to know whether the circumstances in the United States regarding Resolution VI—which for years delayed take-up there of all the Washington recommendations—would have been different had a change in the political administration not hindered the adoption of the conference's resolutions. Decisive power over that question lay not with politicians, or even with leading astronomers, but with naval officials. Even as this is true, there was a preparedness within the ranks of those who opposed the unification of the civil and astronomical days to recognize such a thing in the future, provided other nations would do so. The adoption of Resolution VI may be characterized as continued postponement in the absence of the one means, an international conference, that several nations' representatives pressed for but did nothing to make happen. In the case of Resolution II, the only serious alternative to the primacy of Greenwich that was raised after 1884 was the Jerusalem option. British diplomats did not extend to the Italian government the possibility of another international meeting to debate a prime meridian whose purpose related only to universal time, not geodesy, astronomy, or navigation. For the British, Washington was the affirmation of existing practice, not a license to subvert it.

Rather than consider Washington's Diplomatic Hall as the place and November 1, 1884, as the date at which metrological uniformity was visited upon the world, we should see Washington 1884 as a key moment but not a defining event. As Chapter 7 shows, the prime meridian had an even more varied afterlife well after the world's time and space was measured from Greenwich.

7

Ruling Space, Fixing Time

WRITING FIFTY YEARS AFTER the Washington meeting recommended Greenwich as the prime meridian, one Australian newspaper
reporter did not think much of what he saw of the world's ruling line, which
had been marked years earlier near the Royal Observatory. Beneath the stra-
pline "Nothing Much to Look At," he observed that it was "just a diagonal
band of stone across a path with a score along its centre—and most people
would pass it by without a second glance." Appearances, however, were de-
ceptive: "It is really one of the most interesting things in the world . . . .
A notice board beside it labels it as the Prime Meridian, and it is accepted as
such for the reckoning of longitude and time by the whole world."[1]

This book has examined the prime meridian as a question of geography:
one concerned with the authority of science and accuracy; with metrology,
politics, longitude, and timekeeping; and, in the late nineteenth century
especially, with modernity, internationalism, and the idea that there are universal common goods—even global needs—that transcend the established
practices of individual nations and professional communities. It began with
the "solution" to the prime meridian question before examining the 1884
Washington International Meridian Conference in some detail and identifying the reasons why that meeting did not resolve the question. The book
has explored the "afterlife" of the 1884 meeting, but its principal concern has
been with the "prime meridian problem"—the presence, use, and multiplicity
of different national prime meridians before and after the global primacy

of Greenwich 0°. This problem occasioned the Washington meeting and in various ways postdated it.

Although in the first section I briefly outline the main features of the prime meridian in the principal period covered by this book, 1634 to 1884, this concluding chapter does not rehearse at length the evidence already examined. Nor does it bring this geographical narrative of the prime meridian to a definitive end. My concern overall is to continue the theme of the afterlife of Greenwich 1884 by looking at events and actions that sought in later years to mark that occasion in time and in space. The second section presents a story of Greenwich's commemoration and of the memorialization of prime meridians in several ways—not just in bands of stone but in monuments; postage stamps; millennial maps; tree plots; and in the case of the Paris meridian, in comic heroes and mass picnics. If this is to illustrate something of different prime meridians' varied representational status, it is in closing also to address the Greenwich prime meridian's significance for modernity. This was underlined in dramatic fashion in February 1894. In that month the Royal Observatory at Greenwich, and by implication the prime meridian, was the intended target of a bomb attack by a French anarchist, Martial Bourdin. What contemporary newspapers termed the "Greenwich Bomb Outrage" was fictionalized by Joseph Conrad in his 1907 novel *The Secret Agent*. For Conrad's protagonist, the prime meridian was the very emblem of modernist authority and of science's overarching global reach: one line and point in space to "rule" the world.

More than a century later, as I show in this chapter's final section, the world's prime meridian has indeed moved. This, as we shall see, is the result not of French anarchism, British imprecision, or global politics arrived at through scientific consensus but of instrumental accuracy, time's passage, and geographical changes in the world.

Prime Meridians / The Prime Meridian

The central elements of the prime meridian problem as a matter of geographical difference were clear by the late eighteenth century. For contemporaries, where you were and who you were influenced which prime meridian you used and why. Differences existed across and within nations. Users worked from observed prime meridians (astronomers and authors of navigational texts and ephemerides), from measured prime meridians (geographers, mapmakers, and topographical surveyors), or from one or the other and sometimes from neither (sailors at sea). For modern researchers, where one looks matters not only in understanding that such geographical confusion existed but also in recognizing its constituent features.

Books matter. Astronomical ephemerides such as the *Connaissance des temps* in France (1679), *The Nautical Almanac and Astronomical Ephemeris* in Britain (1767), and the *Almanaque náutico* in Spain (1792) established a textual correspondence between navigational practice, astronomical prediction, and observed prime meridians as individual nations' first meridians—here for Paris, Greenwich, and Cadiz, respectively. These works could incorporate for reasons of comparison the 0° longitude used in other countries. In books of geography and in some navigational works, certainly in an English-language context, different prime meridians were cited in teaching what a prime meridian was or in documenting its use in practice. Ships' logs indicated the variations in which prime meridian, if any, was employed.

Maps matter. Maps indicated the prime meridian selected by different mapmakers and perhaps by different users. In that sense some maps had symbolic value beyond their practical utility (as Morse recognized: see Figure 2.1). It is for these reasons that the term "geographical confusion" is so apt, as Faden, Gibbon, and their contemporaries knew. And it is for these reasons—differences within and between nations in terms of user communities and in material forms and practices—that viewing the prime meridian question by the late eighteenth century simply at the scale of the nation has limited interpretive value.

From the late eighteenth century, the prime meridian problem required a solution beyond the interests of any one nation or scientific community. The cross-Channel triangulation projects of the British and French between 1787 and 1790 and again in 1821 provide evidence of international collaboration, even if the aim was to more accurately fix each of the two observed prime meridians relative to one another instead of replace one with the other. By contrast, the words of Pierre-Simon Laplace in 1806 and of Benjamin Vaughan in 1810 share a distinct transnational emphasis. Their rhetoric stressed the intellectual and practical exchange value of uniformity and universality—in currency, metrology, language, and mapping—as well as in modes of reckoning longitude: what Vaughan in 1810 presciently termed that "grand change always to be kept sight of . . . that of considering the whole globe as an unit, & having one prime meridian for the whole."[2]

This nascent recognition of a prime meridian solution was facilitated within and across national boundaries by the intellectual power vested in shared practices of measurement and the language of mathematics and by the political and institutional authority that came from accuracy and claims of even greater

accuracy in the future. Considerable evidence points to contemporaries' attention to the political significance of accuracy: in Maskelyne's responses to the
claims of Cassini IV; in Bowditch's words regarding Maskelyne's calculations
and in disputes with William Lambert over the adequacy and number of
Lambert's observations; in Roy's misplaced faith in the authority of transnational triangulation; in Kater's adjustments of Roy; in Airy's trust in "galvanic
connexions"; in the Greenwich and Paris observatories' telegraphic positioning; and, during the Washington meeting and even after, in some people's
views that the world's prime meridian could be precisely sited not in and from
an established observatory but by the distances between several. What hindered moves to a solution before the late nineteenth century were differences
in metrology in terms of linear measurement and in timekeeping: specifically,
between the nautical and the astronomical communities; broadly, in civil life;
and pressingly, in railway time and in telegraphy.

Even as nineteenth-century contemporaries worked with the discrepancies in terms of prime meridians, timekeeping, and metrology, views were
being articulated that something could be done about these issues. From the
early 1870s, the desire to find a prime meridian solution was lent purpose and
direction by several factors. The Greenwich prime meridian had by then come
to assume a position of calculative authority. From the 1790s it was Britain's
established observed first meridian for national and continental measurement, a common first meridian for Britain's geographers, and increasingly
the 0° of Britain's seagoing communities and those of other nations. From
1848 it was the basis for Greenwich mean time and the regulation of civic
time in Britain. *The Nautical Almanac and Astronomical Ephemeris,* known
for its accuracy and annual sales volume, was based on the Greenwich prime
meridian. The meridian also gained status because in 1850 it was formalized
as one of two prime meridians for the United States; because in 1867 it was at
the center of a telegraphic network that further unified Britain's timekeeping
and geographically and astronomically brought Europe, the Americas, and
Britain and its empire into communicative alignment; because the great proportion of the world's maritime commerce used Greenwich (even if they also
used other national prime meridians); and because from 1883 Greenwich was
the baseline for standardizing the timing of America's railroads.

The fact of different prime meridians became the focus of discussion
within scientific and geographical bodies in 1871. The prime meridian problem
was increasingly articulated in a language of transnationalism: in timekeeping
in the work of Sandford Fleming especially but also more widely as the ter-
minology of "global," "universal," or "cosmopolitan" was used in search of a

shared solution for a modernizing world. Between 1871 and 1881, the International Geographical Congress (IGC) provided key sites of deliberation and recommendation over a single prime meridian for the world. The resolution resulting from Antwerp in 1871 was weak, was of service only to the navigational community, and was not sanctioned by the members as a whole—but it was a start. The 1875 Paris IGC, shaped by French geographer Adrien Germain, was less internationally collaborative in tone. The third IGC, in Venice in 1881, was more international in scope than previous studies have supposed.

The Venice IGC was distinguished by the presence of North American delegates who together had been proselytizing on the advantages of universal and standard time. Many had the support of influential institutions within the United States, where the U.S. Congress was by then in no doubt over the need for temporal alignment of the nation's rail network. Sandford Fleming had the support of the Dominion government of Canada and the Canadian Institute and used these connections to disseminate a series of pamphlets in which he urged a universal system of timekeeping—with a single prime meridian as the key first step. Fleming's resolutions in Venice toward "the Unification of Initial Meridians" were influential more because they called upon governments to act upon the communications and less because of his and others' view over which of several possibilities—Greenwich, Ferro, a Bering Strait *mediateur,* or Greenwich's antimeridian—the world's single prime meridian might be.

The 1883 Rome International Geodetic Association (IGA) meeting was an important deliberative moment on the road to the prime meridian's solution for several reasons. The authority of an observed prime meridian was accepted. On this basis, the scientific credentials of the four leading candidate sites—Berlin, Greenwich, Paris, and Washington—had nothing to distinguish between them: other practical considerations, such as the extent to which the world's nations used each in navigation and in works of geography, needed to be considered. Also accepted was the view that a global system of universal timekeeping should be based upon the initial meridian chosen. In Rome, delegates were free to vote as individuals upon these principles. Further action upon them, at a later meeting, would be on behalf of governments. Rome 1883 shaped what was to be possible in Washington 1884.

The International Meridian Conference was not the first scientific meeting to make recommendations concerning a single prime meridian for the world, but it was the first at which delegates had the authority to vote on behalf of nations. Many countries already used Greenwich as the baseline employed in navigation. For years after Washington, different national prime meridians continued to be used on large-scale topographic maps even when legislation directed otherwise—as for France, which agreed in 1896 in principle to adopt the Greenwich prime meridian over that of Paris as the world's baseline. The IGC affirmed Greenwich as the world's prime meridian in relation to time in 1904. The French followed in 1911. Together, these variations in timing between countries, the drawn-out debates over Resolution VI, and the imprecations of Italian scientists over the Bologna proposal and an altogether different prime meridian for use with universal time show that the prime meridian solution was not felt equally everywhere or everywhere the same.

Marking the Prime Meridian

In *The Cosmic Time of Empire*, Adam Barrows discusses the representation and significance of Greenwich mean time (much less, directly, that of the prime meridian itself) in relation to several canonical modernist texts. One of these, Joseph Conrad's *The Secret Agent*, is set in 1886 (the year Japan became the first country to adopt the recommendations of the Washington conference), but its central incident is based on an actual occurrence in February 1894. The facts behind the Greenwich Bomb Outrage are briefly these: after the sound of an explosion was heard in Greenwich Park, a French anarchist, Martial Bourdin, was found there "terribly mutilated" after mistakenly having detonated the bomb he was carrying. No clear motive was ever determined. Although comments in the *Times* (London) suggested that the French had long objected to the preeminence of Greenwich and its worldwide regulatory position, there is no firm evidence that this influenced Bourdin or sparked any violent antipathy among French anarchists or astronomers to a rival prime meridian. Bourdin died of his injuries in a Greenwich hospital. In the novel, Conrad takes the target to be the observatory, the building, astronomy, and the prime meridian itself—all symbols of the power and authority of science. In the words of the eponymous protagonist "Mr. Vladimir," "The blowing up of the first meridian is bound to raise a howl of execration." As he later remarks, "You don't know the middle classes as well as I do. Their sensibilities are jaded. The first meridian. Nothing better, and nothing easier, I should think."[3]

To Barrows, Greenwich by the early twentieth century "had entered modernist consciousness as a powerful symbol of authoritarian control from a distance and of the management of diverse populations." The Greenwich prime meridian was more than the world's baseline for the measurement of time and of space: it was, almost archetypically, an apolitical site, associated, as one character in Conrad's novel has it, with the "sacrosanct fetish" of science. Toward the end of his life in his essay "Geography and Some Explorers," Conrad outlined the shift over time from what he called "Geography Fabulous," the portrayal on maps of mythical creatures for want of certain information; to "Geography Militant," the conquest of the earth by explorers and governments and depicted in maps and other documents of territorial control; to "Geography Triumphant," Conrad's own age, where virtually the whole world, including the polar regions, is under the authoritative gaze of the geographer.

It is tempting to argue that the events depicted in *The Secret Agent* should be read similarly, as an expression of protest against Geography Triumphant and Greenwich's temporal, metrological, and geographical rule, locally inscriptive and globally imperative. In James Joyce's *Ulysses* (1922), Barrows argues that the disorientation felt by one main character, Leopold Bloom, over the difference in 1904 between local Dublin time and Greenwich mean time (twenty-five minutes and twenty-two seconds) may be used to challenge "the complicity of Greenwich standard time with imperial designs to restructure and redirect the spaces and rhythms of a colonized Ireland." Sandford Fleming emphasized how the facts of modernity required a single global prime meridian and agreed standards of time. As Conrad, Joyce, and others have noted and as study of the afterlife of Greenwich's primacy from 1884 has revealed, however, the meaning of the resolution to adopt this single base point for the world and the prime meridian's later meaning are different things. Considered in relation to his later "Geography and Some Explorers," Conrad's *Secret Agent* may be read as part of the contemporary disquiet over national identity, geographical difference, and the organization of time and space "plotted" against the world-defining line that is the Greenwich prime meridian. This is not the same thing as stating, as have others, that the "1884 congress on the prime meridian in Washington represents a culminating episode in the long story of longitude."[4] Greenwich's afterlife is testament to its continuing significance.

The Greenwich prime meridian has been marked in several other ways in the twentieth century. Higgitt and Dolan identify ten acts of what we might term "prime meridian commemoration" between 1910 and 1959. These ranged from a line in the path on the north boundary of the observatory in 1910 to similar strips in other places through which the 0° line passes (the small towns of Cleethorpes and Louth in Lincolnshire, for example) to obelisks in Lewes and Peacehaven in Sussex. A stone strip—the one that in an earlier and less distinguished form had prompted the journalist for

FIGURE 7.1 Continuing confusion over the differences between the prime meridians of Greenwich and of Paris was an element in a mystery concerning missing pirate treasure solved by Tintin, the reporter and explorer. *Source:* Hergé's *The Adventures of Tintin: Red Rackham's Treasure* (London: Methuen, 1959), 23. © Hergé / Moulinsart 2016. Reproduced by permission from Moulinsart SA.

the *Mercury* (Hobart, Tasmania) in 1934 to deem the world's ruling line not worth looking at—was relaid between November 1934 and early March 1935, and a descriptive plaque was added. In 1953 the line was replaced with a bronze strip inset into larger stones. At much the same time, children's storybooks continued to hint that different prime meridians were still at work in the world, if only in fiction (Figure 7.1). British postage stamps issued on June 26, 1984, showed the world encircled by Greenwich's line of reference. In 2000, Britain's Ordnance Survey issued a special edition of its *Explorer* series of maps, marking where "the zero line" ran with a green line on the maps' covers. In France the new millennium was marked by a mass picnic held across the country on July 14, 2000, as near as possible to the line of the Paris meridian. This event of gastronomic popular geography was part of a more general nationwide project, La Méridienne Verte (The Green Meridian). In-

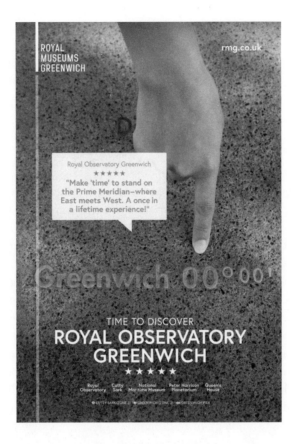

FIGURE 7.2 The importance of Greenwich as the world's 0° is clear from this poster, widely distributed throughout London, inviting people to discover the Royal Observatory. © National Maritime Museum, Greenwich, London.

stigated in 1999 by the architect Paul Chemotov, the plan was to involve the administrations of eight regions, twenty departments, and 336 communes in a line north–south across France, from Dunkirk to Perpignan. It would mark the nation's first meridian by planting trees that in time would be visible from space: a living "definition" of one nation's former reference point.[5]

This evidence in stone, paper, and trees illustrates for the prime meridian what one scholar has considered history in the guise of its representation. This is as true, I contend, for geography as an act of memory—situated acts of memorialization designed to mark events in space and time in space and time. The Greenwich prime meridian may not have been debated in the twentieth century with the same vigor as at the nineteenth-century IGCs or with the insistence of Cesare Tondini de Quarenghi or Sandford Fleming, but in addition to its regulatory function, its legacy endures in these commemorative circumstances.[6] In the summer of 2016, the Royal Observatory, Greenwich, even advertised the prime meridian as a tourist destination—the world's 0° as the start point of global longitude, the measurement of time, and the place "where East meets West" (Figure 7.2).

Moving the Prime Meridian

The global significance of the prime meridian is not in doubt even when the alignment of the prime meridian in space changes over time—that is, when the placement of the world's ruling line commemorates its own former authoritative position. Put simply, the Greenwich prime meridian is no longer where it once was.

The world's prime meridian is still at Greenwich, but it is today 102.5 meters (or 336 feet 4 inches) east of where the Airy Transit Circle, the nineteenth-century telescopic instrument at the Royal Observatory, positioned it "at the centre of the transit instrument at the Observatory at Greenwich as the initial meridian for longitude" according to Resolution II (see Table 5.1). The Greenwich prime meridian is now positioned in relation to the International Terrestrial Reference Frame (ITRF) and the World Geodetic System (WGS 84). This movement can be explained by changes in the technology of measurement. George Biddell Airy, William Christie, and others calculated the position of the Greenwich prime meridian using a telescope pointing toward the heavens to a so-called clock star. Ideally, this method of measurement should be exactly perpendicular to the earth's plane of measurement, but it was not: both local terrain and the shape of the earth cause local distortions to gravity. A decade or so after completing the sea trials of Harrison's timepiece and of Mayer's lunar tables, which in combination underlay his production of *The Nautical Almanac and Astronomical Ephemeris*, Nevil Maskelyne had noted just this phenomenon when making gravitational calculations on the slopes of Schiehallion in Highland Perthshire.[7] The Airy Transit Circle used from 1851 is very slightly off perpendicular. This deflection from the vertical, modern researchers have shown, accounts for the entire recorded longitude shift in Greenwich. From 1984, a century after Greenwich's ruling authority was proposed for the world, the earth's geocentric measurement has been managed through the Bureau International de l'Heure (BIH), which coordinates the results derived from satellite navigation systems—forms of space-based measurement unaffected by the earth's gravity. Technological change has prompted great accuracy. With greater accuracy come claims to scientific authority. Nathaniel Bowditch, William Roy, Henry Kater, and others would all have understood this.[8]

There is no need to adjust systems of timekeeping. Nor should geographers redraw their maps or astronomers recalculate their ephemerides. As the Royal Observatory's public astronomer, Marek Kukula, noted in August 2015 in one newspaper article covering this "displacement" of the Greenwich prime meridian, "The great thing about meridian lines is that it doesn't really matter—as long as everybody is using the same one."[9] This book has been about the problems that arose when everybody did not use the same one.

NOTES

1. William Parker Snow, *An International Prime Meridian: A Circular Letter* (privately printed, c. 1883). Snow's two-page letter was in the form of a printed circular. It began with "To the President and Council of the . . . Geographical Society or whomsoever it may concern," the blank being completed presumably by Parker Snow himself. It has not proved possible to know how many he circulated or exactly to whom: records of the Royal Geographical Society and the American Geographical Society confirm that they at least received the letter (although no trace of it now survives).

2. The claim that Parker Snow was psychic was made by journalist and newspaper editor W. T. Stead, "Mr. W. Parker Snow—Sailor, Explorer, and Author," *Review of Reviews* 7 (1893): 371–386, from which the phrase about the "Fourth Dimension" is taken (p. 376). Parker Snow's long-held (and erroneous) belief that Sir John Franklin and his crew were still alive and would be found in the Arctic is evident in his address to the British Association for the Advancement of Science at its 1860 Oxford meeting. This was later published in pamphlet form: W. P. Snow, *A Paper on the Lost Polar Expedition and Possible Recovery of Its Scientific Documents, Read on Thursday June 28 before the Geographical and Ethnological Section* (London: Edward Stanford, 1860). For a summary, see Captain Parker Snow, "On the Lost Polar Expeditions and Possible Recovery of Its Scientific Documents," *Report of the Thirtieth Meeting of the British Association for the Advancement of Science* (London: John Murray, 1861), 180–181. Parker Snow's insistence upon Franklin's survival, despite mounting evidence to the contrary and the opposing views of many British mariners and scientists, several of whom heard his Oxford talk, led to a fund under his direction, Snow's Renewed Arctic Search, to renew efforts to find the lost ships.

    Snow's earlier failed exploratory attempts in this regard are recounted in his *Voyage of the* Prince Albert *in Search of Sir John Franklin: A Narrative of Every-Day Life in the Arctic Seas* (London: Longman, Brown, Green, and Longmans, 1851). Parker Snow's work on the military leaders of the Confederacy in the American Civil War was published as *Southern Generals: Who They Are, and What They Have Done* (New York: C. B. Richardson, 1865) and later reissued under a different title. Parker Snow has not been well served at any length by modern scholars: see Ian Stone, "William Parker Stone, 1817–1895," *Polar Record* 19 (1978): 163–165; William Barr, "Searching for Franklin from Australia: William Parker Snow's Initiative of 1853," *Polar Record* 33 (1985): 145–150. In his brief obituary of Parker Snow in 1895, Clements Markham, president of the Royal Geographical Society, made no mention of his 1883 views over a single prime meridian, referring to him only as "a brother Arctic": the two had sailed together on the *Prince Albert*. Clements Markham, "Obituary: Captain William Parker Snow," *Geographical Journal* 5 (1895): 500–501.

3. "Meridian," *The London Encyclopaedia, or Universal Dictionary of Science, Art, Literature, and Practical Mechanics*, 22 vols. (London: Thomas Tegg, 1825), 14, 302–303.

4. My use of "Britain" throughout refers to Great Britain, made up of England, Scotland, Wales, and Northern Ireland. Where reference is later made to Ireland, it is to the country of that name, not as part of the British Isles.

5. Parker Snow, *An International Prime Meridian*, 2.

6. Ibid., 3 (original emphasis).

7. William Parker Snow, "Ocean Relief Depots," *Chambers's Journal of Popular Literature, Science, and Art*, Saturday, November 27, 1880, 753–755, quote from p. 754.

INTRODUCTION

1. The quotes are taken from the printed proceedings of the Washington International Meridian Conference. This is titled *International Conference Held at Washington for the Purpose of Fixing a Prime Meridian and a Universal Day. Protocols of the Proceedings* (Washington, DC: Gibson Brothers, 1884). The first words cited are those of the conference chair, Rear Adm. C. R. P. Rodgers, on October 1, 1884, from p. 6. The selected summary of the resolutions is from pp. 199 and 201.

2. Among many contemporary discussions, see William Ellis, "The Prime Meridian Conference," *Nature* 31 (1884): 7–10. Among modern studies of the 1884 conference and the key metrological role of Greenwich, see Ian R. Bartky, *Selling the True Time: Nineteenth-Century Timekeeping in America* (Stanford, CA: Stanford University Press, 2000); Bartky, *One Time Fits All: The Campaign for Global Uniformity* (Stanford, CA: Stanford University Press, 2007); Stuart R. Malin, "The International Prime Meridian Conference, Washington, October 1984 [1884]," *Journal of Navigation* 30 (1985): 203–206; Stuart Malin and Carole Stott, *The Greenwich Meridian* (London: Her Majesty's Stationery Office [HMSO], 1984); Derek Howse, *Greenwich Time and the Longitude* (London: Philip Wilson, 1997); Charles Jennings, *Greenwich the Place Where Days Begin and End* (London: Abacus, 2001). The journal *Vistas in Astronomy* 20 (1986) carries a themed set of papers from a commemorative conference held to mark the centenary of Greenwich and 1884. The view of Greenwich 1884 as "the world's touchstone" is from Chet Raymo, *Walking Zero: Discovering Cosmic Space and Time along the Prime Meridian* (New York: Walker, 2006), xi. The significance of Greenwich and universal time in motivating new conceptions of time and modernism in English literature is addressed by Adam Barrows, *The Cosmic Time of Empire: Modern Britain and World Literature* (Berkeley: University of California Press, 2011).

3. This is the tone of Graham Dolan, *On the Line: The Story of the Greenwich Meridian* (London: HMSO, 2003) and even of Howse, *Greenwich Time and the Longitude*. Although Howse's detailed account is aware of the wider historical context, it is largely a story of Greenwich alone, not of Greenwich in relation to other prime or first meridians.

4. On these points, see Matthew Edney, "Cartographic Confusion and Nationalism: The Washington Meridian in the Early Nineteenth Century," *Mapline: The Quarterly Newsletter Published by the Hermon Dunlap Center for the History of Cartography at the Newberry Library* 69–70 (1993): 3–8, quote from p. 4.

5. Matthew Edney, "Meridians, Local and Prime," in *Cartography in the European Enlightenment: Volume 4, History of Cartography*, ed. Matthew Edney and Mary Pedley (Chicago: University of Chicago Press, forthcoming). I am grateful to Professor Edney for granting me access to his contribution to this forthcoming work.

6. John Senex, *A New General Atlas containing a Geographical and Historical Account of all the Empires, Kingdoms, and other Dominions of the World* (London: printed for Daniel Browne, Thomas Taylor, John Darby, John Senex, William Taylor, and four others, 1721), 6.

7. Several of these works appear in Sandford Fleming, *Papers on Time-Reckoning and the Selection of a Prime Meridian to Be Common to All Nations* (Toronto: Wilson, 1880), 12. On Fleming and universal time, see Clark Blaise, *Time Lord: Sandford Fleming and the Creation of Standard Time* (London: Weidenfeld and Nicholson, 2000).

8. On this, see Donald Janelle, "Global Interdependence and Its Consequences," in *Collapsing Space and Time: Geographic Aspects of Communications and Information*, ed. Stanley Brunn and Thomas Leinbach (London: HarperCollins, 1991), 49–81; Barney Warf, *Time-Space Compression: Historical Geographies* (Abingdon, UK: Routledge, 2008). On telegraphy and globalization in the nineteenth century, see Roland Wenzlhuemer, *Connecting the Nineteenth-Century World: The Telegraph and Globalization* (Cambridge: Cambridge University Press, 2012). For a discussion of

the prime meridian in relation to late twentieth-century postmodernism and national space, see Ronald R. Thomas, "The Home of Time: The Prime Meridian, the Dome of the Millennium, and Postnational Space," in *Nineteenth-Century Geographies: The Transformation of Space from the Victorian Age to the American Century*, ed. Helena Mitchie and Ronald R. Thomas (New Brunswick, NJ: Rutgers University Press, 2003), 23–39.

9. Fleming, *Papers on Time-Reckoning*, 52.

10. Edney, "Meridians, Local and Prime."

11. Don Juan Pastorin, "Remarks on a Universal Prime Meridian, by Don Juan Pastorin, Lieut-Commander of the Spanish Navy," *Proceedings of the Canadian Institute* 2 (1885): 49–51, quotes from pp. 49 and 51. Pastorin was expressing disbelief at a proposal then being brought forward by the Geographical Society of Madrid for yet another prime meridian within Spain: "One more Meridian, when there were so many already!" It was not adopted.

12. On this point, see Ernst Mayer, "Die Geschichte des ersten Meridians und die Zählung der Geographischen Länge," *Mittheilungen aus dem Gebiete des Seewesens* 6 (1878): 49–61 and particularly Heinrich Haag, *Die Geschicht des Nullmeridians (Mit Einer Karte)* (Leipzig, Germany: Otto Wigand, 1913). Haag's 1913 work was presented as a doctoral thesis to the philosophy faculty in Leipzig in 1912. He traces the prime meridian from its expression in the work of classical geographers, such as Eratosthenes, through the early modern period, to discussions during several international geographical congresses, beginning with the first in 1871 (the subject of Chapter 4 in this book). He closes by briefly examining the various prime meridians that cartographers of different countries used at the end of the nineteenth century. There is then a gap of some years before others address the prime meridian as an object of historiographical inquiry: see, for example, W. G. Perrin, "The Prime Meridian," *Mariner's Mirror* 13 (1927): 109–124. For modern reviews of the topic, see André Gotteland, *Les méridiennes du monde et leur histoire*, 2 vols. (Paris: Éditions Le Manuscrit, 2008); Edney, "Meridians, Local and Prime."

13. The fullest discussion from a British perspective of the establishment of the Greenwich Royal Observatory and of Maskelyne's influential role in helping establish Greenwich as Britain's prime meridian from the later eighteenth century is Howse, *Greenwich Time and the Longitude*. See also Rebekah Higgitt and Graham Dolan, "Greenwich, Time and 'The Line,'" *Endeavour* 34 (2009): 35–39. For France, the history of the prime meridian in the longer term is the subject of Jean-Pierre Martin, *Une histoire de la méridienne: Textes, enjeux, débats et passions autour du méridien de Paris 1666–1827* (Cherbourg, France: Editions Isoète, 2000). The international dimensions of the prime meridian question, between Britain and France in particular, are the subject of Jean-Pierre Martin and Anita McConnell, "Joining the Observatories of Paris and Greenwich," *Notes and Records of the Royal Society* 62 (2008): 355–372 and, at fuller length, of McConnell and Martin, *Méridien, méridienne: Textes, enjeux, debats et passions autour des méridiens de Paris et de Greenwich (1783–2000)* (Cherbourg, France: Editions Isoète, 2013). On the American prime meridian in general terms, see Joseph Hyde Pratt, "American Prime Meridians," *Geographical Review* 32 (1942): 233–244; Silvio A. Bedini, *The Jefferson Stone: Demarcation of the First United Meridian of the United States* (Frederick, MD: Professional Surveyors, 1999); Matthew Edney, "Cartographic Culture and Nationalism in the Early United States: Benjamin Vaughan and the Choice for a Prime Meridian, 1811," *Journal of Historical Geography* 20 (1994): 384–395.

14. Elisabeth Crawford, Terry Shinn, and Sverker Sörlin, "The Nationalisation and Denationalization of the Sciences: An Introductory Essay," in *Denationalizing Science: The Contexts of International Scientific Practice*, ed. Elisabeth Crawford, Terry Shinn, and Sverker Sörlin (Dordrecht, Netherlands: Kluwer, 1993), 1–42, quote from p. 3; Josep Simon and Néstor Harran, *Beyond Borders: Fresh Perspectives in History of Science*, with the editorial assistance of Tayra Lanuza-Navarro, Pedro Luiz Castell, and Xino-Guillem Llobat (Newcastle upon Tyne, UK: Cambridge Scholars, 2008). The 1884 Washington meeting is read in these terms by Allen W. Palmer, "Negotiation and Resistance

in Global Networks: The 1884 International Meridian Conference," *Mass Communication and Society* 5 (2002): 7–24.

15. Peter Galison, *Einstein's Clocks and Poincaré's Maps: Empires of Time* (London: Hodder and Stoughton, 2003), quote from p. 38; Barrows, *Cosmic Time of Empire*. The prime meridian, surprisingly, does not feature in Stephen Kern, *The Culture of Time and Space, 1880–1918* (1983; repr., Cambridge, MA: Harvard University Press, 2003).

16. On these issues, see M. Norton Wise, "Introduction," in *The Values of Precision*, ed. M. Norton Wise (Princeton, NJ: Princeton University Press, 1994), 3–13. In dealing with questions of precision, accuracy, and trust as I do throughout the book, I have also drawn on Theodore Porter, *Trust in Numbers: Objectivity in Science and Public Life* (Princeton, NJ: Princeton University Press, 1995); Steven Shapin, *Never Pure: Historical Studies of Science as If It Was Produced by People with Bodies, Situated in Time and Space, Culture, and Society, and Struggling for Credibility and Authority* (Baltimore: Johns Hopkins University Press, 2010).

17. On the meter, see Ken Alder, *The Measure of All Things: The Seven-Year Odyssey That Transformed the World* (London: Little, Brown, 2002); Alder, "A Revolution to Measure: The Political Economy of the Metric System," in *The Values of Precision*, ed. M. Norton Wise (Princeton, NJ: Princeton University Press, 1994), 39–71. The history and variability of measures is the subject of Witold Kula, *Measures and Men*, trans. R. Szreter (Princeton, NJ: Princeton University Press, 1986). For a general discussion of the prime meridian as one element in the measuring of the globe, see Raymo, *Walking Zero*; Avraham Ariel and Nora Ariel Berger, *Plotting the Globe: Stories of Meridians, Parallels, and the International Date Line* (Westport, CN: Praeger, 2006); Paul Murdin, *Full Meridian of Glory: Perilous Adventures in the Competition to Measure the Earth* (New York: Springer, 2009).

18. Simon Schaffer, "Metrology, Metrication, and Victorian Values," in *Victorian Science in Context*, ed. Bernard Lightman (Chicago: University of Chicago Press, 1997), 438–474. See also Schaffer, "Modernity and Metrology," in *Science and Power: The Historical Foundations of Research Policies in Europe*, ed. Luca Guzzetti (Luxembourg City, Luxembourg: European Communities, 2000), 71–91; Andrew Barry, "The History of Measurement and the Engineers of Space," *British Journal for the History of Science* 26 (1993): 459–468.

19. Elisabeth Crawford, "The Universe of International science, 1880–1939," in *Solomon's House Revisited: The Organization and Institutionalization of Science*, ed. Tore Frängsmyr (Canton, MA: Science History, 1990), 251–269; Maurice Crosland, "Aspects of International Scientific Collaboration and Organization before 1900," in *Proceedings of the XVth International Congress of the History of Science*, ed. Eric Forbes (Edinburgh: Scottish Academic Press, 1977), 114–125. The history and the geography of early scientific internationalization is usefully discussed in Ken Alder, "Scientific Conventions: International Assemblies and Technical Standards from the Republic of Letters to Global Science," in *Nature Engaged: Science in Practice from the Renaissance to the Present*, ed. Mario Biagioli and Jessica Riskin (New York: Palgrave Macmillan, 2012), 19–39.

20. Crawford, Shinn, and Sörlin, "Nationalisation and Denationalization of the Sciences," 1–42. The essays in this book cover these topics with reference to vegetation science, molecular biology, ecology, and physics, among other subjects. On these issues for the social sciences, see Johan Heilbron, Nicolaas Guilhot, and Laurent Jeanpierre, "Toward a Transnational History of the Human Sciences," *Journal of the History of the Behavioural Sciences* 44 (2008): 146–160; Ann Curthoys and Merilyn Lake, eds., *Connected Worlds—History in Transnational Perspective* (Canberra: Australian National University English Press, 2008). For an elaboration of these points, see Martin Geyer and Johannes Paulman, eds., *The Mechanics of Internationalism: Culture, Society, and Politics from the 1840s to the First World War* (Oxford: Oxford University Press, 2001), quote from pp. 8–9. On the questions of scientific standards, see the essays in Oliver Schlaudt and Lara Huber, eds., *Standardization in Measurement: Philosophical, Historical and Sociological Issues* (London: Pickering and Chatto, 2015) and Chapter 3 of this book.

21. Walter Nash, *At Prime Meridian* (Shrewsbury, UK: Feather Books, 2000) is a book of poetry and does not mention the prime meridian. Linda Emma, *Prime Meridian* (Deadwood, OR: Wyatt-Mackenzie, 2009) is a novel about the first Gulf War and the memories of women in a town called Meridian as they mourn soldiers killed in that conflict. Christopher Bayly, *The Imperial Meridian: The British Empire and the World 1780–1830* (London: Longman, 1989) uses the term to reflect on British imperialism in this period. For a discussion of the utility of this in a French context, see David Todd, "A French Imperial Meridian, 1814–1870," *Past and Present* 210 (2011): 155–186.

22. On these points, see Diarmid A. Finnegan and Jonathan Jeffrey Wright, "Introduction: Placing Global Knowledge in the Nineteenth Century," in *Spaces of Global Knowledge: Exhibition, Encounter and Exchange in an Age of Empire*, ed. Diarmid A. Finnegan and Jonathan Jeffrey Wright (Farnham, UK: Ashgate, 2015), 1–15; Simone Turchetti, Néstor Herran, and Soraya Boudia, "Introduction: Have We Ever Been 'Transnational'? Towards a History of Science across and beyond Borders," *British Society for the History of Science* 45 (2012): 319–336. This introduces a set of essays on the topic of transnational perspectives in the history of science.

23. In a wide literature, see David N. Livingstone, "The Spaces of Knowledge: Contributions towards a Historical Geography of Science," *Environment and Planning D: Society and Space* 13 (1995): 5–34; Crosbie Smith and Jon Agar, eds., *Making Space for Science: Territorial Themes in the Shaping of Knowledge* (Basingstoke, UK: Macmillan, 1998); David N. Livingstone, *Putting Science in Its Place: Geographies of Scientific Knowledge* (Chicago: University of Chicago Press, 2003); Ana Simões, Ana Carneiro, and Maria Paula Diogo, eds., *Travels of Learning: A Geography of Science in Europe* (Dordrecht, Netherlands: Kluwer, 2003); Richard Powell, "Geographies of Science: Histories, Localities, Practices, Futures," *Progress in Human Geography* 31 (2007): 309–330; Peter Meusburger, David N. Livingstone, and Heike Jöns, eds., *Geographies of Science* (Dordrecht, Netherlands: Kluwer, 2010); Charles W. J. Withers and David N. Livingstone, "Thinking Geographically about Nineteenth-Century Science," in *Geographies of Nineteenth-Century Science*, ed. David N. Livingstone and Charles W. J. Withers (Chicago: University of Chicago Press, 2011), 1–19. On the nominalist and realist positions invoked in this work, see Robert J. Mayhew, "A Tale of Three Scales: Ways of Malthusian Worldmaking," in *The Uses of Space in Early Modern History*, ed. Paul Stock (New York: Palgrave Macmillan, 2015), 197–226, especially pp. 197–200, 219–220.

CHAPTER 1 ○ "ABSURD VANITY"

1. Anton-Friedrich Büsching, *A New System of Geography*, 6 vols. (London: printed for A. Millar in the Strand, 1762), 1:21; Ebenezer MacFait, *A New System of General Geography, in which the Principles of that Science are Explained* (Edinburgh: printed for the author, 1780), 53; Edward Gibbon, "The Circumnavigation of Africa," in *The English Essays of Edward Gibbon*, ed. Patricia C. Craddock (Oxford: Clarendon Press, 1972), 375; William Faden, *Geographical Exercises; Calculated to facilitate the Study of Geography; and, by an expeditious Method, to imprint a Knowledge of the Science on the Minds of Youth* (London: printed for the proprietor, 1777), 1.

2. On prime meridians in non-Western cartographic cultures, see Gerald R. Tibbetts, "The Beginnings of a Cartographic Tradition," in *The History of Cartography Volume Two, Book One: Cartography in the Traditional Islamic and South Asian Societies*, ed. J. B. Harley and David Woodward (Chicago: University of Chicago Press, 1992), 90–107, especially pp. 100–104; David A. King and Richard P. Lorch, "Qibla Charts, Qibla Maps, and Related Instruments," in Harley and Woodward, *History of Cartography Volume Two, Book One*, 189–205, quote from pp. 196–197. On Uijain, where the prime meridian in Hindu culture was based since the mid-second century, see David Pingree, "Astronomy and Astrology in India and Iran," *Isis* 54 (1963): 229–246. On Kyoto in Japan, see Kazutaka Unno, "Cartography in Japan," in *The History of Cartography Volume Two, Book Two:*

Cartography in the Traditional Islamic and South Asian Societies, ed. J. B. Harley and David Woodward (Chicago: University of Chicago Press, 1994), 346–477. On Beijing in China, see Cordell D. K. Yee, "Traditional Chinese Cartography and the Myth of Westernization," in Harley and Woodward, *History of Cartography Volume Two, Book Two,* 170–202.

3. This and the following paragraphs on the prime meridian in the classical and early modern European context is taken from W. G. Perrin, "The Prime Meridian," *Mariner's Mirror* 13 (1927): 109–124. On the "correction" of Ptolemy's terrestrial measurements, see the work of Christian Marx, "Investigations of the Coordinates in Ptolemy's *Geographike Hyphegesis* Book 8," *Archive for History of Exact Sciences* 66 (2012): 531–555. For a late eighteenth-century commentary on Ptolemy's first meridian, see William Vincent, *The Commerce and Navigation of the Ancients in the Indian Ocean* (London: T. Cadell and W. Davies, 1797). In the second (1807) edition of this work, Vincent has a short "Dissertation II: On the First Meridian of Ptolemy" as an appendix, pp. 576–578.

4. Tycho Brahe positioned his own astronomical observations on the island of Hven in relation to two different prime meridians, those used by Ptolemy and by Copernicus: see Hans Raeder, Elis Strömgren, and Beryl Strömgren, eds., *Tycho Brahe's Description of His Instruments and Scientific Works as Given in* Astronomine Instauratae Mechanica *(Wandburgi, 1598)* (Copenhagen: Det Kongelige Danske Videnskabernes Skelskab, 1976), 139. I am grateful to Prof. Michael Jones for this information.

5. Jerry Brotton, "Toleration: Gerard Mercator, World Map, 1569," in *A History of the World in Twelve Maps* (London: Allen Lane, 2012), 218–259; Perrin, "Prime Meridian," 116.

6. The extract from John Davis is taken from Albert Hastings Markham, ed., *Voyages and Works of John Davis, the Navigator,* 2 vols. (London: Hakluyt Society, 1880), 1:284. On Blundeville and others, see Horace E. Ware, "A Forgotten Prime Meridian," *Publications of the Colonial Society of Massachusetts* 12 (1908–1909): 382–398; Ware, "Supplement to a Forgotten Prime Meridian," *Publications of the Colonial Society of Massachusetts* 13 (1911): 226–234.

7. Art R. T. Jonkers, "Parallel Meridian: Diffusion and Change in Early-Modern Oceanic Reckoning," in *Noord-zuid in Oostindisch perspectief,* ed. J. Parmentier (The Hague: Walburg), 17–42, especially p. 17. Jonkers also provides a summary history of "The Era of the Agonic Meridian (1500–1650)," 2–6. For fuller discussions of the agonic meridian and of the intellectual history of geomagnetism, see Jonkers, *North by Northwest: Seafaring, Science, and the Earth's Magnetic Field 1600–1800,* 2 vols. (Göttingen, Germany: Cuvillier, 2000); Jonkers, *Earth's Magnetism in the Age of Sail* (Baltimore: Johns Hopkins University Press, 2003). These issues of credibility and trust at a distance from different informants, of which the case of the agonic prime meridian is a good example, are the central themes of Steven Shapin's *A Social History of Truth: Civility and Science in Seventeenth-Century England* (Chicago: University of Chicago Press, 1994). For consistency I have used the term "magnetic variation" throughout, even though the phenomenon it describes is sometimes termed "magnetic declination": magnetic variation is the angle between magnetic north and true north.

8. Samuel E. Dawson, "The Lines of Demarcation of Pope Alexander VI and the Treaty of Tordesillas AD 1493 and 1494," *Transactions of the Royal Society of Canada* 5, 2nd ser. (1899): 468–546. Humboldt's mistaken views over papal knowledge of the phenomenon of terrestrial magnetism are discussed on p. 488.

9. Perrin, "Prime Meridian," 118.

10. Jonkers describes the sixteenth century as "a veritable cauldron of different geomagnetic ideas, thrown together in conflicting and mutually adapting arrangements." Jonkers, *Earth's Magnetism in the Age of Sail,* 45. While this is true, Jonkers's work shows that the main features of the agonic meridian as an intellectual and practical concern from c. 1508 to about 1650 form a distinct element within the intellectual history of the prime meridian as well as of geomagnetism.

11. Lucie Lagarde, "Historique du probléme du méridien origine en France," *Revue d'histoire des sciences* 32 (1979): 289–304; Perrin, "Prime Meridian," 119–120. On García de Céspedes, see María Portuondo, "An Astronomical Observatory for the Escorial of Philip II: An Exercise in Historical Inference," *Colorado Review of Hispanic Studies* 7 (2009): 101–117. On Grotius's arguments in relation to conceptions of territory on land, see Stuart Elden, *The Birth of Territory* (Chicago: University of Chicago Press, 2013), 239–241.

12. Jonkers, "Parallel Meridian," 6.

13. Ibid., 10.

14. Ibid., 10–11. The view of Perrin that "whatever the real origin of the 1634 decree, it certainly had the effect of encouraging the use of Ferro" must now be qualified in the light of Jonkers's work. Perrin, "Prime Meridian," 119. On King Philip II of Spain's 1573 edict over Toledo as the prime meridian for all Spanish maps, see James J. Parsons, "Before Greenwich: The Canary Islands, El Hierro and the Dilemma of the Prime Meridian," in *Person, Place, and Things: Interpretative and Empirical Essays in Cultural Geography,* ed. Shue T. Wong (Baton Rouge: Louisiana State University Geoscience, 1995), 61–78.

15. Larry Stewart, "Other Centres of Calculation, or, Where the Royal Society Didn't Count: Commerce, Coffee-Houses and Natural Philosophy in Early Modern London," *British Journal for the History of Science* 32 (1999): 133–153; Katy Barrett, "The Wanton Line: Hogarth and the Public Life of Longitude" (PhD diss., University of Cambridge, 2014).

16. On these issues, see Steven Shapin, *The Scientific Revolution* (Chicago: University of Chicago Press, 1996); Daniel R. Headrick, *When Information Came of Age: Technologies of Knowledge in the Age of Reason and Revolution, 1700–1850* (Oxford: Oxford University Press, 2000). The idea of the fiscal-military state and of the control of space as a means to the control of society and empire is central to John Brewer, *The Sinews of Power: War, Money, and the English State, 1688–1783* (Cambridge, MA: Harvard University Press, 1990). The point about "modes" of cartographic inquiry is from Matthew Edney, "Cartography: Disciplinary History," in *Sciences of the Earth: An Encyclopedia of Events, People, and Phenomena,* 2 vols., ed. Gregory A. Good (New York: Garland, 1998), 1:81–85.

17. For fuller accounts, see Joseph W. Konvitz, *Cartography in France, 1660–1848: Science, Engineering and Statecraft* (Chicago: University of Chicago Press, 1987); Anne M. C. Godlewska, *Geography Unbound: French Geographic Science from Cassini to Humboldt* (Chicago: University of Chicago Press, 1999); Brotton, *History of the World in Twelve Maps,* 294–336.

18. These paragraphs are based on Lagarde, "Historique du probléme du méridien origine en France"; Jean-Pierre Martin, *Une histoire de la méridienne: Textes, enjeux, débats et passions autour du méridien de Paris 1666–1827* (Cherbourg, France: Editions Isoète, 2000). Unfortunately, Martin's otherwise excellent account does not fully reference the sources on which it is dependent: among the several key texts cited here, see M. l'Abbe de la Caille, "Extrait de la relation du voyage fait en 1724, aux Isles Canaries, par le P.Feuillée Nimine, pour determiner la vraie position du premier méridien," *Histoire de l'Academie Royale des Sciences* (hereafter cited as *HRAS*), 1746, 129–151; Guillaume Delisle, "Determination géographique de la situation and d'l'étendue des differentes parties de la terre," *HRAS,* 1720, 365–383; Cassini de Thury, *La meridienne de l'Observatoire Royal de Paris* (Paris: Hippolyte-Louis Guerin et Jacques Guerin, 1744). The notion of the center of calculation is from Bruno Latour, *Science in Action: How to Follow Scientists and Engineers through Society* (Cambridge, MA: Harvard University Press, 1987). For a discussion of French navigational science in the late seventeenth-century Atlantic, see Nicholas Dew, "Scientific Travel in the Atlantic World: The French Expedition to Gorée and the Antilles, 1681–1683," *British Journal for the History of Science* 43 (2010): 1–17. On the geographers and astronomers in the French court and in the Academy of Sciences at this period, see Michael Heffernan, "Geography and the Paris Academy of Sciences: Politics and Patronage in Early 18th-Century France," *Transactions of the Institute of British Geographers* 39 (2014): 62–75.

19. On terrestrial measurement and the shape of the earth debate in the 1730s, see Mary Terrall, *The Man Who Flattened the Earth: Maupertuis and the Sciences in the Enlightenment* (Chicago: University of Chicago Press, 2003); Rob Iliffe, "'Aplatisseur du Monde et de Cassini': Maupertuis, Precision Measurement, and the Shape of the Earth in the 1730s," *History of Science* 31 (1993): 335–375. The view of 1744 as a monumental year in Enlightenment geodesy is from Martin, *Histoire de la méridienne*, 88.

20. Dava Sobel, *Longitude: The True Story of a Lone Genius Who Solved the Greatest Scientific Problem of His Time* (New York: Walker, 1995).

21. The Latin originals of the Cruzado-Oldenburg-Flamsteed correspondence are in the Royal Society, January 20, 1675, MSS EL/C2/1; December 10, 1675, EL/C2/2; September 15, 1675, EL/O2/159); February 3, 1676, EL/O2/161. For transcriptions, see A. Rupert Hall and Marie Boas Hall, eds., *The Correspondence of Henry Oldenburg Volume 11* (London: Taylor and Francis, 1986), 444, 501; Hall and Hall, eds., *The Correspondence of Henry Oldenburg Volume 12* (London: Taylor and Francis, 1986), 69, 150, 172. See also "An Extract to the Publisher from a Spanish Professour of the Mathematicks, Proposing a New Place for the First Meridian, and Pretending to Evince the Equality of all Natural Daies, as Also to Shew a Way of Knowing the True Place of the Moon," *Philosophical Transactions of the Royal Society of London* 10 (1675): 425–432.

22. Alexi Baker, ed., *The Board of Longitude 1714–1828: Science, Innovation and Empire* (London: Palgrave, 2016); Richard Dunn and Rebekah Higgitt, *Ships, Clocks and Stars: The Quest for Longitude* (London: Collins; Greenwich: Royal Museums Greenwich, 2014).

23. William Whiston and Humphry Ditton, *A New Method for Discovering the Longitude Both at Sea and Land* (London: printed for John Phillips at the Black Bull in Cornhill, 1714), quotes from pp. 23, 38, 52, 54, 79. Howse, *Greenwich Time and the Longitude* terms this a "bizarre" scheme (p. 56) but also discusses the popular reaction—disbelieving and scathing—to Whiston and Ditton's 1714 work. Some contemporaries presented their own (similarly unworkable) schemes concerning longitude with reference to Whiston and Ditton's work: see, for example, R. B. [pseud.], *Longitude To be found with a new Invented Instrument, both by Sea and Land*, which advertised in its subtitle *With a better Method for discovering Longitude, than that lately proposed by Mr. Whiston and Mr. Ditton* (London: printed for F. Burleigh in Amen Corner, 1714).

24. *An Essay Towards a New Method to Shew the Longitude at Sea; especially near the Dangerous Shores* (London: printed for E. Place, at Furnival's-Inn-Gate in Holborn, 1714), quotes from pp. 16, 19.

25. Jane Squire, *A Proposal to Determine our Longitude*, 2nd ed. (London, 1731; repr., London: printed for the author, 1743). The quote regarding Bethlehem is from the second edition, p. 65. For one commentary upon Squire and her work, see Eva G. R. Taylor, *The Mathematical Practitioners of Hanoverian England 1714–1840* (Cambridge: Cambridge University Press, 1966), 19–20, 193. A more recent summary is provided by Alexi Baker, "Squire, Jane," *Oxford Dictionary of National Biography Online*, accessed October 14, 2015, www.oxforddnb.com/public/dnb/45826-article.html.

26. This paragraph is based upon the first three chapters of Derek Howse, *Greenwich Time and the Longitude* (Greenwich: Philip Wilson/National Maritime Museum in association with ATKearney, 1997), itself based on Howse, *Greenwich Time and the Discovery of the Longitude* (Oxford: Oxford University Press, 1990). See also Howse, "Nevil Maskelyne, the Nautical Almanac, and G. M. T.," *Journal of Navigation* 38 (1985): 159–177; William J. H. Andrewes, ed., *The Quest for Longitude* (Cambridge, MA: Harvard University Press, 1996); Dunn and Higgitt, *Ships, Clocks and Stars*. The quote from Maskelyne is in Nevil Maskelyne, *The British Mariner's Guide* (London: printed for the author, 1763), i.

27. Howse puts it thus, speaking of the adoption of Greenwich at the Washington meeting: "And it was the publication in 1766 of *The Nautical Almanac* which had started the chain of events described above." Howse, *Greenwich Time and the Longitude*, 71. In "Nevil Maskelyne," he writes

"that publication . . . was the principal reason why we are this year [1984] celebrating the centenary of the international adoption of Greenwich as the world's prime meridian" (p. 165). D. H. Sadler and G. A. Wilkins, "Astronomical Background to the International Meridian Conference of 1884," *Journal of Navigation* 38 (1985): 191–199 refer, a little guardedly perhaps, to "the contributions made by the Royal Observatory at Greenwich that led, almost inevitably, to the choice of the Airy transit circle to define the zero meridian for the measurement of longitude and the beginning of the universal day" (p. 191). They also note, "It was almost inevitable that, with its widespread astronomical and nautical usage and its civil status, G. M. T. and the Greenwich meridian should be chosen in 1884" (p. 197).

28. Jonkers, *Earth's Magnetism in the Age of Sail,* 29–30; Jonkers, "Parallel Meridian," 11–12. The extract from "Of the First Meridian" is in *Sayling by the True Sea Chart,* Sloane MSS 3143, fol. 65r, British Library; Howse, *Greenwich Time and the Discovery of the Longitude,* 71. See also H. Harries, "Pre-Greenwich Sea Longitudes," *Observatory* 50 (1927): 315–319. The quote from John Brisbane is cited in Simon Schaffer, "In Trans: European Cosmologies in the Pacific," in *The Atlantic World in the Antipodes: Effects and Transformations since the Eighteenth Century,* ed. Kate Fullagar (Newcastle upon Tyne, UK: Cambridge Scholars Press, 2012), 70–93, quote from p. 84.

29. On the nonappearance of *The Nautical Almanac and Astronomical Ephemeris* in British texts of practical navigation in the 1770s and 1780s, see Jane Wess, "Navigation and Mathematics: A Match Made in the Heavens?," in *Navigational Enterprises in Europe and Its Empires, 1730–1850,* ed. Richard Dunn and Rebekah Higgitt (London: Palgrave Macmillan, 2015), 201–222, quote from Ewing from p. 207. On French discussions over longitude and the testing of claims at sea, see Guy Boistel, "From Lacaille to Lalande: French Work on Lunar Distances, Nautical Ephemerides and Lunar Tables, 1742–1785," in Dunn and Higgitt, *Navigational Enterprises,* 47–64; Martina Schiavon, "The Bureau des Longitudes: An Institutional Study," in Dunn and Higgitt, *Navigational Enterprises,* 65–88. Boistel is particularly cautious about the relationship of *The Nautical Almanac and Astronomical Ephemeris* to Lacaille's earlier work: "The *Nautical Almanac* appears, therefore, to be an adaptation of the French ephemeris and a realization of Lacaille's earlier proposals for a nautical almanac. The full extent of what Maskelyne owed to Lacaille, however, has yet to be fully explored" (p. 60).

30. On the Russian example, see Endel Varep, *The Prime Meridian of Dagö and Ösel* (Tartu, USSR: Tartu State University, 1975). For Varep, this westernmost echo of Ptolemy was, for Russia, "but a chance episode in the history of Russian cartography" (p. 14). On Spain see Parsons, "Before Greenwich," 67–68; Juan Pimentel, "A Southern Meridian: Astronomical Undertakings in the Eighteenth-Century Spanish Empire," in Dunn and Higgitt, *Navigational Enterprises,* 13–31. On the Netherlands, see Karel Davids, "The Longitude Committee and the Practice of Navigation in the Netherlands, c. 1750–1850," in Dunn and Higgitt, *Navigational Enterprises,* 32–46.

31. The claim that the first known (British) chart with longitude zero through Greenwich dates from 1738 is made by G. E. W. Gosnell, "Greenwich Prime Meridian Marks," *Journal of the British Astronomical Association* 63 (1953): 104–107, from p. 105. The quotes from Jefferys's atlas from his editor, Robert Sayer, are in Thomas Jefferys, *The West-India Atlas* (London: printed for Robert Sayer and John Bennett, 1780), ii. The evidence relating to Samuel Dunn is in his *Nautical Propositions and Institutes* (London: printed for the author at Boar's Head Court, Fleet Street, 1781), section 22; Dunn, *The Theory and Practice of Longitude at Sea,* 2nd ed. (London: printed for the author at No. 1, Boar's-Head Court, Fleet Street, 1786): section 74, 35–36; Dunn, *A New Atlas of the Mundane System; or, Of Geography and Cosmography,* 2nd ed. (London: printed for Robert Sayer, No. 53, Fleet Street, 1788), 11.

32. On this, see Martin, *Une histoire de la méridienne,* 87–97; Brotton, *History of the World in Twelve Maps,* 294–336, quotes from pp. 315, 322; Godlewska, *Geography Unbound,* 57–86, in which Godlewska deals with Cassini IV; the decline of "terrain geography;" and Cassini's banishment from science including, after 1795, his exclusion from the Bureau des Longitudes. These are features of

what Godlewska terms geography's "demotion" by the end of the eighteenth century in France. I do not share her interpretation.

33. These paragraphs are based upon Nevil Maskelyne, "Concerning the Latitude and Longitude of the Royal Observatory at Greenwich: With Remarks on a Memorial of the Late M. Cassini de Thury. By the Rev. Nevil Maskelyne, D. D., F. R. S. and Astronomer Royal," *Philosophical Transactions of the Royal Society of London* 77 (1787): 151–187, quotes from pp. 180, 181.

34. William Roy and Isaac Dalby, "An Account of the Trigonometrical Operation, Whereby the Distance between the Meridians of the Royal Observatories of Greenwich and Paris Has Been Determined. By Major-General William Roy, F. R. S. and A. S.," *Philosophical Transactions of the Royal Society of London* 80 (1790): 111–614. For a French perspective, see Martin, *Histoire de la méridienne,* 97–107. On Roy's role in founding the Ordnance Survey and earlier information on the Military Survey of Scotland, 1747–1755, and the resultant "Great Map," see Rachel Hewitt, *Map of a Nation: A Biography of the Ordnance Survey* (London: Granta Books, 2010).

35. MM. Cassini, Méchain, and Le Gendre, *Exposé des opérations faites en France en 1787 pour la jonction des Observattoires de Paris et de Greenwich* (Paris: L'Imprimerie de L'Institution des Sourds-Muet, 1790).

36. Roy and Dalby, "An Account of the Trigonometrical Operation," 183, 186.

37. These paragraphs are from Martin and McConnell, "Joining the Observatories of Paris and Greenwich." See also Martin, *Une histoire de la méridienne,* especially pp. 97–107; McConnell and Martin, *Méridien, méridienne.* On the different toise in use in French topographic science at this time and their part in the Greenwich-Paris measurements of the 1780s, see Michael Kershaw, "A Different Kind of Longitude: The Metrology and Conventions of Location by Geodesy," in Dunn and Higgitt, *Navigational Enterprises,* 134–158.

CHAPTER 2 ○ DECLARATIONS OF INDEPENDENCE

1. Joseph Hyde Pratt, "American Prime Meridians," *Geographical Review* 32 (1942): 233–244.

2. Silvio A. Bedini, *The Jefferson Stone: Demarcation of the First United Meridian of the United States* (Frederick, MD: Professional Surveyors, 1999); Matthew Edney, "Cartographic Culture and Nationalism in the Early United States: Benjamin Vaughan and the Choice for a Prime Meridian, 1811," *Journal of Historical Geography* 20 (1994): 384–395.

3. This point is made by Matthew Edney, "Cartographic Confusion and Nationalism: The Washington Meridian in the Early Nineteenth Century," *Mapline: The Quarterly Newsletter Published by the Hermon Dunlap Center for the History of Cartography at the Newberry Library* 69–70 (1993): 3–8.

4. The emergence in late eighteenth-century America of this new geographic culture of letters is the subject of Martin Brückner, *The Geographic Revolution in Early America: Maps, Literacy and National Identity* (Chapel Hill: University of North Carolina Press, 2006).

5. On Morse, see Ralph Brown, "The American Geographies of Jedidiah Morse," *Annals of the Association of American Geographers* 31 (1941): 145–217; David N. Livingstone, "'Risen into Empire': Moral Geographies of the American Republic," in *Geography and Revolution,* ed. David N. Livingstone and Charles W. J. Withers (Chicago: University of Chicago Press, 2005), 304–335; Brückner, *Geographic Revolution in Early America,* 113–120, 151–158, 163–170. As Susan Schulten puts it, "Jedediah Morse envisioned a body of geographical knowledge that would become a foundation for civic virtue and national identity." Susan Schulten, *Mapping the Nation: History and Cartography in Nineteenth-Century America* (Chicago: University of Chicago Press, 2012), 75.

6. Jedidiah Morse, *The American Geography; or, A View of the Present Situation of the United States of America,* 2nd ed. (London: printed for John Stockdale, 1792), vi–vii, 5. Morse's reference to "London" may mean Greenwich, or it may refer to St. Paul's Cathedral in London, which was used as a prime meridian before the late 1760s and even by a few mapmakers after that time.

7. Lewis Evans, *Geographical, Historical, Political, Philosophical and Mechanical Essays. The First, Containing an Analysis of a General Map of the Middle British Colonies in America* (Philadelphia: printed by B. Franklin; Philadelphia: D. Hall, 1755), 1. The twentieth-century commentator who does not distinguish between the two types is Pratt, "American Prime Meridians." The critic of Morse's scheme for Philadelphia and his inaccuracy is James Freeman, *Remarks on The American Universal Geography* (Boston: Belknap and Hall, 1793), 6. Morse's textual emphasis upon a Philadelphia prime meridian was mirrored by its use as America's first meridian in late eighteenth-century embroidered maps, where manual dexterity was used to delineate national identity. Judith A. Tyner, *Stitching the World: Embroidered Maps and Women's Geographical Education* (Farnham, UK: Ashgate, 2015), 76.

8. Edney, "Cartographic Confusion and Nationalism"; Bedini, *Jefferson Stone,* chaps. 1–2; Pratt, "American Prime Meridians," 235.

9. Bedini, *Jefferson Stone,* especially chaps. 1–2. The reasons for Jefferson letting his earlier interest in the prime meridian drop remain unclear. Bedini notes that Lambert continued to send Jefferson his astronomical observations until 1825, the year before his death, and that Jefferson forwarded them to the American Philosophical Society in Philadelphia. As Bedini further notes, "No mention has been found in Jefferson's papers and correspondence indicating that he had communicated his plan for the establishment of an American prime meridian to anyone, neither to the Congress, nor to his particular associates in his scientific endeavors.... Nor does there seem to have been any effort to develop the project after the markers had been installed. It is probable that the pressure of current state affairs preoccupied him so fully that he then had no time to do so, and that later it may have appeared to be no longer practical" (p. 39).

10. William Lambert, *Calculations for Ascertaining the Latitude North of the Equator, and the Longitude West of Greenwich Observatory, in England, of the Capitol, at the City of Washington, in the United States of America* (Washington, DC: printed for author, A. and G. Way, 1805).

11. *American State Papers: Documents, Legislative and Executive, of the Congress of the United States, from the First Session of the Eleventh to the Second Session of the Seventeenth Congress, Inclusive: Commencing May 22, 1809, and Ending March 3, 1823* (Washington, DC: Gales and Seaton, 1834), II, 54.

12. Ibid., 53.

13. Ibid., 53–54.

14. Bedini, *Jefferson Stone,* 44. On the development of the U.S. Naval Observatory, see Steven J. Dick, *Sky and Ocean Joined: The U.S. Naval Observatory 1830–2000* (Cambridge: Cambridge University Press, 2003).

15. Benjamin Vaughan, "An Account of Some Late Proceedings in Congress Respecting the Project for Establishing a First Meridian, with Remarks," MS B/V46p, Benjamin Vaughan Papers, American Philosophical Society. This manuscript has been edited and commented upon by the leading historian of cartography, Matthew Edney, in an unpublished typescript of December 1992, also in the American Philosophical Society's collections. Vaughan intended his work to constitute a fairly substantial pamphlet, complete with appendices and supporting citations, but he never brought it to a conclusion. The original source of this manuscript is, as Edney notes, a collection of unfinished drafts, notes, and the remnants of extensive cut-and-pasting found in a file in Vaughan's papers cataloged as "Concerning the Establishment of a Universal Prime Meridian, in Reply to William Lambert's Proposal of a U.S. One." Edney has assembled the several

parts in order, expanded some abbreviations, modernized the spelling and punctuation for the sake of clarity, numbered Vaughan's footnotes, and included some interpolations of a few words within the original papers. In order, the papers are: 1) the most advanced and most complete portion of the MS, comprising the first twenty pages; 2) a fairly neat folio of four MS pages; 3) and 4) variants of item 2 that present a digression concerning the state of contemporary maps; 5) a rough draft of two MS pages; 6) a two-sided MS sheet made from two separate glued-together pages, presumably from earlier drafts and notes; this item continues from item 5; 7) and 8) odd scraps; 9) a nine-page rough draft on the question of how to determine longitude, given contemporary interests in and problems with measuring it; and 10) and 11) a single page and four pages, respectively, located at the end of the file. I cite here from the original Vaughan MS, using folio references where they exist and making reference to Edney's 1992 transcript as necessary.

16. Ibid., 2, 3, 4, 6, 11, 14.

17. Edney, "Cartographic Culture and Nationalism," 391; Vaughan, "An Account of Some Late Proceedings in Congress," fol. 21: Edney 1992 transcript, 10.

18. Vaughan, "An Account of Some Late Proceedings in Congress," fol. 6: Edney 1992 transcript, 3.

19. [Nathaniel Bowditch], "Report of the Committee to Whom Was Referred, on the 25th of January, 1810, the Memorial of William Lambert, Accompanied with Sundry Papers Relating to the Establishment of a First Meridian for the United States, at the Permanent Seat of Their Government," *Monthly Anthology and Boston Review* 9 (1810): 246–265, quotes from pp. 250, 252, 257, 265. Bowditch here is referring to Lambert's use of these terms in Lambert, *Calculations for Ascertaining the Latitude North of the Equator* (see text of Chapter 2 and Figure 2.2).

20. [William Lambert], "To the Critical Reviewers of Boston, in the State of Massachusetts," *Independent Chronicle*, December 10, 1810.

21. [Nathaniel Bowditch], "Defence of the Review of Mr. Lambert's Memorial," *Monthly Anthology and Boston Review* 10 (1811): 40–48, quotes from p. 48 (original emphasis).

22. Nathaniel Bowditch, *The New American Practical Navigator* (Newburyport, MA: Edward M. Blunt, 1802), v, vi; Nevil Maskelyne, *Tables Requisite to be used with the Astronomical and Nautical Ephemeris* (London: printed by W. Richardson and S. Clark, 1766). This work usually appears as pt. 2 of *The Nautical Almanac and Astronomical Ephemeris* (1767) and in all later editions of that work. The relationship between Moore and Bowditch and their respective navigational texts, as well as historians' later unjustified dismissal of Moore, is addressed by Charles H. Cotter, "John Hamilton Moore and Nathaniel Bowditch," *Journal of Navigation* 30 (1977): 323–326.

23. [Pierre-Simon Laplace], *Mécanique céleste*, trans. Nathaniel Bowditch, 4 vols. (Boston: 1829–1839), 4, 54. With a memoir of the translator by N. I. Bowditch. Laplace's work in its original French was in five volumes: Bowditch never completed his translation of the fifth.

24. *American State Papers*, 195.

25. Ibid., 546, 759.

26. Ibid., 753–796, quote from p. 792; Bedini, *Jefferson Stone*, 48.

27. Henry S. Tanner, *New American Atlas Containing Maps of the Several States of North American Union* (Philadelphia: H. S. Tanner, 1823), 1.

28. Ferdinand Hassler, *Principal Documents Relating to the Survey of the Coast of the United States, since 1816* (New York: William van Norden, 1834), 74 (original emphasis).

29. Richard Stachurski, *Finding North America: Longitude by Wire* (Columbia: University of South Carolina Press, 2009), 49. On the persistent theme of the longitudinal "fixing" of the observatories of Paris and Greenwich, see the discussions in Chapters 2, 4, and 7 of this book.

30. Charles H. Davis, "Upon the Prime Meridian," *Proceedings of the American Association for the Advancement of Science* 2 (1849): 78–85, quotes from pp. 78, 79.

31. Craig B. Waff, "Charles Henry Davis, the Foundation of the *American Nautical Almanac*, and the Establishment of an American Prime Meridian," *Vistas in Astronomy* 28 (1985): 61–66, quote from p. 63.

32. The members of this committee were Prof. A. D. Bache (superintendent of the U.S. Coastal Survey), Lt. M. F. Maury (superintendent of the National Observatory), Prof. Frederick A. P. Barnard (University of Alabama), Prof. Lewis R. Gibbes (Charleston, South Carolina), Prof. Edward Courtenay (University of Virginia), Prof. Stephen Alexander (Princeton University), Prof. John Frazer (University of Pennsylvania), Prof. H. J. Anderson (New York), Prof. O. M. Mitchel (Cincinnati), Prof. A. D. Stanley (Yale University), the Hon. William Mitchell of Nantucket, Prof. Joseph Lovering (Harvard University), Prof. William Smyth (Bowdoin College), Prof. George Coakley (College of St. James, Maryland), Professor Curley (Georgetown College), Prof. J. Smith Fowler (Franklin College, Tennessee), Prof. James Phillips (University of North Carolina), Prof. William H. C. Bartlett (West Point), Prof. Ebenezer S. Snell (Amherst College), Prof. Alexis Caswell (Brown University), and Lieutenant Davis himself. The letters of evidence are presented in several of the volumes of the *Proceedings of the American Association for the Advancement of Science* (see notes following). A fuller account, together with the politicians' brief summary of the legislative history concerning the prime meridian in the United States before 1850, is contained in Committee on Naval Affairs, American Prime Meridian, H.R. Rep. No. 286-31, at 1–70 (May 2, 1850) (hereafter cited as H.R. Rep. No. 286-31, with the pagination, date, and author following). For a discussion of Davis's prime meridian paper in relation to his organization of the Nautical Almanac Office, see Dick, *Sky and Ocean Joined,* 124–127.

33. H.R. Rep. No. 286-31, at 14–17 (November 1849) (copy of the "remonstrance" referred to). The total of 732 persons is calculated from the numbers given on p. 17. The quote from Burne of the Baltimore Board of Trade is from November 5, 1849, p. 13.

34. H.R. Rep. No. 286-31, at 24 (January 10, 1850) (Smyth); H.R. Rep. No. 286-31, at 44 (n.d.) (Lovering); H.R. Rep. No. 286-31, at 34 (October 16, 1849) (Coakley). The final quote here is from I. F. Holton, "On an American Prime Meridian," *Proceedings of the American Association for the Advancement of Science* 2 (1849): 381–383, quote from p. 381.

35. H.R. Rep. No. 286-31, at 50–68, quote from p. 52. (n.d. but probably January 1850) (Charles H. Davis, "Remarks upon the Establishment of an American Prime Meridian").

36. Charles H. Davis, *Remarks upon the Establishment of an American Prime Meridian* (Cambridge, MA: Metcalf, 1849), 7, 30.

37. In a commentary on his father's achievements, Davis's son, also C. H. Davis, remarked that his father's work on the American *Nautical Almanac* was his greatest achievement, "a monument . . . more enduring than brass or marble." C. H. Davis, "Memoir of Charles Henry Davis, 1807–1877" (paper read before the National Academy, April 1896, Washington, DC, privately printed, 1896), 31.

38. Joseph Lovering, "On the American Prime Meridian," *American Journal of Science and Arts* 9 (1850): 1–15, quote from p. 11.

39. "Report on the Committee on the Prime Meridian—Appointed at the Meeting of the American Association for the Advancement of Science, Held at Cambridge, August 14, 1849," *Proceedings of the American Association for the Advancement of Science* 4 (1851): 155–157, quote from p. 155.

40. Lovering, "On the American Prime Meridian," 1.

41. "Report on the Committee on the Prime Meridian," 155.

42. Bedini, *Jefferson Stone,* 58.

43. Stachurski, *Finding North America*, 165. The contemporary cited was Alexander Dallas Bache, director of the United States Coastal Survey from 1843 to 1867 (and successor to Hassler). On further details on *The American Ephemeris and Nautical Almanac* as a hybrid text, see Dick, *Sky and Ocean Joined*, 127.

44. Ian R. Bartky, *Selling the True Time: Nineteenth-Century Timekeeping in America* (Stanford, CA: Stanford University Press, 2000); Bartky, *One Time Fits All: The Campaign for Global Uniformity* (Stanford, CA: Stanford University Press, 2007); Carlene Stephens, *On Time: How America Has Learned to Live by the Clock* (Washington, DC: Smithsonian and Bullfinch Press, 2002); Peter Galison, *Einstein's Clocks and Poincaré's Maps: Empires of Time* (New York: Hodder and Stoughton, 2003).

45. Sandford Fleming, *Papers on Time-Reckoning and the Selection of a Prime Meridian to Be Common to All Nations* (Toronto, 1880), 52.

46. Sandford Fleming, *Longitude and Time Reckoning: A Few Words on the Selection of a Prime Meridian to Be Common to All Nations, in Connection with Time-Reckoning* (Toronto, 1883), 55.

47. Ibid., 61.

48. Charles P. Daly, "Annual Address. Geographical Work of the World in 1878 & 1879," *Journal of the American Geographical Society of New York* 12 (1880): 1–107, quote from p. 7.

49. Smyth to Latimer, "Letter from C. Piazzi Smyth to Charles B. Latimer," *Appeal to the Earnest and Thoughtful* 2: 178–179, quote from p. 178.

50. *What Shall Be the Prime Meridian for the World?*, report from the Committee on Standard Time and Prime Meridian, 1884, International Institute for Preserving and Perfecting Weights and Measures, Cleveland. Professor C. Piazzi Smyth's report is on pp. 10–13. Charles Piazzi Smyth, *Memorandum Requested by the Committee on Kosmic Time and Prime Meridian, Appointed by the International Institute for Preserving and Perfecting Weights and Measures* (Edinburgh: privately printed, 1883). In separate copies of this memorandum, the date "5 June 1883" is given, perhaps indicating that the work was performed in 1883 but not published until 1884. The membership of this "Committee on Standard Time" included the Rev. H. G. Wood of Sharon, Pennsylvania; Piazzi Smyth; Msr. L'Abbe F. Moigno, canon of Saint Denis in Paris; Sandford Fleming; William H. Searles of Beech Creek, Pennsylvania; Jacob Clark, civil engineer, New York; Professor Stockwell, the astronomer, from Cleveland; and Charles Latimer, president of the institute, from Cleveland.

51. Frederick A. P. Barnard, "The Imaginary Metrological System of the Great Pyramid," *School of Mines Quarterly* 5 (1884): 97–127, 193–217, 289–329, quote from p. 300.

52. Fleming, *What Shall Be the Prime Meridian for the World?*, 1.

53. H.R. Rep. No. 286-31, at 39 (October 15, 1849) (William H. C. Bartlett).

CHAPTER 3 ◦ INTERNATIONAL STANDARDS?

1. For a summary, see M. E. Himbert, "A Brief History of Measurement," *European Physical Journal Special Topics* 172 (2009): 25–35. The definitive study of the different terminologies and systems of measurement in Europe is Witold Kula, *Measures and Men*, trans. R. Szreter (Princeton, NJ: Princeton University Press, 1986). Robert P. Crease, *World in the Balance: The Historic Quest for an Absolute System of Measurement* (New York: W. W. Norton, 2011) is less detailed in its historical evidence than Kula but offers non-European examples. On Britain's weights and measures, see R. D. Connor, *The Weights and Measures of England* (London: Her Majesty's Stationery Office, 1987); Julian Hoppit, "Reforming Britain's Weights and Measures, 1660–1824," *English Historical Review* 108 (1993): 82–104; R. D. Connor and Allan D. C. Simpson, *Weights and Measures in*

Scotland: A European Perspective (Edinburgh: National Museums of Scotland in association with Tuckwell Press, 2004). The example of Arthur Young is taken from John Heilbron, "The Measure of Enlightenment," in *The Quantifying Spirit in the Eighteenth Century,* ed. Tore Frängsmyr, John Heilbron, and Robin Rider (Berkeley: University of California Press, 1990), 207–242. The term "metrology" was first coined in 1816: see Patrick Kelly, *Metrology; or, An Exposition of Weights and Measures, Chiefly Those of Great Britain and France* (London: printed for the author by J. Whiting, 1816).

2. The quotes from Pierre-Simon Laplace are from his *The System of the World,* 2 vols., trans. J. Pond (London: Richard Phillips, 1809), 1:37, 135, 152. They are also cited in Juan Pastorin, "Remarks on a Universal Prime Meridian," *Proceedings of the Canadian Institute* 2 (1885): 49–51, from p. 50.

3. Jean-Alexandre Carney, "Mémoire sur un premier méridien universel, et sur une ère universelle, a laquelle il se lierait," *Recueil des bulletins, Publiés par la Société libre des Sciences et Belles-Lettres de Montpellier* 1, no. 11 [1803]: 19–24. This first volume contains bulletins 1 to 14: Carney's paper appears in bulletin 1 and, with further discussion of it as part of a section devoted to questions of geography, in bulletin 12, 280–284.

4. Jean-Pierre Martin and Anita McConnell, "Joining the Observatories of Paris and Greenwich," *Notes and Records of the Royal Society* 62 (2008): 355–372; Michael Kershaw, "A Different Kind of Longitude: The Metrology and Conventions of Location by Geodesy," in *Navigational Enterprises in Europe and its Empires, 1730–1850,* ed. Richard Dunn and Rebekah Higgitt (London: Palgrave Macmillan, 2015), 134–158.

5. Henry Kater, "An Account of Trigonometrical Operations in the Years 1821, 1822 and 1823, for Determining the Difference of Longitude between the Royal Observatories of Paris and Greenwich," *Philosophical Transactions of the Royal Society of London* (hereafter cited as *Philosophical Transactions*) 118 (1828): 153–239, quotes from pp. 153, 154, 156, 157, 158, 160–161. On Kater's work in India, see Matthew H. Edney, *Mapping an Empire: The Geographical Construction of British India, 1765–1843* (Chicago: University of Chicago Press, 1998), 182, 183, 247. The 1821 paper referred to is Henry Kater, "An Account of the Comparison of Various British Standards of Linear Measure," *Philosophical Transactions* 111 (1821): 75–94.

6. Kater, "An Account of Trigonometrical Operations," 184, 192–193. Kater was incorrect in noting that the work of Arago and his French colleagues had not been published by then: it had been, but its appearance among British intellectual circles was delayed. See Jean-Baptiste Biot, François Arago, and Veuve de Louis Courcier, *Recueil d'observations géodésiques, astronomiques et physiques, exécutées par ordre du Bureau des Longitudes de France, en Espagne, en France, en Angleterre et en Ecosse, pour déterminer la variation de la pesanteur et des degrés terrestres sur le prolongement du méridien de Paris, faisant suite au troisième volume de la base du système métrique* (Paris: Mme Ve Courcier, 1821).

7. J. F. W. Herschel, "Account of a Series of Observations Made in the Summer of the Year 1825, for the Purpose of Determining the Difference of Meridians of the Royal Observatories of Greenwich and Paris," *Philosophical Transactions* 116 (1826): 77–126, quotes from pp. 81, 126. For a summary, see Christopher Wood, "The Determination of the Difference in Meridians of the Paris and Greenwich Observatories," *Antiquarian Horology* 24 (1998): 234–236.

8. Thomas Henderson, "On the Difference of Meridians of the Royal Observatories of Greenwich and Paris," *Philosophical Transactions* 117 (1827): 286–296, quotes from pp. 286, 287, 295. Thomas Henderson (1798–1844) was the first Astronomer Royal for Scotland. John Pond was also engaged in work on positioning the Royal Observatory, of which he was in charge. It is unclear how this work was connected to the error noted by Thomas Henderson and transmitted to Herschel: see John Pond, "On the Latitude of the Royal Observatory of Greenwich," *Memoirs of the Astronomical Society of London* 2 (1826): 317–319.

9. Edward J. Dent, "On the Difference of Longitude between the Greenwich and Paris Observatories," *Memoirs of the Royal Astronomical Society* 11 (1840): 69–72, quotes from p. 70.

10. There is a considerable literature on metrology, its social dimensions, and its intellectual and institutional history. This paragraph is drawn from Andrew Barry, "The History of Measurement and the Engineers of Space," *British Journal for the History of Science* 26 (1993): 459–468; Simon Schaffer, "Modernity and Metrology," in *Science and Power: The Historical Foundations of Research Policies in Europe,* ed. Luca Guzzetti (Luxembourg City, Luxembourg: European Communities, 2000), 71–91; Schaffer, "Metrology, Metrication, and Victorian Values," in *Victorian Science in Context,* ed. Bernard Lightman (Chicago: University of Chicago Press, 1997), 438–474; Schaffer, "Late Victorian Metrology and Its Instrumentation: A Manufactory of Ohms," in *Invisible Connexions: Instruments, Institutions and Science,* ed. Robert Bud and Susan Cozzens (Bellingham, WA: SPIE Press, 1995), 23–56; Joseph O'Connell, "The Creation of Universality by the Circulation of Particulars," *Social Studies of Science* 23 (1993): 129–173; Martha Lampland and Susan Leigh Star, eds., *Standards and Their Stories: How Quantifying, Classifying and Formalizing Shape Everyday Life* (Cornell, NY: Cornell University Press, 2008). The quote is from Schaffer, "Modernity and Metrology," 91.

11. Connor, *Weights and Measures of England,* 345. Connor notes that it is uncertain whether this invitation was ever sent.

12. Hoppit, "Reforming Britain's Weights and Measures" provides details of the variation in standards of weights and measures in the early modern period. For discussions in the Royal Society in the 1740s over English standard units and English and French standards, see Anonymous, "An Account of the Proportions of the English and French Measures and Weights, from the Standards of the Same, kept at the Royal Society," *Philosophical Transactions* 42 (1742–1743): 185–188; Anonymous, "An Account of a Comparison Lately Made by Some Gentlemen of the Royal Society, of the Standard of a Yard, and the Several Weights Lately Made for Their Use; With the Original Standards of Measures and Weights in the Exchequer, and Some Others Kept for Public Use, at Guild-Hall, Founders-Hall, the Tower, &c," *Philosophical Transactions* 42 (1742–1743): 541–556. Thomas Williams's discussion of the earth's dimensions and a possible new metrology is in his *Method to Discover the Difference of the Earth's Diameters, Likewise a Method for Fixing an Universal Standard for Weights and Measures* (London: John Stockdale, 1788), quote from p. 73. On the debates in the Royal Society at the end of the eighteenth century, see George Shuckburgh Evelyn, "An Account of Some Endeavours to Ascertain a Standard of Weights and Measures," *Philosophical Transactions* 88 (1798): 133–182. Sir James Steuart's remarks are from his "A Plan for Introducing an Uniformity of Weights and Measures over the World and for Facilitating the More Speedy Accomplishment of Such a Scheme within the Limits of the British Empire," in *The Works, Political, Metaphysical and Chronological; Collected by His Son General Sir James Steuart,* 6 vols. (London: T. Cadell and W. Davies, 1805), 5:379–415, quotes from pp. 382, 387, 397.

13. Kater's principal works in this respect and upon which these paragraphs are based are as follows: Henry Kater, "An Account of Experiments for Determining the Length of the Pendulum Vibrating Seconds in the Latitude of London," *Philosophical Transactions* 108 (1818): 33–102; "On the Length of the French Meter Estimated in Parts of the English Standard," *Philosophical Transactions* 108 (1818): 103–109; "An Account of Experiments for Determining the Variation in the Length of the Pendulum Vibrating Seconds, at the Principal Stations of the Trigonometrical Survey of Great Britain," *Philosophical Transactions* 109 (1819): 337–508; "An Account of the Comparison of Various British Standards of Linear Measure," *Philosophical Transactions* 111 (1821): 75–94; "An Account of the Re-Measurement of the Cube, Cylinder, and Sphere, Used by the Late Sir George Shuckburgh Evelyn, in His Enquiries Respecting a Standard of Weights and Measures," *Philosophical Transactions* 111 (1821): 316–326; "An Account of the Construction and Adjustment of the New Standards of Weights and Measures of the United Kingdom of Great Britain and Ireland," *Philosophical Transactions* 116 (1826): 1–52; "On the Error in Standards in Linear

Measure, Arising from the Thickness of the Bar on Which They are Traced," *Philosophical Transactions* 120 (1830): 359–381; "An Account of the Construction and Verification of a Copy of the Imperial Standard Yard Made for the Royal Society," *Philosophical Transactions* 121 (1831): 345–347. On Kater's royal commission and later work, see Connor, *Weights and Measures of England*, 253–261.

14. George Biddell Airy, "Account of the Construction of the New National Standard of Length, and of Its Principal Copies," *Philosophical Transactions* 147 (1857): 621–705. Airy provided an abstract of his fuller account for the Royal Society in May 1857 (see *Philosophical Transactions* 147 (1857): 530–534; Connor, *Weights and Measures of England*, 261–272.

15. The quote from Herschel is from his *Outlines of Astronomy* (London: Longman, Brown, Green and Longmans, and John Taylor, 1849), 87. On the legislative history of metrication in nineteenth-century Britain, see Connor, *Weights and Measures of England*, 279–288. As Connor notes (p. 284), the Metric (Weights and Measures) Act of 1864 "only legalized the use of metric terms in contracts but not the use of metric units for trading purposes. The metre, kilogram, and litre could be written about in commercial undertakings but metre bars could not be used to measure anything for sale, nor could kilogram weights be used to weigh anything in the market-place. They could be used for scientific purposes but not for business."

16. This and the following paragraphs are based on: Connor, *Weights and Measures of England*, app. C, "A Brief Account of the Metric System," 344–357 (from whom I take the words "utterly chaotic" to describe metrology in pre-Revolutionary France); Kula, *Measures and Men*, chaps. 23–24; Ken Alder, "A Revolution to Measure: The Political Economy of the Metric System in France," in *The Values of Precision*, ed. M. Norton Wise (Princeton, NJ: Princeton University Press, 1995), 39–71; Alder, *The Measure of All Things: The Seven-Year Odyssey That Transformed the World* (London: Little, Brown, 2002); Paul Murdin, *Full Meridian of Glory: Perilous Adventures in the Competition to Measure the Earth* (New York: Springer, 2009), chaps. 5–6. Also useful is Heilbron, "Measure of Enlightenment."

17. On the significance of different toise in French metrology and terrestrial measurement, see Kershaw, "A Different Kind of Longitude."

18. The quote is from Connor, *Weights and Measure of England*, 348.

19. There is a large literature on this topic from a variety of disciplinary perspectives. For a good summary and a discussion of the "annihilation of space and time" that addresses the geographical dimensions and technological bases of the impact of telegraphy in the nineteenth century, see Roland Wenzlhuemer, *Connecting the Nineteenth-Century World: The Telegraph and Globalization* (Cambridge: Cambridge University Press, 2013).

20. Herschel, "Account of a Series of Observations," 107. In a footnote, Herschel asked, "Might not telegraphs be employed to ascertain the difference of longitudes of the stations between which they are established?"

21. Reports to the Board of Visitors: The Printed Reports of George Airy, Astronomer Royal, to the Board of Visitors, 1836–1857, June 5, 1852, p. 5, RGO 17/1/1, Cambridge University Library.

22. Ibid.; Printed Reports of George Airy, 1836–1857, June 3, 1854, p. 14, RGO 17/1/1.

23. Printed Reports of George Airy, 1858–1870, June 2, 1860, p. 19, RGO 17/1/2.

24. This evidence is taken from Alexander R. Clarke, *Comparisons of the Standards of Length of England, France, Belgium, Prussia, India, Australia, Made at the Ordnance Survey Office, Southampton* (London: Eyre and Spottiswoode, 1866), quotes from pp. vii and viii. The work is attributed to Clarke, but the words are James's and the date is August 10, 1866. For a full account, see *Account of the Observations and Calculations, of the Principal Triangulation; and of the Figure, Dimensions and Mean Specific Gravity, of the Earth as Derived Therefrom* [Drawn up by Captain Alexander Ross

Clarke under the direction of Lt. Col. H. James] (London: Eyre and Spottiswoode, 1858). The different units of length measured in the Ordnance Survey's "Bar Room" were the Russian standard double toise; Prussian standard toise; Belgian standard toise; platinum meter of the Royal Society, compared with the standard meter of France, by M. Arago; English standard yard [eight different specimens of the same standard being measured]; Ordnance Survey ten-feet standard bar; Indian ten-feet standard bar, new and old; Australian ten-feet standard bar; and the ten-feet standard bar of the Cape of Good Hope.

25. Printed Reports of George Airy, 1858–1870, June 1, 1867, p. 21, RGO 17/1/2.

26. Printed Reports of George Airy, 1871–1881, June 3, 1871 (p. 19) to June 5, 1880 (p. 18), RGO 17/1/3. The quote from Bache is from Richard Stachurski, *Finding North America: Longitude by Wire* (Columbia: University of South Carolina Press, 2009), 159.

27. On metrology and time's standardization, I have drawn from Elisa Arias, "The Metrology of Time," *Philosophical Transactions* 363 (2005): 2289–2305; Eviatar Zerubavel, "The Standardization of Time: A Sociohistorical Perspective," *American Journal of Sociology* 88 (1982): 1–23. On timekeeping in early modern Britain, see Paul Glennie and Nigel Thrift, *Shaping the Day: A History of Timekeeping in England and Wales 1300–1800* (Oxford: Oxford University Press, 2009). On the geographies and cultural practices of time, see the essays in John Hassard, ed., *The Sociology of Time* (Basingstoke, UK: MacMillan, 1990); Robert Levine, *A Geography of Time: The Temporal Misadventures of a Social Psychologist, Or How Every Culture Keeps Time Just a Little Bit Differently* (New York: Basic Books, 1997). My point on the effect of railways upon conceptions of time and space in the nineteenth century draw from Wolfgang Schivelbusch, *The Railway Journey: The Industrialization of Time and Space in the Nineteenth Century* (Leamington Spa, UK: Berg, 1986); Wenzlhuemer, *Connecting the Nineteenth-Century World,* 31–34, 59–62. On industrial capitalism and new time-based work routines, see E. P. Thompson, "Time, Work-Discipline, and Industrial Capitalism," *Past and Present* 38 (1967): 59–97. On the regulation of time in the nineteenth century, I am indebted to Ian R. Bartky, *Selling the True Time: Nineteenth-Century Timekeeping in America* (Stanford, CA: Stanford University Press, 2000); Bartky, *One Time Fits All: The Campaigns for Global Uniformity* (Stanford, CA: Stanford University Press, 2007); Peter Galison, *Einstein's Clocks, Poincaré's Maps* (London: Hodder and Stoughton, 2003).

28. Anonymous, "Greenwich Time," *Blackwood's Edinburgh Magazine* 63 (March 1848): 354–361, quotes from p. 355. The observation of Greenwich's time signal in 1874 is from Printed Reports of George Airy, 1871–1881, June 6, 1874, p. 17, RGO 17/1/3. On Greenwich's coordinating role with respect to time, see Howse, *Greenwich Time and the Longitude,* 91–94.

29. Zerubavel, "Standardization of Time," 7–10; Bartky, *Selling the True Time,* passim; Bartky, *One Time Fits All,* passim.

30. On these matters, see Schaffer, "Metrology, Metrication, and Victorian Values," 438–474; Crease, *World in the Balance,* 128–142.

31. This is the claim made by H. A. Brück and M. T. Brück, *The Peripatetic Astronomer: The Life of Charles Piazzi Smyth* (Philadelphia: Hilger, 1988). They note: "The immediate reason for Piazzi Smyth's conversion to Taylor's ideas was his reading of a second pamphlet of John Taylor's called 'The battle of the standards,' published in 1864, which dealt with the problem of the use of proper weights and measures, claiming the superiority of the ancient British standards and their link with the sacred system of the Bible" (p. 100).

32. Charles Piazzi Smyth, *Present State of the Longitude Question in Navigation* (Edinburgh: Chamber of Commerce, 1859), 61. He referred to changes in chronometry (his own time gun of 1861, which was (and still is) fired every day at 1:00 p.m. from the ramparts of Edinburgh Castle) as "the fourth age of Marine Longitude" (pp. 56–57).

33. C. P. Smyth correspondence, MS. A13.59, Royal Observatory Edinburgh. Both letters (as do most others) date from 1877 and so probably relate to the third edition of *Our Inheritance in the Great Pyramid*, which was published in that year.

34. J. F. W. Herschel, pamphlet, "Two Letters to the Editor of *The Athenaeum* on a British Modular Standard of Length," n.d. The copy I examined of this pamphlet, in the archives of the Royal Observatory Edinburgh, has marginalia in Piazzi Smyth's hand. The quote, "It is simply this . . ." has the word "Italic" alongside, indicating Piazzi Smyth's endorsement of the point. See also Brück and Brück, *Peripatetic Astronomer,* 100–101.

35. Charles Piazzi Smyth, *Life and Work at the Great Pyramid,* 3 vols. (Edinburgh: Edmonston and Douglas, 1867), 1:xii. The critic was Frederick A. P. Barnard, "The Imaginary Metrological System of the Great Pyramid," *School of Mines Quarterly* 5 (1884) 1:97–127, quote from p. 103. Barnard was president of the American Metrological Society and offered his lengthy dismissal of the claims of Smyth and the pyramidologists in a paper read before that society on December 27, 1883. Pt. 2 and pt. 3 of his paper appear in this journal and volume, pp. 193–217 and 289–329, respectively. Barnard was an advocate of the metric system: see his *The Metric System of Weights and Measures* (Boston: American Metrics Bureau, 1879).

36. The reports of what was, from the outset, a strongly prometric lobby in the British Association for the Advancement of Science feature in reports of its annual meetings from 1864 to 1869; for example, "Report on the Best Means of Providing for a Uniformity of Weights and Measures, with Reference to the Interests of Science," *Report of the Thirty-Fourth Meeting of the British Association for the Advancement of Science* (London: John Murray, 1865), 375–378; Charles K. Davies, *The Metric System: Considered with Reference to Its Introduction into the United States; Embracing the Reports of the Hon. John Quincy Adams, and the Lecture of Sir John Herschel* (New York: A. S. Barnes, 1874); Crease, *World in the Balance,* 132–142.

37. Charles Piazzi Smyth, *Memorandum Requested by the Committee on Kosmic Time and Prime Meridian, Appointed by the International Institute for Preserving and Perfecting Weights and Measures* (Edinburgh: privately printed, 1883).

38. Anonymous, "Professor C. Piazzi Smyth's Report," *What Shall Be the Prime Meridian for the World?* [Report of Committee on Standard Time and Prime Meridian, International Institute for Preserving and Perfecting Weights and Measures] (Cleveland, 1884), pp. 10–13.

39. [International Institute for Preserving Weights and Measures], *Appeal to the Earnest and Thoughtful, and Especially to the Members of the International Institute for Preserving Weights and Measures* (Cleveland, 188[8]), 2.

40. Schaffer, "Metrology, Metrication and Victorian Values," 457–459; William M. F. Petrie, *Inductive Metrology; or, the Recovery of Ancient Measures from the Monuments* (London: H. Saunders, 1877); Melancthon W. H. Lombe Brooke, *The Great Pyramid of Gizeh: Its Riddle Read, It Secret Metrology Fully Revealed as the Origin of British Measures* (London: H. Banks and Son, 1908).

CHAPTER 4 ∘ GLOBALIZING SPACE AND TIME

1. Msr. Roux de Rochelle, "Mémoire sur la fixation d'un premier méridien, lu à la Société de Géographie, le 4 Octobre 1844," *Bulletin de la Société de Géographie,* troisième série 3 (1845): 145–153; L. E. Sédillot, "Longitude, latitude; premiers méridiens," *Bulletin de la Société de Géographie,* quatrième série 1 (1851): 167–172; L. E. Sédillot, "Appel aux gouvernements des principaux États de L'Europe et de l'Amérique pour l'adoption d'un premier méridien commun dans l'énonciation des longitudes terrestres," *Bulletin de la Société de Géographie,* quatrième série 1 (1851): 193–205; Edmé-François Jomard, "Lettre de M. Jomard sur la même sujet (méridien universel)," *Bulletin*

de la Société de Géographie, quatrième série 1 (1851): 206–209; Antoine d'Abbadie, "Lettre de M. Antoine d'Abbadie sur le même sujet (méridien universel)," *Bulletin de la Société de Géographie,* quatrième série 1 (1851): 210–211; L. E. Sédillot, "Lettre de M. A. Sédillot," *Bulletin de la Société de Géographie,* cinquième série 12 (1866): 408–410; Anonymous, "Meridién origine des longitudes." *Bulletin de la Société de Géographie,* sixième serie 8 (1874): 241–246.

2. Henry James, *On the Rectangular Tangential Projection of the Sphere and Spheroid . . . for a Map of the World on the Scale of Ten Miles to an Inch* (Southampton: Ordnance Survey Office, 1868), 3.

3. This is a central theme of Ian Bartky's discussion of the prime meridian and particularly of the emergence of standard time in America and globally in the later nineteenth century. Ian R. Bartky, *One Time Fits All: The Campaigns for Global Uniformity* (Stanford, CA: Stanford University Press, 2007), especially chaps. 2–5. Although I draw on Bartky's analysis, my interpretation differs from his in several ways, as is clear from my text.

4. Elisabeth Crawford, "The Universe of International Science, 1880–1939," in *Solomon's House Revisited: The Organisation and Institutionalization of Science,* ed. Tore Frängsmyr (Canton, MA: Science History, 1990), 251–269.

5. This assessment of Struve's 1870 paper is based on Otto W. Struve, "Du premier méridien," *Bulletin de la Société de Géographie* 9 (1875): 46–64. This is a translation of Struve's Russian original "O pervom meridiane," *Izvestiya Imperatorskogo Russkogo Obschestva* 6 (1870): 14–34. See also Bartky, *One Time Fits All,* 37–40.

6. Bartky, *One Time Fits All,* 40.

7. V. de Saint-Martin, "Premier Méridien," in *L'année géographique revue annuelle, neuvième et dixième années (1870–1871),* ed. V. de Saint-Martin (Paris: Libraire Hachette, 1872), 442–444.

8. The literature on the history of geography in these terms is voluminous. For this summary I have made use of David Livingstone, *The Geographical Tradition: Episodes in the History of a Contested Enterprise* (Oxford: Blackwell, 1992); Gary S. Dunbar, ed., *Geography: Discipline, Profession and Subject since 1870* (Dordrecht, Netherlands: Kluwer Academic, 2000); Helene Blais and Isabelle Laboulais, eds., *Géographies plurielles: Les sciences géographiques au moment de l'émergence des sciences humaines (1750–1850)* (Paris: L'Harmattan, 2006); Karin Morin, *Civic Discipline: Geography in America, 1860–1890* (Farnham, UK: Ashgate, 2011). For a more detailed account of geography's institutionalized development as a civic and an academic science in nineteenth-century Britain, see Charles W. J. Withers, *Geography and Science in Britain 1831–1939: A Study of the British Association for the Advancement of Science* (Manchester, UK: Manchester University Press, 2010). The figures given for the foundation of geographical societies before and after 1871 are from J. S. Keltie and H. R. Mill, *Report of the Sixth International Geographical Congress* (London: John Murray, 1895), xiii.

9. The groups or sections at Antwerp were Geography; Cosmography; Ethnography; and the all-embracing Navigation, Voyages, Commerce, Meteorology, and Statistics. Discussion also took place across these groups. For the six meetings in Antwerp in 1871 that discussed the prime meridian, see *Compte-rendu du Congrès des Sciences Géographiques, Cosmographiques et Commerciales tenu a Anvers du 14 au 22 Aout 1871. Tome second* (Antwerp, Belgium: Gerrits and Van Merlen, 1872), 1:176, 183, 184, 206–209, 381; 2:234, 254–257.

10. [E. Ommaney], "Additional Notices," *Proceedings of the Royal Geographical Society* 16 (1871–1872): 134.

11. I have found no trace of manuscript records of the meeting and so take the printed record— inevitably a summary—as the only record of proceedings. On the discussion following Omanney's question, see *Compte-rendu du Congrès des Sciences Géographiques,* 1:206–208; *Compte-rendu du Congrès des Sciences Géographiques,* 2:254–255.

12. Ibid., 2:255–256.

13. G. Visconti, "Du premier méridien, par Otto Struve," *Bulletin de la Société de Géographie* 9 (1875): 46–64; A. Germain, "Le premier meridien et la *Connaissance des temps,*" *Bulletin de la Société de Géographie* 9 (1875): 504–521; Bartky, *One Time Fits All,* 43–44.

14. Committee Minute Book, September 1872–October 1877, December 17, 1874, fol. 136, Royal Geographical Society (with the Institute of British Geographers).

15. This summary and that of the preceding paragraph is based on *Congrès International des Sciences Géographiques tenu à Paris du Ier au 11 Août 1875: Compte rendu des séances,* 2 vols. (Paris: Sociéte de Géographie, 1880), 1:26–27, 29, 30; 2:400–402. See also Bartky, *One Time Fits All,* 45–46.

16. A. Salomon, F. de Morsier, and L.-H. de Laharpe, "Mémoire sur la fixation d'un premier méridien," *Mémoires de la Société de Géographie de Genève* 40 (1875): 87–94.

17. This paragraph is based on the several papers on this issue by Henri Bouthillier de Beaumont: "Le méridien unique," *L'exploration* 1 (1877): 131–132; "Choix d'un méridien initial," *L'exploration* 7 (1879): 132–136; "Note [d'un méridien initial]," *Le globe* 18 (1879): 202–208; and, most important, his 1880 pamphlet *Choix d'un méridien initial unique* (Geneva: Libraire Desrogis, 1880). See also Bartky, *One Time Fits All,* 47; H. M. Smith, "Greenwich Time and the Prime Meridian," *Vistas in Astronomy* 20 (1976): 219–229. On de Beaumont's own discussion of the 1875 Paris congress, see de Beaumont, "Quelques mots sur l'exposition Géographique de Paris," *Bulletin de la Société de Géographie de Genève* 40 (1875): 210–226, especially pp. 210–216.

18. On the 1876 Brussels meeting and its colonial originating contact and legacy, see Sandford Bederman, "The 1876 Brussels Geographical Conference and the Charade of European Cooperation in African Cooperation," *Terrae Incognitae* 21 (1989): 63–73. That de Beaumont was intending to give his paper on the Bering Strait prime meridian was reported upon in *Le globe,* the journal of the Geographical Society of Geneva, at a meeting in February 1876: *Le globe* 15 (1876): 22. The society discussed the issue the previous year, reporting that a common initial meridian and "point of [longitudinal] departure" was in the best interests of all nations: *Le globe* 14 (1875): 216–217. The discussion of delegates to the 1879 International Congress on Commercial Geography over a single prime meridian for commercial reasons and as an advantage in maritime affairs is summarized in *Bulletin de la Société Belge de Géographie* 3 (1879): 592–596; E. Cortambert, "Selecting a First Meridian," *Popular Science Monthly* 15 (June 1879): 156–159, quote from p. 159. That this article was translated from the French periodical *La nature* attests to the shared interest in the question across national boundaries but within the same type of scientific periodical.

19. C. P. Daly, "Annual Address. Geographical Work of the World in 1878 and 1879," *Journal of the American Geographical Society of New York* 12 (1880): 1–107, quotes from pp. 7, 8. Mayor Daly's geographical work more generally is the subject of Morin, *Civic Discipline.* I am grateful to Professor Morin for drawing this reference to my attention.

20. Bartky, *One Times Fits All,* 59.

21. G. M. Wheeler, *Report upon the Third International Geographical Congress and Exhibition at Venice, Italy, 1881* (Washington, DC: Government Printing Office, 1883), 30–31. In his listing of the prime meridians used in topographical mapping, Wheeler has a footnote to the effect that "Spain at different epochs has counted longitude from no less than eleven distinct and separate meridians" (p. 30). Further evidence of the many prime meridians then in use can be found in different genres of topographic mapping. In the scientific cartography that informed his popular journal, *Geographische mitteilungen,* August Petermann used either the Paris or the Greenwich prime meridian before favoring Greenwich after 1885. See Jan Smits, *Petermann's Maps: Carto-bibliography of the Maps in* Petermann's Geographische Mitteilungen *1885–1945* (t'Goy, Utrecht: HES and De Graaf, 2004). By contrast, in the limited number of topographical maps of the region available to

British military authorities in the Crimean War (1853–1856), the prime meridian employed was commonly that of Ferro. See Daniel Foliard, *Dislocating the Orient: British Maps and the Making of the Middle East, 1854–1921* (Chicago: University of Chicago Press, 2017), 35.

22. S. Fleming, *The Adoption of a Prime Meridian to Be Common to All Nations. The Establishment of Standard Meridians for the Regulation of Time, Read before the International Geographical Congress at Venice, September, 1881* (London: Waterlow and Sons, 1881), quotes from pp. 4, 6, 7, 13.

23. Wheeler, *Report upon the Third International Geographical Congress*, 26. For the three papers in full, see *Terzo Congresso Geografico Internazionale tenuto a Venezia dal 15 al 22 Settembre 1881, notizie e rendiconti*, vol. 1 (Rome: Società Geografica Italiana, 1882); *Terzo Congresso Geografico Internazionale tenuto a Venezia dal 15 al 22 Settembre 1881, communicazioni e memori*, vol. 2 (Rome: Società Geografica Italiana, 1884). The Barnard-Daly, Hazen-Wheeler, and Fleming papers appear in vol. 2, pp. 7–9, 10–18, 18–19.

24. Fleming, *Adoption of a Common Prime Meridian*, 15.

25. Otto Struve to Sandford Fleming, January 23, 1881, MG 29-B1, fol. 332, Sandford Fleming Papers, Library and Archives Canada.

26. This is taken from the "List of Views Issued by Each Group and Not Presented at the General Sessions," in *Terzo Congresso Geografico Internazionale tenuto a Venezia* 1:392; Wheeler, *Report upon the Third International Geographical Congress*, 23. See also Bartky, *One Time Fits All*, 66–67. For the first expression of A.-E. Béguyer de Chancourtois's proposal, see his "Programme d'un système de géographie," *Bulletin de la Société de Géographie*, series VI 8 (1874): 270–336. His paper to Venice in 1881 appears as "L'adoption d'un méridien initial international," *Terzo Congresso Geografico Internazionale tenuto a Venezia* 1 (1882): 20–22. See also his *Observation au sujet de la circulaire du Gouvernement des États Unis, concernant l'adoption d'un méridien initial commun et d'une heure universelle* (Paris: privately printed, 1883).

27. George Wheeler to Sandford Fleming, March 2, 1882, vol. 53, fol. 365, ff. 1–3, MG 29-B1, Sandford Fleming Papers.

28. Wheeler, *Report upon the Third International Geographical Congress*, 27–28; Oscar Meyer, ed., *The Third International Geographical Congress of Venice: A Short Account by Oscar Meyer, Commissioner for New South Wales* (Florence, 1882), 38. The links between Wheeler and from Teano to the Italian government are clear from Wheeler's letter to Fleming (see note 27), though there is no suggestion that Wheeler was exerting undue influence upon the Italians. A rather uncomplimentary view of the Venice meeting is provided by the Austrian explorer Gustav Kreitner, who, although awarded a medal for his Central Asian explorations, considered the meeting badly organized, the academic elements overtaken by protocol at the expense of substance, and the trade element far too dominant: G. Kreitner, *Report of the Third International Geographical Congress, Venice* (Venice, 1882), 8–9.

29. Bartky, *One Time Fits All*, 50.

30. The claim by Fleming that his interest in the irregularities of timekeeping was prompted by personal experience in Ireland (missing a train owing to incorrect scheduling and errors in printed timetables) is dismissed as "myth" by Bartky, *One Time Fits All*, 50–51. Where it is listed at all, *Terrestrial Time* is sometimes dated as 1876, but for the reasons explored by Bartky, it dates to 1878. As Bartky notes, Fleming published three almost identical pamphlets to this with a similar title: *Terrestrial Time* (March 1878), *Temps terrestre: Mémoire* (August 1878), and *Uniform Non-Local Time (Terrestrial Time)* (November 1878). The variations in this pamphlet are not crucial to my argument and so are not discussed here. See Bartky, *One Time Fits All*, 51–56, 221n11. See also M. Creet, "Sandford Fleming and Universal Time," *Scientia Canadensis* 14 (1990): 66–89; Sandford Fleming, *Terrestrial Time* (London, 1876), 36.

31. The two pamphlets were published together, with continuous pagination, as S. Fleming, *Papers on Time-Reckoning and the Selection of a Prime Meridian to Be Common to All Nations: Transmitted to the British Government by His Excellency the Governor-General of Canada* (Toronto: Copp, Clark, 1879). In order to identify the pamphlet in question, I have here used the relevant separate title but retained the continuous pagination of the composite work.

32. Sandford Fleming, *Time-Reckoning and the Selection of a Prime Meridian to Be Common to All Nations* (Toronto: Copp, Clark, 1879), quotes from pp. 12, 28, 29.

33. Sandford Fleming, *Longitude and Time-Reckoning* (Toronto: Copp, Clark, 1879), quotes from pp. 55, 56.

34. Fleming, *Longitude and Time-Reckoning,* 60. It is not clear with respect to the data presented in Tables 4.2 and 4.3 how Fleming distinguished between ships or tonnage from countries that used more than one prime meridian (see, in Table 4.2, Germany, for example, whose mariners used three different first meridians) and how such distinctions were then aggregated to give the relative percentages by different prime meridian (Table 4.3). The claim that the illustration represents the first integration of Fleming's time and longitude proposals is taken from Bartky, *One Time Fits All,* 57.

35. Ian R. Bartky, *Selling the True Time: Nineteenth-Century Timekeeping in America* (Stanford, CA: Stanford University Press, 2000); Bartky, *One Time Fits All,* 60–61.

36. Fleming, *Papers on Time-Reckoning and the Selection of a Prime Meridian.* See also Fleming, *Longitude and Time-Reckoning.* The memorial from the Canadian Institute is paginated as pp. 1–7 in this cumulative work; quotes are from pp. 6, 7.

37. On the initial request, sent from Downing Street to the Royal Geographical Society, see Colonial Office Covering Letters, Edward Wingfield, August 1879, MSS CB6/511, Royal Geographical Society (with Institute of British Geographers) Archives. On the response, see Sandford Fleming, "Sir G. B. Airy, Astronomer Royal, Greenwich, to the Secretary of State for the Colonies (18 June 1879)," in *Universal or Cosmic Time: Together with Other Papers, Communications and Reports in the Possession of the Canadian Institute Respecting the Movement for Reforming the Time-System of the World, and Establishing a Prime Meridian as a Zero Common to All Nations* (Toronto: Copp, Clark, 1885), 32–34. This publication is a collection of twenty-nine different documents on the prime meridian and universal time, dated between May 1879 and March 1885 and prepared by the Canadian Institute. The composite nature of this 1885 work is important in demonstrating the role of the Canadian Institute in promoting the issues, in bringing together many of Fleming's shorter papers and lectures (including his 1881 Venice paper, pp. 56–63), and in allowing us to see how Fleming's involvement in the 1884 Washington conference was colored by others' interests (see Chapter 5 of this book).

38. The papers were sent to scientific bodies and organizations in France, Germany, Italy, Norway, Sweden, the United States, and Russia (each received eight copies), as well as to Austria, Belgium, Brazil, Denmark, Japan, the Netherlands, Portugal, Spain, Switzerland, Turkey, Greece, and China (each received four copies). In Great Britain the papers were sent to the Admiralty; to Airy as Astronomer Royal; to Charles Piazzi Smyth, the Astronomer Royal for Scotland; to the Royal Astronomical Society; to the Royal Geographical Society; to the Royal United Service; and to the Royal Society.

39. The responses of the British bodies are given in the "Supplementary Papers, Communications and Reports" that make up much of Fleming, *Universal or Cosmic Time*: the Admiralty, 38; the Royal Society, 39; Piazzi Smyth, 35. See Gen. Sir J. H. Lefroy's response, November 19, 1879, MSS CB8/1377, Royal Geographical Society (with the Institute of British Geographers) Archives; Bartky, *One Time Fits All,* 64.

40. The societies in Europe to whom Wilson's second memorial, Fleming's 1879 papers, and Cleveland Abbe's "Report on Standard Time" on behalf of the American Metrological Society were

NOTES TO PAGES 171-175

sent in May 1880 were: The Institut de France and the Société de Géographie (both in Paris); the Société Belge de Géographie (in Brussels); the Königliché de Preussische Akademie der Wissenschaften and the Gesellschaft für Erdkunde (both in Berlin); the Kaiserliche Akademie der Wissenschaften and the Kaiserliche Geographische Gessellschaft (both in Vienna); the Nicolaevskaia Glavania Observatoria (in Pulkova); the Imperial Rousskae Geograticheskoe Obsehestov and the Imperial Akademia Nauk (both in Saint Petersburg); and the Société de Géographie in Geneva.

41. Sandford Fleming, "Notes from His Excellency the Governor-General of Canada, Transmitting Mr. Sandford Fleming's Papers Together with the Report of the American Metrological Society, to Various Scientific Societies in Europe," in Fleming, *Universal or Cosmic Time*, 43–44, quote from p. 43.

42. Struve's address to the Imperial Academy of Sciences in Saint Petersburg on the Canadian Institute's questions was given on September 30, 1880. It was later published, in translation, in O. Struve, "Report on Universal Time and on the Choice for That Purpose of a Prime Meridian; Made to the Imperial Academy of Sciences, St. Petersburg, by M. Otto Struve, Member of the Academy and Director of the Observatory at Pulkova," *Proceedings of the Canadian Institute* 2 (July 1885): 45–49, quotes from pp. 46–47, 48–49, 48.

43. On the rather muted paper offered to the Berlin Geographical Society (on July 2, 1881) in discussing the Abbe-Fleming papers, see G. V. Boguslawski, "Remarks upon a Normal Time to Be Common to the Whole Earth, and a Prime Meridian, to Be Accepted by All Nations," *Proceedings of the Canadian Institute* 2 (1885): 52–55. For the Belgian discussion, see E. Adan, "De L'heure universelle," *Société Belge de Géographie* 4 (1880): 403–411. Colonel Adan was the leading voice in the Belgian Geographical Society pushing for adoption of a single prime meridian. On the Spanish evidence, see J. Pastorin, "Remarks on a Universal Prime Meridian, by Don Juan Pastorin, Lieut.-Commander of the Spanish Navy," *Proceedings of the Canadian Institute* 2 (1885): 49–51.

44. Creet, "Sandford Fleming and Universal Time," 66–89; C. Blaise, *Time Lord: Sir Sandford Fleming and the Creation of Universal Time* (London: Weidenfeld and Nicolson, 2000).

45. For a fuller discussion of the term and what it signifies, see Simone Turchetti, Nèstor Herran, and Soraya Boudia, "Introduction: Have We Ever Been 'Transnational'? Towards a History of Science across and beyond Borders," *British Journal for the History of Science* 45 (2012): 319–336.

46. Fleming's quote is from his "Letter to the President of the American Society for the Advancement of Science on the Subject of Standard Time for the United States of America, Canada and Mexico, by Sandford Fleming, C. E.," Montreal Conference of the AAAS, August 1882, 1.

47. Ibid., 5. On the prime meridian issue in the U.S. Congress in June 1882, see "To Fix a Common Prime Meridian," H.R. Rep. No. 1519-47, at 1–2 (June 24, 1884), quotes from p. 2. On the role of this varied institutional involvement in leading the United States toward the adoption of standard time in 1883, see Bartky, *One Time Fits All*, 68–73. The Newcomb quote is his response in its entirety: see Simon Newcomb, "Remarks on the Cosmopolitan Scheme for Regulating Time," in Fleming, *Universal or Cosmic Time*, 64. Newcomb's position with regard to timekeeping in America and his role in the Naval Observatory is explored in S. J. Dick, *Sky and Ocean Joined: The U.S. Naval Observatory 1830–2000* (Cambridge: Cambridge University Press, 2003). In most nineteenth-century Mexican maps, the prime meridian was the easternmost tower of the Metropolitan Cathedral in Mexico City (although some topographical maps of Mexico used the Greenwich meridian). I am grateful to Prof. Luz Maria Tamayo for this information. The work of the American Geographical Society is clear from the Minutes of Council 1879–1885, February 4, 1882, vol. 9, box 4, 109–111; Correspondence between Hazen, General Cullum, and Charles Daly, 1879–1885, AC 1, box 79, vol. 9, American Geographical Society Papers, Letterpress Books.

48. J. H. Lefroy to Clements Markham and Douglas Freshfield, November 15, 1882, MSS JMS/21/49, Royal Geographical Society. Lefroy suggested that some notice of the proposal (that is, for the international meeting) should appear in the published proceedings of the society, but this was not taken up. The council minutes of November 24, 1879, simply note that "Sir J. H. Lefroy's report of Mr. Sandford Fleming's proposals regarding a Prime Meridian was adopted." In his *One Time Fits All*, Bartky (p. 67, 225n28) conflates Lefroy's 1879 and 1882 responses. Made separately, they also offer separate reasons for not accepting a prime meridian other than Greenwich. For a summary of the ten replies received overall, see *Terzo Congresso Geografico Internazionale tenuto a Venezia dal 15 al 22 Settembre 1881, Communicazioni e memori*, vol. 2, xv–xviii.

49. A. Hirsch to C. W. Siemens, January 13, 1883, MSS Eur F127/188, files 348–353, British Library, quotes from file 351.

50. Col. Donnelly to W. H. M. Christie, August 9, 1883, RGO 7/142, Papers of William Christie, Cambridge University Library.

51. "Report by the Astronomer Royal and Colonel Clarke FE CB FRS Delegates to the Geodetic Conference at Rome, December 10, 1883," RGO 7/142, Papers of William Christie. The Swedish response and Gylden's proposal (for universal time based on 240 ten-minute intervals) is discussed in Hugo Gyldén, "Om eqvidistanta lokaltider," *Ymer tidskrift utgifven af Svenska Sällskapet för Antropologi oct Geografi* 3 (1883): 40–48. On the Belgian evidence, see E. Hennequin, "Le premier méridien et l'heure universelle a la Septième Conférence Géodésique Internationale," *Bulletin de la Société Royale Belge de Géographie* 7 (1883): 782–805.

52. Bartky, *One Time Fits All*, 81. While I am indebted to Bartky's discussion of the Rome 1883 meeting (pp. 75–81), I have also drawn from H. Smith, "Greenwich Time and the Prime Meridian," *Vistas in Astronomy* 20 (1976): 219–229; W. Lambert, "The International Geodetic Association and Its Predecessors," *Bulletin Géodésique* 17 (1963): 299–324; *Resolutions de l'Association Geodesique Internationale concernant l'unification des longitudes et des heures* (Geneva, 1883).

CHAPTER 5 ○ GREENWICH ASCENDANT

1. This is based on a detailed assessment of the printed record of the conference, specifically the printed protocols of its proceedings: "International Conference Held at Washington for the Purpose of Fixing a Prime Meridian and a Universal Day. October, 1884," *Protocols of the Proceedings* (Washington, DC: George Brothers, 1884) (hereafter cited as *Protocols*).

2. *Protocols*, November 1, 1884, 206.

3. The 1884 International Meridian Conference is discussed in different ways by those who have studied the prime meridian question. Derek Howse, *Greenwich Time and the Longitude* (Greenwich: Philip Wilson, in association with the National Maritime Museum, 1997) itemizes each resolution and presents a useful table of the voting patterns on each (pp. 133–143). Peter Galison, *Einstein's Clocks, Poincaré's Maps: Empires of Time* (London: Sceptre, 2003) offers a summary of the main arguments (pp. 144–155). Adam Barrows, *The Cosmic Time of Empire: Modern Britain and World Literature* (Berkeley: University of California Press, 2011) places the Washington meeting alongside the near-contemporary Berlin conference on West Africa as a harbinger of imperial modernity and documents the main features of the debates around the single prime meridian (pp. 36–46). For a brief history, see S. R. Malin, "The International Prime Meridian Conference, Washington, October 1984 [1884]," *Vistas in Astronomy* 38 (1985): 203–206. Allen W. Palmer, "Negotiation and Resistance in Global Networks: The 1884 International Meridian Conference," *Mass Communication and Society* 5 (2002): 7–24 errs in several matters of fact but stresses how the 1884 conference should be understood in an age of scientific and political transnationalism. The fullest account of the meeting, and

of its uneven consequences, is offered by Ian R. Bartky, *One Time Fits All: The Campaigns for Global Uniformity* (Stanford, CA: Stanford University Press, 2007), chaps. 6–7, pp. 82–119.

4. I take the idea of the "truth spot" from Thomas Gieryn, *Cultural Boundaries of Science: Credibility on the Line* (Chicago: University of Chicago Press, 1999); Gieryn, "Three Truth-Spots," *Journal of the History of the Behavioural Sciences* 38 (2002): 113–132; Gieryn, "City as Truth-Spot: Laboratories and Field-Sites in Urban Studies," *Social Studies of Science* 36 (2006): 5–38. The idea of "speech space" is taken from Martin Hewitt, "Aspects of Platform Culture in Nineteenth-Century Britain," *Nineteenth Century Prose* 29 (2002): 1–32; David N. Livingstone, "Science, Site and Speech: Scientific Knowledge and the Spaces of Rhetoric," *History of the Human Sciences* 20 (2007): 71–98; Diarmid A. Finnegan, "Placing Science in an Age of Oratory: Spaces of Scientific Speech in Mid-Victorian Edinburgh," in *Geographies of Nineteenth-Century Science,* ed. David N. Livingstone and Charles W. J. Withers (Chicago: University of Chicago Press, 2011), 153–177.

5. By country and in the order in which they are listed, the nominated delegates were: Baron Ignatz von Schlaeffer (on behalf of Austria-Hungary); Dr. Luiz Cruls (Brazil); Cdre. S. R. Franklin of the U.S. Navy (Colombia); Mr. Juan Francisco Echeverria (Costa Rica); Mr. A. Lefaivre and Mr. J. Janssen (France); Baron H. von Alvensleben (Germany); Capt. Sir Frederick J. O. Evans, Prof. John C. Adams, Lt. Gen. Richard Strachey, and Mr. Sandford Fleming (Great Britain); Mr. Miles Rock (Guatemala); the Hon. W. D. Alexander and the Hon. Luther Aholo (Hawaii); Count Albert de Foresta (Italy); Professor Kikuchi (Japan); Mr. Leandro Fernandez and Mr. Angel Anguliano (Mexico); Capt. John Stewart (Paraguay); Mr. C. de Struve, Major General Stebnitzki, and Mr. J. de Kologrivoff (Russia); Mr. M. de J. Galvan (San Domingo); Mr. Antonio Batres (Salvador); Mr. Juan Valera, Mr. Emilio Ruiz del Arrol, and Mr. Juan Pastorin (Spain); Count Carl Lewenhaupt (Sweden); Col. Emile Frey (Switzerland); Rear Adm. C. R. P. Rodgers, Mr. Lewis M. Rutherfurd, Mr. W. F. Allen, Comm. W. T. Sampson, and Prof. Cleveland Abbe (United States); and Señor Dr. A. M. Soteldo (Venezuela). Delegates not present (they arrived later) were: Mr. Francisco Vidal Gormas and Mr. Alvaro Bianchi Tupper (Chile); Mr. Carl Steen Andersen de Bille (Denmark); Mr. Hinckeldeyn (Germany); Mr. William Coppinger (Liberia); Mr. G. de Weckherlin (the Netherlands); and Rustum Effendi (Turkey). "Mr C. de Struve," one of the Russian delegates, was Karl von Struve, half-brother to the astronomer Otto Wilhelm von Struve. Known in America as Karl or Carl de Struve, he was Russian ambassador to the United States between 1882 and 1892. Germany might reasonably have expected to have been represented by astronomer and geodesist Heinrich Georg von Boguslawski, who had spoken on the prime meridian question to the Berlin Geographical Society in 1881, but Boguslawski died in March 1884 before the German delegation was finalized. See W. T. Lynn, "Boguslawski, Father and Son," *Observatory* 22 (1899): 124–125.

6. The other members of the Russian delegation to Washington in 1884 were a Mr. Kologruvoff, a member of the Imperial Russian Council of Routes and Communications, and Major General Stebnitzki, a geodesist and surveyor. The proposals for the Russian delegation—that they be guided by the resolutions from Rome in 1883 (see Table 4.4)—were first drawn up by Lieutenant Colonel Rylke of the Russian War Department. See Otto Struve, "The Resolutions of the Washington Meridian Conference," in *Universal or Cosmic Time,* ed. Sandford Fleming (Toronto: Copp, Clark: 1885), 84–101, especially p. 91.

7. "The Seventh International Geodetic Conference," *Times,* October 27, 1883, 9. Hervé Faye's report to the Paris Academy of Sciences appears in *Nature* 29 (November 8, 1883– December 13, 1883): 44, 183; *Monthly Notices of the Royal Astronomical Society* 44 (1884): 206–208.

8. This link between appointment and colonial representation and the suggestion that a further person might be appointed to represent the Australian colonies is made in the letter sent to John Adams in Cambridge confirming his appointment: June 19, 1884, St. John's Library / Adams / 10/30/1. On the presumptive appointment of Fleming by the Canadians and

the exchanges between the branches of government in Britain and the Dominion government in Canada, see Letters of May 31, 1883; June 5, June 7, June 8, June 14, July 21, 1883, between Foreign Office, Colonial Office, and Canadian government officials, RG 25, vol. 31 (A–L), items 1–9, Library and Archives Canada.

9. J. Donnelly to W. H. Christie, February 22, 1884, and Christie to Donnelly, March 28, 1884, RGO 7/142, Papers of William Christie, Cambridge University Library.

10. J. Donnelly to R. Strachey, July 7, 1884, MS Eur F127/188, ff. 424–426, British Library.

11. Édouard Caspari, *Rapport fait au nom de la Commission de l'Unification des Longitudes et des Heures* (Paris: Ministère de l'Instruction Publique et des Beaux-Arts, Imprimerie Nationale, 1884). The first page of this eighteen-page document lists the members of the commission. See also Bartky, *One Time Fits All,* 85–86.

12. Caspari, *Rapport fait au nom de la Commission de l'Unification des Longitudes et des Heures,* 16–18; Bartky, *One Time Fits All,* 86.

13. Struve, "Resolutions of the Washington Meridian Conference," 84–101, quote from pp. 94–95.

14. The document to which delegates had access was *Unification des longitudes par l'adoption d'un meridien initial unique, et introduction d'une heure universelle [Extrait des comptes rendus de la Septième Conférénce Generale de L'Association Géodésique Internationale réunie à Rome, en Octobre 1883, rédigé par les secretaries A.* [Adolphe] *Hirsch* [and] *Th.* [Theodore] *v. Oppolzer publié par le Bureau Central de l'Association Géodésique Internationale.*

15. *Protocols,* October 2, 1884, quotes from pp. 10, 20.

16. Ibid., quotes from pp. 24, 25.

17. Ibid., quotes from pp. 29, 30, 32.

18. "A Common Prime Meridian. The International Conference Not Yet Able to Agree," *New York Times,* October 3, 1884; *Times,* October 2, 1884, 9; *Times,* October 3 and 4, 1884.

19. *Protocols,* October 2, 1884, quotes from pp. 37, 38, 40, 41.

20. *Protocols,* October 6, 1884, quotes from pp. 43, 44, 50, 51.

21. Ibid., quotes from pp. 52, 53, 54, 55, 56.

22. Ibid., quotes from pp. 60, 67. John Adams's judgment of Newcomb's view appears in a draft version of Adams's contribution to the protocols in Adams's personal papers: St. John's Library / Adams / 27/4/5; no date but 1884 from internal evidence. Adams had written to Newcomb in July 1884 in advance of coming to Washington, expressing his desire to meet with him and hoping that the Washington meeting would be "strictly confined to the question of the Prime Meridian, & the time from which the Astronomical day shall be reckoned, and not allowed to be mixed in any way with other questions." In this letter Adams made reference to Laplace's views, expressed in his *Mecanique céleste,* over the importance of a single prime meridian and of metrological uniformity. He also noted Professor Forster's argument at Rome on these questions, the German's argument there being judged "very weak and inconclusive." See July 14, 1884, St. John's Library / Adams / 37/21/4. This letter, with those of Struve to Fleming and Wheeler to Fleming (see Chapter 4 in this book), is one of the strongest pieces of evidence that the prime meridian question was discussed among individual scientists outside the context of the IGC or the Washington conference.

23. *Times,* October 8, 1884, 5; C.H. Mastin to R. Strachey, Letters to Sir Richard Strachey, 1883–1889, October 11, 1884, file 124, MS Eur 127/151, British Library; Jules Janssen to Henriette Janssen, October 13, 1884, MSS 4133–278, Bibliothéque Institut de Français, cited in Françoise Launay, *The Astronomer Jules Janssen A Globetrotter of Celestial Physics,* trans. Storm Dunlop (New York:

Springer, 2012), 132. Interestingly, with respect to this communication to others over the prime meridian conference, Britain's John Adams kept a diary of his North American work and travels in 1884 (he traveled to the 1884 Montreal meeting of the British Association for the Advancement of Science before heading to Washington). His diary has full entries to mid-August but is then blank until mid-November, at which point full entries resume. There is frustratingly no mention at all of the Washington conference: St. John's College / Adams 21/17. It may be of course that he had neither the time nor the desire to record the events of the conference in his diary, having spent all day debating the issues.

24. *Times,* October 11, 1884, 5; *Protocols,* October 13, 1884, quotes from pp. 75, 76–77.

25. *Protocols,* October 13, 1884, quotes from pp. 90, 92.

26. The total sales of British Admiralty charts in the period 1877 to the first quarter of 1884 to each of the seven leading purchasing nations was as follows: France: 35,744; Germany: 36,679; United States: 23,867; Italy: 14,440; Russia: 52,930; Turkey: 4591; Austria: 6544. The annual sales of *The Nautical Almanac and Astronomical Ephemeris* in the period 1877 to 1883 (not given by Evans in terms of by-nation sales) was in 1877: 18,439 copies; 1878: 16,408; 1879: 16,290; 1880: 14,561; 1881: 15,870; 1882: 15,071; 1883: 15,535. See *Protocols,* October 13, 1884, 97–98. The claim by Howse that it was Sandford Fleming's table of tonnage statistics and the use of different prime meridians that broke the deadlock in Washington is unfounded: there is no evidence to this effect in the printed *Protocols.* The scientific delegates would almost certainly have known of this evidence before they came; diplomatic delegates made no remark on it. Fleming's evidence was part of the material that suggested a primacy for Greenwich on practical grounds, although Fleming advocated a prime meridian 180° from Greenwich on the grounds of neutrality. Derek Howse, "1884 and Longitude Zero," *Vistas in Astronomy* 28 (1985): 11–19.

27. *Times,* October 15, 1884, 9.

28. "The Meridian Conference," *Science* 4 (October 17, 1884): 376–378, quote from p. 378.

29. *Protocols,* October 14, 1884, quotes from pp. 130, 131.

30. [Sandford Fleming], *The International Prime Meridian Conference Washington, October 1884: Recommendations Suggested by Sandford Fleming Respectfully Submitted,* October 1884.

31. *Protocols,* October 13, 1884, 105, 106, 109; *Protocols,* October 14, 1884, 117–118, 123, 124.

32. *Protocols,* October 14, 1884, 132–133.

33. Ibid., 134.

34. Ibid., 147, 149.

35. *Protocols,* October 20, 1884, 158, 162, 163.

36. Ibid., 167–168.

37. Ibid., 170, 171, 173, 181–182.

38. To judge from Rodgers's letter (to Fleming), the conference was already breaking up. Rodgers wrote how "many delegates have gone or are impatient to go": C. R. P. Rodgers to Sandford Fleming, October 31, 1884, MG 29-B1, vol. 41, fol. 295, Library and Archives Canada.

CHAPTER 6 ○ WASHINGTON'S "AFTERLIFE"

1. Derek Howse, "1884 and Longitude Zero," *Vistas in Astronomy* 28 (1985): 11–19, quote from p. 18; Howse, *Greenwich Time and the Longitude* (London: Philip Wilson, 1997), chap. 6.

2. *Australian Town and Country Journal,* November 22, 1884, 25; Rebekah Higgitt and Graham Dolan, "Greenwich, Time and 'The Line,'" *Endeavour* 34 (2009): 34–39; *Philadelphia Inquirer,* January 2, 1885, 2.

3. Janssen's remarks are cited in Françoise Launay, *The Astronomer Jules Janssen A Globetrotter of Celestial Physics,* trans. Storm Dunlop (New York: Springer, 2012), 134. For Janssen's report in full, see Jules Janssen, "Rapport sur le Congrès de Washington et sur les propositions qui y ont été adoptées, touchant le premier méridien, l'heure universelle et l'extension du système décimal à la mesure des angles et à celle du temps," *Comptes rendus de l'Académie des Sciences* 100 (March 9, 1885): 706–726. On Adams's reflections, see John Adams to F. Bashforth, November 21, 1884, St. John's Library / Adams / 4/26/4, St. John's College Library, Cambridge (original emphasis); [Sandford Fleming], "Report on the Washington International Conference, to the Canadian Government, by Mr. Sandford Fleming, Delegate of Great Britain, Representing Canada. (31st December, 1884)," in *Universal or Cosmic Time, Together with Other Papers, Communications and Reports in the Possession of the Canadian Institute Respecting the Movement for Reforming the Time-System of the World, and Establishing a Prime Meridian as a Zero Common to All Nations* (Toronto: Copp, Clark: 1885), 67–73, quotes from pp. 69, 71; George Wheeler to Sandford Fleming, June 23, 1885, MG 29–B1, vol. 53, fol. 365, Sandford Fleming Papers, Library and Archives Canada (LAC); f. 1; C. R. P. Rodgers to Sandford Fleming, December 15, 1887, and Rodgers to Fleming, March 21, 1889, MG 29–B1, vol. 41, fol. 295, Sandford Fleming Papers.

4. U.S. Congress. *Message from the President of the United States Transmitting a Communication from the Secretary of State, Relative to International Meridian Conference Held at Washington,* 48th Cong., 2nd sess., December 4, 1884; Senate Committee on Foreign Relations, *Concurrent Resolution Authorizing the President to Communicate to the Governments of All Nations the Resolutions Adopted by the International [Meridian] Conference . . . for Fixing a Prime Meridian,* 48th Cong., 2nd sess., February 7, 1885.

5. Rodgers to Fleming, December 15, 1887, MG 29–B1, vol. 41, fol. 295, Sandford Fleming Papers.

6. Letter from the secretary of the navy transmitting communications concerning the proposed change in the time for beginning the astronomical day, February 17, 1885, RGO 7/146, Papers of William Christie, Cambridge University Library, Washington, DC.

7. Ian R. Bartky, *One Time Fits All: The Campaigns for Global Uniformity* (Stanford, CA: Stanford University Press, 2007), 100–119, supplemented by additional primary material.

8. Ibid., 107–109; *Thirty-Fourth Report of the Department of Science and Art of the Committee of Council on Education, with Appendices* (London: Eyre and Spottiswoode, 1887), enclosure 11, p. 21.

9. Bartky, *One Time Fits All,* 106; Otto Struve, "The Resolutions of the Washington Meridian Conference," in Fleming, *Universal or Cosmic Time,* 84–101, quotes from pp. 93, 97, 99.

10. D. Kikuchi to J. Adams, December 12, 1884, St. John's Library / Adams / 24/16/2. John Tennant wrote to Adams offering the view that "I don't believe we shall soon see universal time in general uses while quite seeing that the American plan is good and practical there for Railway purposes." J. Tennant to J. Adams, February 23, 1885, St. John's Library / Adams / 14/41/3.

11. "Minutes and Correspondence: Washington Prime Meridian Conference, 10–12" [Mr. Duncombe, Under Secretary of State at the Foreign Office, January 30, 1886], *Thirty-Fourth Report of the Department of Science and Art of the Committee of Council on Education,* app. A. On the Royal Society's Prime Meridian Committee, see Royal Society, June 25, 1885, MSS CMB / 3/8, where the committee recommended to the Royal Society's council the approval of Resolution VI and its uptake in "the Nautical Almanacs for all Nations" from 1890. See also [Minutes of June 18, 1885], *Minutes of Council of the Royal Society from October 30th 1884 to June 30th 1892* (London: Harrison and Son, 1893), 46.

12. "Minutes and Correspondence: Washington Prime Meridian Conference," app. A, 10.

13. Ibid.

14. RGO 7/146, items 1–15, Papers of William Christie; June 25, 1885, CMB / 3/8, Royal Society of London; Bartky, *One Time Fits All*, 109–113.

15. [Sandford Fleming], "Report on the Washington International Conference, to the Canadian Government," 67–73, quotes from p. 73; Bartky, *One Time Fits All*, 113, where he notes "Why he [Fleming] did so at this particular time [1892] is not known." Sandford Fleming's *Time-Reckoning for the Twentieth Century* (Washington, DC: Adams, 1889) was first published as part of the *Report of the Smithsonian Institution for 1886* and takes its pagination (pp. 346–361) from that form of publication. Bartky does not cite this work.

16. Fleming, *Time-Reckoning for the Twentieth Century*, 359. Sandford Fleming was also receiving private correspondence on his proposals from interested parties at this time. In a letter from Lieutenant General Lefroy in September 1885, Lefroy complimented Fleming on his publications advocating universal time and noted that while he [Lefroy] was surprised by Newcomb's reactions, he was not at all taken aback by Airy's seeming rejection of Greenwich, "which is very characteristic of the man." J. H. Lefroy to Sandford Fleming, September 2, 1885, MG 29 B-1, vol. 28, fol. 199, Sandford Fleming Papers.

17. "Correspondence Relative to a Universal Prime Meridian and to a Proposed Reform in Time Reckoning," *Thirty-Eighth Report of the Department of Art and Science of the Committee of Council on Education, with Appendices* (London: Eyre and Spottiswoode, 1890), app. A, 16–32, quotes from pp. 25 and 27. Fleming's "Memorandum," dated November 20, 1890, is on pp. 16–20.

18. Sandford Fleming to W. Christie, July 4, 1982, RGO 7/146, Papers of William Christie. Fleming clearly got much of his information over the matter from this meeting with Donnelly: "He [Donnelly] very kindly showed me all the papers which had been received on the question of Time-reckoning . . . I find he has a considerable number of interesting communications."

19. The printed summary of returns to the Canadian joint committee's circular of April 1893 allows the identification of individual respondents within each country. In Great Britain, for example, one of the four respondents who was against altering the astronomical day in the ways and for the date proposed was Lt. Gen. John Tennant, who had written to John Adams in February 1885 inquiring about the consequences of the Washington meeting and the issue of universal time (see Note 10). The implication of his remarks about the advantage of universal time for the American railways but not in other respects is consistent with later evidence of his disapproval over the 1893 proposal. See *Forty-Second Report of the Department of Art and Science of the Committee of Council on Education, with Appendices* (London: Her Majesty's Stationery Office, 1895), app. A, 19. Sandford Fleming's remarks over the Belknap-Newcomb report are not cited in this printed report but appear in a manuscript in RGO 7/146, Papers of William Christie, 14. The third figure referred to here by Fleming was William C. Whitney, secretary of state to the U.S. Navy: Bartky, *One Time Fits All*, 116 incorrectly dates this remark to 1895.

20. *Forty-Second Report of the Department of Art and Science of the Committee of Council on Education*, app. A, 21; Bartky, *One Time Fits All*, 21; RGO 7/146, August 3, 1894, Papers of William Christie.

21. [Replies from Foreign Governments], RGO 7/146, Papers of William Christie.

22. C. Abbe to Sandford Fleming, May 21, 1895, MG 29-B1, vol. 2, fol. 1, Sandford Fleming Papers. The indented quote is from Abbe to Fleming, May 24, 1895.

23. [Questions put to Shipmasters], RGO 7/146, Papers of William Christie.

24. Fleming to Christie, October 10, 1895, RGO 7/146, Papers of William Christie. On the Christie-Fleming correspondence in 1896, see Christie to Fleming, March 7, 1896, RG vol. 31, A–L, item 4, LAC and the reply, Fleming to Christie, October 10, 1896, MG-29 B1, fol. 65, LAC.

25. Bartky, *One Time Fits All*, 112.

26. [Replies from Foreign Governments], November 27, 1896 and January 19, 1897 [for the confirmation from French authorities of the intention to adopt Greenwich as the prime meridian], RGO 7/146, Papers of William Christie; RGO 16/2, Papers on the Astronomical Day, Cambridge University Library; Bartky, *One Time Fits All*, 155–156.

27. Otto Struve, "The Resolutions of the Washington Meridian Conference," in Fleming, *Universal or Cosmic Time*, 98.

28. Henri Bouthillier de Beaumont, "Les projections dans la cartographie et présentation d'une nouvelle projection de la sphere comme planisphère," *Le globe* 27 (1888): 216–217. The report upon de Beaumont's delivery to the congress of Swiss geographers in August 1888, from which I cite the comments upon its reception, appears in *Le globe* 28 (1889): 13. In his *One Time Fits All* (p. 97), Bartky suggests that de Beaumont's work in April 1888 was a substantial paper (Bartky erroneously cites it as pp. 1–26 of the 1888 issue of *Le globe*) in which de Beaumont was disparaging of attempts at uniformity since 1871 and in which he offered comments about the unrepresentative nature of the IGA meeting at Rome and the 1884 Washington meeting. This is not so: de Beaumont makes only passing reference to earlier work and offers no comments of substance upon the two earlier meetings.

29. Cesare Tondini de Quarenghi, "A Suggestion from the Bologna Academy of Science towards an Agreement on the Initial Meridian for the Universal Hour," *Report of the Fifty-Eighth Meeting of the British Association for the Advancement of Science* (London: John Murray, 1889), [transactions of section A], 618–619.

30. The work in the early 1870s of the BAAS committee charged with reporting upon "the best means for a uniformity of weights and measures; with reference to the interest of science" is discussed in the annual report of the BAAS for 1871 (p. 197), 1872 (p. 217), and 1874 (p. 359). On Fleming's intended paper to the Glasgow BAAS meeting in 1876, see Sandford Fleming, "Observations on the Conventional Division of Time Now in Use, and Its Disadvantages in Connexion with Steam Communications in Different Parts of the World; with Remarks on the Desirability of Adopting Common Time over the Globe for Railways and Steam-Ships," *Report of the Forty-Sixth Meeting of the British Association for the Advancement of Science* (London: John Murray, 1887), 182. The mention of the title without a description of the paper's content was BAAS convention, showing the paper was offered but not delivered. On the Ravenstein-Fleming correspondence over the invitation to speak to the 1886 Birmingham BAAS meeting, see E. G. Ravenstein to Sandford Fleming, July 30, 1886, and the reply, August 5, 1886, vol. 6, fol. 37, MG–29 B1, LAC. On Ravenstein's endorsement of the metric system, see E. G. Ravenstein, "A Plea for the Metre," *Report of the Fifty-Seventh Meeting of the British Association for the Advancement of Science* (London: John Murray, 1888), 805.

31. "Report of the Committee, Consisting of Dr. Glaisher, Mr. W. H. M. Christie, Sir R. S. Ball, and Dr. Longstaff, Appointed to Consider the Proposals of M. Tondini de Quarenghi Relative to the Unification of Time, and the Adoption of a Universal Prime Meridian, Which Have Been Brought before the Committee by a Letter from the Academy of Sciences of Bologna," *Report of the Fifty-Eighth Meeting of the British Association for the Advancement of Science*, 49.

32. R. P. Tondini de Quarenghi, "Du choix du méridien de Jérusalem pour fixer l'heure universelle," *IVᵉ Congres Internationale des Sciences Geographiques tenu a Paris en 1889*, 2 vols. (Paris: Bibliotheque des Annales Économiques, 1890), 1:193–204, quote from p. 204. The Jerusalem proposal is also discussed on pp. 41–42 and 164–166 of vol. 1 of the *IVᵉ Congres*. The initials R. P. for Tondini de Quarenghi (used also by Bartky in his mention of the Italian) signal here an abbreviation of his status as a Catholic priest and missionary—Révérend Père. His Christian name was Cesare (see Note 29).

33. The voting totals, but not the names of those who voted either for or against, are given in *IVᵉ Congres Internationale des Sciences Geographiques tenu a Paris en 1889*, 1:166. In his *One Time Fits All*, Bartky

(p. 97) does not mention that the Bologna proposal over Jerusalem was first presented to the 1888 BAAS meeting in Bath, assuming rather that the committee's rejection of it in Newcastle in 1889 was the first time the association had dealt with the matter and that the BAAS only entertained the issue then because of its rejection by delegates in Paris at the fourth IGC. As the evidence discussed shows, the BAAS Newcastle meeting formally rejected what had been drawn to their attention a year before.

34. "Correspondence Relative to a Universal Prime Meridian and to a Proposed Reform in Time Reckoning," app. A, 29.

35. Ibid., app. A, pp. 29–30 (Tornielli), p. 30 (Donnelly).

36. Ibid., app. A, 32, contains Donnelly's response to the 1891 Italian proposal. The IGA's rejection of the Bologna proposal is briefly noted in Bartky, *One Time Fits All*, 98. On the papers discussed at the Berne meeting in 1891, see Henri Bouthillier de Beaumont, "L'expression de la longitude par l'heure"; de Beaumont, "Presentation, avec cartes nouvelles, d'une cartographié générale pour le meilleur enseignement de la géographie"; [C.] Tondini de Quarenghi, "Le statu quo dans le marine, l'astronomie et la topographie et le méridien de Jerusalem-Nyanza pour fixer l'heure eniverselle," *Compte rendu du Vme Congrès International des Sciences Geographiques tenu à Berne du 10 au 14 Aout 1891* (Berne: Schmid, Francke, and CIE, 1892), 222–224, 351–354, 229–248. For notice of Tondini de Quarenghi's projected talk to the 1891 BAAS Leeds meeting, see C. Tondini de Quarenghi, "The Actual State of the Question of the Initial Meridian for the Universal Hour," *Report of the Sixty-First Meeting of the British Association for the Advancement of Science* (London: John Murray, 1892), 897. De Quarenghi's published paper in 1888 in the journal of the Italian Geographical Society on this question is a curious hybrid of his interests in calendar reform, universal time, and the candidacy of Jerusalem as a world prime meridian: [Signor] Tondini de Quarenghi, "Nota sul calendario Gregoriano e sull'era universale del. Sig[lio]," *Bolletino della Societa Geografica Italiana* 20 (1888): 621–623. It is possible that the role of the Italian government in pushing for Jerusalem in 1890 and 1891 was the result of the recent appointment of Francesco Crispi as Italy's prime minister: Crispi was the initiator of Italy's expansionist colonial vision of Italy as a "great power" to match Britain, France, and the United States. I am grateful to Prof. Matteo Proto of the University of Bologna for this suggestion.

37. [T. H. Holdich], "Geographical Notes," *Proceedings of the Royal Geographical Society and Monthly Record of Geography* 13 (1891): 615–616. Bartky does not refer to Holdich by name, referring to this evidence only as that "of the delegate from the Indian government." On Holdich's distinguished career as an imperial surveyor, see Kenneth M. Mason and H. L. Crosthwait, "Colonel Sir Thomas Hungerford Holdich," *Geographical Journal* 75 (1930): 209–217. Holdich's remark here that he was "quite aware of the nature of the reasons which have prevented its adoption hitherto" [referring to the failure before then of other nations to adopt Greenwich] is intriguing: we must suppose he is referring to the issues that are discussed in Chapters 3 and 4 of this book.

38. Bartky, *One Time Fits All*, 98.

39. Norman J. W. Thrower, *Maps and Civilization: Cartography in Culture and Society* (Chicago: University of Chicago Press, 1999), 164–171; Alastair W. Pearson and Michael Heffernan, "The American Geographical Society's Map of Hispanic America: Million-Scale Mapping between the Wars," *Imago Mundi* 61 (2009): 215–243. On the public announcement of the resolution in Berne in 1891 that Greenwich should be universally adopted, see the *Times*, August 14, 1891, 3.

40. D'Italo Enrico Frassi, "On Time-Reform, and a System of Hour Zones," in *Report of the Sixth International Geographical Congress: Held in London, 1895* (London: John Murray, 1896), 261–268, quote from p. 262. On the other papers cited, see M. J. de Rey-Pailhade, "L'application du système décimale a la mesure du temps et des angles," in *Report of the Sixth International Geographical Congress*, 255–256; Henri Bouthillier de Beaumont, "Resolution as to Standard Time," in *Report of the Sixth International Geographical Congress*, 259.

41. *Verhandlungen des siebenten Internationalen Geographen-Kongresses, Berlin, 1899*, 2 vols. (Berlin: W. H. Kuhl, 1901). The recommendation of Greenwich and the metric system for Penck's map project is made in vol. 1, 5. On Mill's advocacy of the metric system generally, see H. R. Mill, "On the Adoption of the Metric System of Units in All Scientific Geographical Work," *Verhandlungen* 2:120–124.

42. *Report of the Eighth International Geographical Congress Held in the United States, 1904* (Washington, DC: International Geographical Union, 1905), 109–110.

43. The most detailed study of "Partitioning the World's Time" is given by Bartky, *One Time Fit All*, chaps. 8–9, from whom I take this term (p. 120). In his *Greenwich Time and the Longitude*, Derek Howse gives a list of the dates at which different countries adopted systems of timekeeping based on the Greenwich meridian: table 3, pp. 148–149. The examples of different countries cited here are taken from material in RGO 7/146, Papers of William Christie. The commentary on different timekeeping in Australia is from the *Sydney Morning Herald*, January 30, 1895, 6.

44. *Sydney Morning Herald*, February 14, 1885, 8.

45. George Biddell Airy, *Account of Observations of the Transit of Venus, 1874, December 8, Made under the Authority of the British Government* (London: Her Majesty's Stationery Office, 1881), 284–285. The Mokattam is the name given to that range of hills southeast of Cairo that formed part of the sites for Airy's triangulation of longitude in the 1874 expedition.

46. *Reports to the Board of Visitors, 1882–1896*, June 2, 1888, June 1, 1889, June 6, 1891, June 4, 1892, June 3, 1893, RGO 17/4, Cambridge University Library; *Reports to the Board of Visitors, 1897–1910*, June 5, 1897, June 3, 1899, June 3, 1905, RGO 17/5.

47. Bartky, *One Time Fits All*, 138–157.

48. *Times*, March 13, 1911, 9.

CHAPTER 7 ○ RULING SPACE, FIXING TIME

1. *Mercury*, January 13, 1934, 11 (Hobart, Tasmania).

2. Benjamin Vaughan, "An Account of Some Late Proceedings in Congress Respecting the Project for Establishing a First Meridian, with Remarks," MS B/V46p, fol. 14, Benjamin Vaughan Papers, American Philosophical Society.

3. Joseph Conrad, *The Secret Agent. Edited and with an Introduction and Notes by Martin-Seymour Smith* (London: Penguin Books, 1984), 68, 70. Conrad took his information about the Greenwich Bomb Outrage from reading the *Anarchist* newspaper and other contemporary sources. For a discussion of these, see M. Kellens Williams, "'Where All Things Sacred and Profane Are Turned into Copy': Flesh, Fact, and Fiction in Joseph Conrad's *The Secret Agent*," *Journal of Narrative Theory* 32 (2002): 32–52; Mary Burgoyne, "Conrad among the Anarchists: Documents on Martial Bourdin and the Greenwich Bombing," *Conradian* 32 (2007): 147–185; Adam Barrows, *The Cosmic Time of Empire: Modern Britain and World Literature* (Berkeley: University of California Press, 2011), chap. 4, 100–128.

4. The extracts quoted in this paragraph are from Barrows, *Cosmic Time of Empire*, 101, 102, 112. On Joseph Conrad's discussion of geography, see Conrad, "Geography and Some Explorers," in *Last Essays*, ed. R. Curle (London: Dent, 1926), 121–134. The claim over Washington and Greenwich 1884 as a culminating episode is made by Ronald L. Thomas, "The Home of Time: The Prime Meridian, the Dome of the Millennium, and Postnational Space," in *Nineteenth-Century Geographies: The Transformation of Space from the Victorian Age to the American Century*, ed. Helena Michie and Ronald L. Thomas (New Brunswick, NJ: Rutgers University Press, 2003), 23–39, quote from p. 26.

As Bartky shows, the United Kingdom of Great Britain and Ireland had two legal times in place between 1880 and 1916, Dublin mean time and Greenwich mean time: Ian R. Bartky, *One Time Fits All: The Campaigns for Global Uniformity* (Stanford, CA: Stanford University Press, 2007), 134–136.

5. Rebekah Higgitt and Graham Dolan, "Greenwich, Time and 'The Line,'" *Endeavour* 34 (2014): 35–39, box 1 on p. 38. As they note, a line of poplars planted in 1903 by John Henry Buxton on his estate in Ware, Hertfordshire, seems not to have had any connection with marking either the prime meridian or the observatory. I have not included it here in noting the ten forms of prime meridian commemoration between 1910 and 1959. For illustrations of several of the memorials, such as the obelisk at Peacehaven in Sussex where the Greenwich prime meridian "leaves" Britain, see Stuart Malin and Carole Stott, *The Greenwich Meridian* (London: Her Majesty's Stationery Office, 1984), 9. On the issue by Ordnance Survey of these millennial maps, see Chet Raymo, *Walking Zero: Discovering Cosmic Space and Time along the Prime Meridian* (New York: Walker, 2006), 19. Produced between 1999 and 2001, the Ordnance Survey Explorer maps so marked, by sheet number, were 122, 135, 147, 148, 161, 162, 174, 194, 209, 225, 227, 235, 249, 261, 273, 283, 284, and 292. I am grateful to Gordon Street, Ordnance Survey Customer Services, for this information. The French millennial picnic and the planting of trees to mark La Méridienne Verte is discussed by Paul Murdin, *Full Meridian of Glory: Perilous Adventures in the Competition to Measure the Earth* (New York: Springer, 2009), 3–5; Claude Teillet, "Mission 2000 en France: La Méridienne Verte," *Comptes-rendus et mémoires de la Société Archeologique et histoire de Clermont en Beauvaisis* 40 (1998–2002): 193–222.

6. This idea is the central focus of Patrick Hutton, *History as an Act of Memory* (Burlington, VT: University of Vermont/University Press of New England, 1993).

7. Nicky Reeves, "'To Demonstrate the Exactness of the Instrument': Mountainside Trails of Precision in Scotland, 1774," *Science in Context* 22 (2009): 323–340.

8. This is distilled from Stephen Malys, John H. Seago, Nikolaos K. Pavlis, P. Kenneth Seidelmann, and George H. Kaplan, "Why the Greenwich Meridian Moved," *Journal of Geodesy* 89 (2015): 1263–1272. I am grateful to Dr. Richard Dunn for drawing this reference to my attention as I was completing this book.

9. Dr Kukula is quoted in Chris Green, "There's No Need to Adjust Your Watch, But the Greenwich Meridian Is on the Move," *Independent*, August 13, 2015, 7. Greenwich has extraterrestrial influence too: legally, the time zone of the International Space Station is specified as Greenwich mean time.

ACKNOWLEDGMENTS

BECAUSE THERE HAVE BEEN so many prime meridians in different national settings and there is no such thing as a single "prime meridian archive" (or a "single prime meridian" archive) this inquiry has depended, more than is usual, upon the assistance of numerous archivists and librarians. This book is very strongly the consequence of following leads from one source or setting to another. The arguments I offer are mine alone, of course, but without the generous assistance of many people, this book would not have been possible. If it is something of a convention, it is also, properly, a pleasure to acknowledge their courtesy, time, and expertise in helping to shape this book.

The book has its origins in a Leverhulme Foundation Major Research Fellowship. I am grateful for this support and to the British Academy and to the University of Edinburgh for additional funding. My debt to numerous institutions and their staffs is heavy: the order of acknowledgment here, in alphabetical order by institution, is not a measure of the debt owed. I am grateful to the staffs of the: American Geographical Society Library and Archives, Milwaukee (particularly to Bob Jaeger and Susan Peschel); American Philosophical Society, Philadelphia (particularly to Roy Goodman and Michael Miller); Bibliothèque nationale de France; Bodleian Library, Oxford; British Library; Cambridge University Library; Centre for Research Collections, University of Edinburgh Library; Library and Archives Canada, Ottawa; Library of Congress, Washington, DC; National Archives, Kew; National Library of Scotland; National Maritime Museum (particularly to Gloria Clifton, Richard Dunn, Gillian Hutchinson, and the staff of the Caird Library); Newberry Library, Chicago; New College Library, Edinburgh; the Paris Observatory, Library and Archives (particularly to Amélia Laurenceau, Emilie Kaftan, and Sandrine Marchal); Royal Astronomical Society (particularly to Sian Prosser of the Library and Archives); Royal Geographical Society, incorporating the Institute of British Geographers (particularly to Eugene Rae of the Foyle Reading Room); Royal Observatory, Edinburgh; Royal Society Library and Archives; and Smithsonian Institution, Washington, DC. It is a pleasure to record my thanks to Kathryn McKee, Archivist, St. John's College, Cambridge. The material cited here from St. John's College Library and Archives is cited by permission of the Master and Fellows of St. John's College, Cambridge.

For additional assistance with sources, I am grateful to Lionel Gaulthier, Department of Geography, University of Geneva; Catherine McIntyre, Library, London School of Economics and Political Science; Matteo Proto, University of Bologna; and Helen Rawson, University of St. Andrews Library.

Research seminars and working papers on aspects of the prime meridian have been presented to academic meetings in Belfast, Chicago, Edinburgh, Leeds, London, Manchester, Paris, Prague, and Valencia, and I acknowledge the many helpful suggestions and criticisms received. I have benefited greatly from the guidance of the two readers appointed by Harvard University Press and thank them for their perceptive remarks on the initial proposal and the near-finished manuscript.

In addition to those mentioned, a number of individuals have at one time or another offered suggestions, thoughts about sources, or feedback and advice on the work. For their courtesies, I am indebted to: Alexi Baker, Sandy Bederman, Paul Bettens, Guy Boistel, Robin Butlin, Catherine Delano Smith, Andrew Dugmore, Matthew Edney, Chris Fleet, Daniel Foliard, Franklin Ginn, Jan de Graeve, Will Hasty, Michael Heffernan, Michael Jones, Bob Karrow, Innes Keighren, William Mackaness, Luz Maria Tamayo, Juan Pimentel, Georgina Rannard, Denis Shaw, Jan Smits, Dan Swanton, Peter van der Krogt, and Iain Woodhouse. There are four people, however, to whom I owe much more than a simple acknowledgment of assistance. Becky Higgitt, formerly of the Royal Museums Greenwich and now of the University of Kent, read drafts of most chapters—sometimes more than once—corrected my errors, urged clarity, and forced me to look up from the prime meridian story to the bigger implications. Anita McConnell read drafts of most of the chapters and made suggestions for improvement. My colleague Fraser MacDonald, himself deeply engaged in a study of the development of rocketry and the politics of American spaceflight, listened patiently while I explained my story. His advice was always apposite and his own writing something to be admired. Most of all I owe a huge amount to Carolyn Anderson, who was a wonderful research assistant: tenacious, imaginative, and obdurate in the face of the many hard-to-find materials, which she did eventually run to ground. Having found many of the materials from which this narrative has been woven, she then had the kindness to read the work in draft form and make suggestions toward its improvement.

The staff at Harvard University Press have been wonderfully supportive—none more so than Ian Malcolm, who first approached me with the idea for this book. His encouragement throughout has been tremendous and his suggestions for improvement, carefully articulated and always helpful. Amanda Peery and, latterly, Anne Zarrella, answered all my queries with charm and efficiency; Stephanie Vyce was patient and helpful with respect to the illustrations and permissions. My colleagues in Edinburgh have likewise encouraged, and at times been a little bemused by, my consuming interest in the geographies of the prime meridian; I can only hope they see this book as a worthwhile outcome. My wife, Anne, has had to live too long with tales from the archives as the research and writing proceeded. This book is for her, my one true line and base point.

INDEX

Magnetic prime meridian. *See* Agonic prime
 meridian
Magnetic variation, 30–31
Mapping, xii, 3, 4, 11, 40, 74, 97, 99, 171
Marsh, Othniel, 227, 232
Maskelyne, Nevil: As Astronomer Royal, 14, 39,
 65–66, 71, 90, 114; *British Mariner's Guide,*
 14, 51, 88; *Nautical Almanac and Astronomical*
 Ephemeris, 14, 39, 46, 51–53, 56–57, 71–72,
 87–88, 93, 141, 169, 180, 203–204, 205, 230,
 232, 265–266, 272; *Tables Requisite,* 88
Mathematics, 17, 31, 40, 110
Mayer, Tobias, 51, 57
Measured prime meridian, 7, 44, 74
Mécanique céleste (Laplace), 89
Méchain, Pierre, 66–67, 110, 114, 121–122, 136
Médiateur, 148, 152, 167
Mercator, Geraldus, 30, 35
"Meridién Fictif," 108
Méridienne vérifiée (Cassini de Thury), 44,
 64–65, 67, 72
Messier, Charles, 66
Mesure de la Terre (Picard), 40
Metre, 121–122, 131–137, 183. *See also* Metric
 system; Metrology
Metric (Weights and Measures) Act (1864),
 120, 131, 291n15
Metric system: Established by law in Britain,
 120; Criticised by Herschel, 132; Metrology
 and, 17, 117–119, 120–122, 131–137, 181–183;
 Supported by Mill, 252–253; Terrestrial
 measurement and, 107–110, 121–122, 178–179,
 183
Metrology: Geography of, 6, 10, 107–137;
 Imperial, 110–112, 116–123, 151; In the work
 of Kater, 110, 117–119; Metric system and, 17,
 117–119, 120–122, 131–137, 181–183; Modernity
 and, 6, 10, 14, 16–18, 85, 107–137, 263–273;
 Politics and, 10, 85, 146; Uniformity in, 85,
 107–108, 146, 213–214, 246–250; Universality
 in, 108, 146, 213, 255–256
Mexico, 196, 298n47
Mill, Hugh Robert, 252–253
Modernity: Conrad on, 268–269; Metrology
 and, 6, 10, 14, 16–18, 85, 107–137, 263–273;
 Prime meridian and, 10, 15–17; Railways
 and, 10, 109, 122–123, 129–131, 162, 177,
 220–222, 258–261; Science and, 6, 14, 15–16,
 18, 137, 139–189, 265–273; Telegraphy and, 10,
 15, 109, 122–123, 129–130, 177; Timekeeping
 and, 13–14, 129–131, 219–242
Monroe, James, 78, 89

Moore, John Hamilton, 88, 96
Morse, Jedidiah, 75, 86; *American Geography,*
 or A Present Situation of the United States
 of America, 75, 86; *Elements of Geography,*
 75; *Geography Made Easy,* 75; Insistence of
 upon an American prime meridian, 75–76
Mr. Blundeville His Exercises, 31
Munich, 129, 151

National Academy of Sciences (USA), 227, 232
National identity: Prime meridian and, 20,
 45–46; Topographical mapping as basis to,
 40–43, 45–46, 110–115, 121
Natural philosophy, 34, 40, 97
Nature, 134, 149
Nautical Almanac and Astronomical Ephemeris
 (Maskelyne): Amended by Bowditch,
 87–89, 93; Claims to accuracy of, 72, 141, 144,
 180–181, 272; Establishes basis to Greenwich prime meridian, 14, 39, 46, 51–53, 72,
 169, 265; Relationship with navigational
 texts, 56–57, 230, 232, 265; Sales of, 180–181,
 203–204, 230
Nautical Propositions (Dunn), 62
Navigation: Importance of longitude in, xii, 1,
 4, 6–7, 10–11, 13–14, 64–65; Importance of
 prime meridian to, 1–3, 7, 14, 19, 64–65, 143,
 171–172, 203–204, 262; Teaching of, 11, 62, 64
Navigators, 1, 3, 7, 10, 28–29, 30, 79, 96, 171
New American Practical Navigator (Maskelyne),
 87–88, 96
New Atlas of the Mundane System (Dunn), 64
Newcastle-upon-Tyne, 134, 246, 247, 250
Newcomb, Simon: As superintendent of the
 American almanac office, 175, 194, 199;
 Opinion of concerning International Me-
 ridian Conference, 199, 226–227, 232–233;
 Rejects Fleming's proposals on cosmo-
 politan time, 175; Resented by Christie,
 240–241
New General Atlas (Senex), 8
New Greenwich, 2–3
New Method for Discovering the Longitude both
 at Sea and Land (Whiston and Ditton),
 49–50
New Orleans, 12, 73, 93, 128; Proposed as
 American prime meridian, 73, 93–98, 103
New Practical Navigator and Daily Assistant
 (Moore), 88
New System of General Geography, in Which
 the Principles of that Science are Explained
 (MacFait), 25